Current Development of Asymmetric Catalysis and Synthesis

Current Development of Asymmetric Catalysis and Synthesis

Guest Editor

Jiaxi Xu

MDPI

Basel • Beijing • Wuhan • Barcelona • Belgrade • Novi Sad • Cluj • Manchester

Guest Editor
Jiaxi Xu
College of Chemistry
Beijing University of
Chemical Technology
Beijing
China

Editorial Office
MDPI AG
Grosspeteranlage 5
4052 Basel, Switzerland

This is a reprint of the Special Issue, published open access by the journal *Molecules* (ISSN 1420-3049), freely accessible at: https://www.mdpi.com/journal/molecules/special_issues/S58ICI5568.

For citation purposes, cite each article independently as indicated on the article page online and as indicated below:

Lastname, A.A.; Lastname, B.B. Article Title. *Journal Name* **Year**, *Volume Number*, Page Range.

ISBN 978-3-7258-6910-7 (Hbk)
ISBN 978-3-7258-6911-4 (PDF)
https://doi.org/10.3390/books978-3-7258-6911-4

Contents

About the Editor

Jiaxi Xu

Jiaxi Xu is a professor in the College of Chemistry at Beijing University of Chemical Technology in China. His research interests include synthetic methodologies and related mechanisms, asymmetric synthesis and catalysis, synthesis of heterocyclic compounds, and unnaturally occurring amino acids and peptides. He has published more than 500 peer-reviewed papers and three chapters in books.

Preface

The following is a Reprint of the Special Issue "Current Development of Asymmetric Catalysis and Synthesis" published in *Molecules*. This Special Issue brings together the following topics: Recent Advances in π-Stacking Interaction-Controlled Asymmetric Synthesis, Recent Advances in the Utilization of Chiral Covalent Organic Frameworks for Asymmetric Photocatalysis, Transition Metal Catalysis for the Asymmetric Synthesis of 2-Arylethylamines: A Review of the New Millennium, Enantioselective Diels–Alder Cycloaddition, Enantioselective Synthesis of the Sex Pheromone, Asymmetric Michael Addition/Cyclization, and Computational Analysis of the Asymmetric Hydrogenation of γ-Ketoacids, among others.

Jiaxi Xu
Guest Editor

molecules

MDPI

Review

Recent Advances in π-Stacking Interaction-Controlled Asymmetric Synthesis

Jiaxi Xu [1,2]

[1] State Key Laboratory of Chemical Resource Engineering, Department of Organic Chemistry, College of Chemistry, Beijing University of Chemical Technology, Beijing 100029, China; jxxu@mail.buct.edu.cn; Tel./Fax: +86-10-6443-5565

[2] College of Science, Henan Agricultural University, Zhengzhou 450002, China

Abstract: The π-stacking interaction is one of the most important intramolecular and intermolecular noncovalent interactions in organic chemistry. It plays an important role in stabilizing some structures and transition states in certain reactions via both intramolecular and intermolecular interactions, facilitating different selectivities, such as chemo-, regio-, and stereoselectivities. This minireview focuses on the recent examples of the π-stacking interaction-controlled asymmetric synthesis, including auxiliary-induced asymmetric synthesis, kinetic resolution, asymmetric synthesis of helicenes and heterohelicenes, and multilayer 3D chiral molecules.

Keywords: asymmetric synthesis; asymmetric induction; diastereoselectivity; π-stacking interaction

1. Introduction

Attractive π-stacking interactions between π-systems (both aromatic ring and other conjugated systems, even double and triple bonds) play various important roles in diverse phenomena, including the stabilization of biological macromolecules, such as the helical structures of DNA and tertiary structures of proteins, even the complexation of biomolecules and small organic compounds [1–3]; the stabilization of the complexation in host–guest systems [4,5]; and controlling selectivities in organic reactions [6–9]. They can not only control chemoselectivity [10–14] and regioselectivity [15–19] but also stereoselectivities, including diastereoselectivity and enantioselectivity, in diverse organic reactions [20,21]. In 1995, Jones and Chapman wrote a comprehensive review on the π-stacking effect in asymmetric synthesis [20]. π-stacking effects in chiral auxiliary-controlled asymmetric synthesis have been summarized. The auxiliaries include cyclohexane-based arenecarbonyl, arylacetyl, *N*-arylcarboxamide, and aryl carboxylate auxiliaries; 4-aryl/arylmethyloxazolidinone-based auxiliaries; axial chirality auxiliaries; natural product-based auxiliaries; and so on [20]. In 2007, Yamada summarized the intramolecular cation–π interaction in organic synthesis in his perspective [22]. In 2010, Xu collected the most important examples of the influence of the electronic effect of catalysts on the enantioselectivity in asymmetric catalytic organic reactions [21]. Attractive noncovalent interactions, especially hydrogen bonding between the ligand and substrate in enantioselective transition metal catalysis, were reviewed in 2020 [23]. Since 1995, some new chiral auxiliaries have been developed and applied in the π-stacking interaction-controlled asymmetric synthesis. Particularly, during the last two decades, the π-stacking interaction has also been applied in the preparation of optically active (hetero)helicenes and multilayer three-dimensional (3D) chiral molecules as potential materials. This minireview focuses on new developments in the π-stacking interaction-controlled asymmetric synthesis, including several newly developed auxiliary-induced asymmetric syntheses, kinetic resolution for the asymmetric syntheses, and asymmetric syntheses of (hetero)helicenes and multilayer 3D chiral molecules as potential organic materials from 1995 to now. All collected examples in this minireview are mentioned or verified by experimental or theoretical calculational results or on the basis of X-ray crystal diffraction analysis.

2. Chiral Auxiliary-Induced Asymmetric Synthesis

2.1. Chiral Arylsulfinyl-Based Auxiliaries in Asymmetric Synthesis

Enantiomerically pure aryl methyl sulfoxides and diaryl sulfoxides are a class of readily available optically active compounds and have been applied as chiral auxiliary starting materials in various asymmetric syntheses, such as syntheses of optically active fluorinated structurally diverse amines, fluorinated α-amino acids and β-amino alcohols [24], and fluorinated and non-fluorinated heterocyclic compounds [25]. The *ortho*-substituted 2-alkylphenyl 4-methylphenyl sulfoxides were first used as precursors of *ortho*-arylsulfinylbenzylcarbanions in the nucleophilic addition with *N*-(4-methylphenylsulfinyl) or *N*-(4-methylphenylsulfonyl) (tosyl) aldimines and ketimines, thereby affording 4-methylphenylsulfinamides and 4-methylphenylsulfonamides as direct addition products [26,27]. The removal of the *N*-tosyl group was actually not a readily available process and was even inefficient in some cases. To make the synthetic strategy more practical and useful in synthetic organic chemistry, both enantiomerically pure aryl methyl sulfoxides and diaryl sulfoxides as readily available optically active compounds have been further applied in various nucleophilic additions of *N*-(4-methoxyphenyl) aldimines and ketimines, giving rise to *N*-(4-methoxyphenyl)-derived amine derivatives, which are readily and efficiently removed via oxidation. Thus, the nucleophilic addition of both enantiomerically pure aryl methyl sulfoxides and diaryl sulfoxides with *N*-(4-methoxyphenyl) aldimines and ketimines has been well investigated and applied in diverse asymmetric syntheses, such as the synthesis of optically active fluorinated structurally diverse amines, fluorinated α-amino acids and β-amino alcohols [24], and fluorinated and non-fluorinated heterocyclic compounds [25].

Peptides have a very wide range of functions in the human body and are a class of widely applied macromolecular medicines. However, they generally survive biodegradation in the human body. To circumvent this biodegradation problem, some structurally similar unnatural peptide analogues are prepared instead of naturally occurring amino acids with non-naturally occurring ones. Fluorinated α-amino acids are one of the choices because fluoro-organic compounds have unique properties, such as lipophilicity, membrane permeability, metabolic stability, and bioavailability. Enantiopure aryl methyl sulfoxides were used in the synthesis of both fluorinated α-amino acids and β-amino alcohols. Enantiomerically pure aryl methyl sulfoxides **1** were first treated with LDA and reacted with fluorinated imidoyl chlorides **2** to form fluorinated chiral arylsulfinyl-derived imines **3**, which were further reduced with tetrabutylammonium borohydride (Bu_4BH_4) in MeOH, affording chiral arylsulfinyl-derived amines **4** in almost quantitative yields with excellent diastereoselectivity. In the reduction step, Bu_4BH_4 nucleophilically attacked the imines **3** from their *Si*-face due to the existence of the π-stacking interaction between the *N*-aryl group of imines and the *S*-aryl group of sulfoxides. The *Re*-face was shielded by the *S*-aryl group of sulfoxides. After the non-oxidative Pummerer reaction, the obtained optically active fluorinated arylsulfinyl-derived amines **4** were further converted into the corresponding fluorinated β-amino alcohols **5**, which were finally transformed into the desired fluorinated α-amino acids **6** in 65–70% yields via the Ru-catalyzed oxidation with $NaIO_4$ as the oxidant (Scheme 1) [24]. The current route is a convenient and useful method for the synthesis of 3,3-difluoro-, 3,3,3-trifluoro-, and 3-chloro-3,3-difluoro-derived alanines.

Enantiomerically pure (*S*)-4-methylphenyl 2-methylphenyl sulfoxide (**7**) was developed as a chiral auxiliary. It was first treated with LDA and reacted with fluorinated aldimines **8** with the *N*-(4-methoxyphenyl) group to form fluorinated chiral arylsulfinyl derived amines (*S*,*S*)-**9** and (*S*,*R*)-**9** in good yields (74–80%) with moderate stereoselectivities (69:31 to 70:30) after workup. The π–stacking interaction between the *N*-(4-methoxyphenyl) group of aldimines **8** and the *S*-aryl group of sulfoxide **7** in both transition states **TS1** and intermediates **A** played a crucial role in controlling the stereoselectivity. However, when (*S*)-4-methylphenyl 2-methylphenyl sulfoxide (**7**) was treated with LDA and then reacted with fluorinated ketimines **10** with the *N*-(4-methoxyphenyl) group to generate the corresponding fluorinated chiral arylsulfinyl derived amines (*S*,*S*)-**11** as major products in satisfactory yields (60–77%) with excellent stereoselectivities ((*S*,*S*)-**11**:(*S*,*R*)-**11** = 90:10 to 96:4). The π–stacking interaction between the *N*-(4-methoxyphenyl) group of ketimines **10** and the *S*-aryl group of sulfoxide **7** in both transition states **TS2** and intermediates **B** also played a key role in controlling the stereoselectivity. The results indicated that ketimines showed better stereoselectivities than aldimines (Scheme 2) [24]. The *ortho*-substituted benzylcarbanion with chiral arylsulfinyl auxiliary improved the stereoselectivity efficiently.

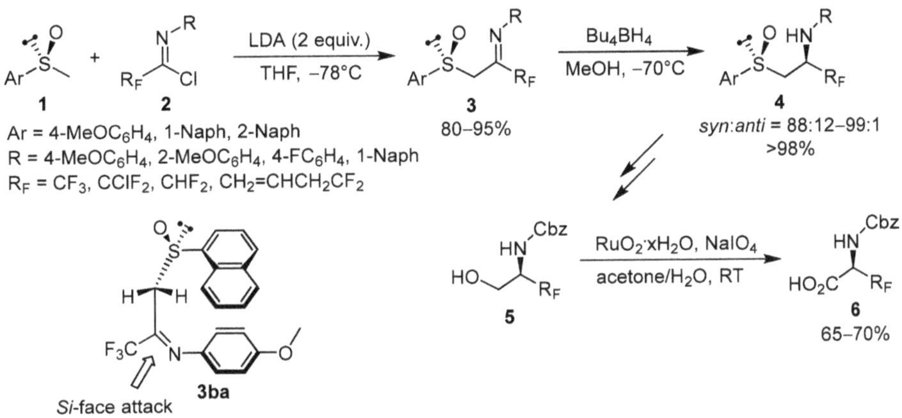

Scheme 1. Synthesis of optically active fluorinated α-amino acids and β-amino alcohols.

To further extend the application of *ortho*-substituted benzylcarbanions with chiral arylsulfinyl auxiliary in the stereoselective nucleophilic addition of various imines, enantiomerically pure (*S*)-4-methylphenyl 2-alkylaryl sulfoxides **12** were also developed as chiral auxiliaries. After the treatment with LDA, they reacted with both fluorinated aldimines **8** and ketimines **10** to give rise to the corresponding fluorinated chiral arylsulfinyl-derived amines (*S*,*S*)-**13** in moderate to good yields (40–86%) with excellent stereoselectivities (>98%) after workup. The similar π-stacking interaction in both transition states **TS3** and intermediates **C** controlled the stereoselectivity almost completely. If fluorinated arylsulfinyl-derived amines (*S*,*S*)-**13** were further treated with KHMDS in THF at 0 °C, they underwent an intramolecular aromatic nucleophilic substitution with arylsulfinyl groups as the leaving groups, affording the corresponding fluorinated indoline derivatives **14** in 60–83% yields. Only trifluoromethyl-derived products (*S*,*S*)-**13** ($R_F = CF_3$) were tested. The results illustrated that indoline derivatives **14** could be generated in one pot. The tandem reaction was attempted. After the nucleophilic addition of 2-(4-methylphenylsulfinyl)benzylcarbanions and aldimines **8** or ketimines **10** at −78 °C, the reaction mixture was warmed to −30 °C and continually stirred for 30 min. Further intramolecular aromatic nucleophilic substitution occurred, producing the desired fluorinated indoline derivatives **14** in 35–71% yields via tandem nucleophilic addition and intramolecular aromatic nucleophilic substitution in one pot, exhibiting high step-economy. In comparison with the step-wise synthetic method, the yields in the tandem fashion were similar to the total yields for two steps in the step-wise route. Furthermore, the one-pot tandem reaction of (*S*)-4-methylphenyl 4-cyano-2-methylphenyl sulfoxide (**16**) with a cyano functional substituent and trifluoromethyl ketimine **10a** was performed with LDA as a base in THF at −78 °C; the desired indoline derivative **16** was obtained in 45% yield after stirring for 30 min. The cyano group survived in the tandem reaction in the presence of *ortho*-arylsulfinylbenzylcarbanion as the strong nucleophile, showing good functional group tolerance (Scheme 3) [25].

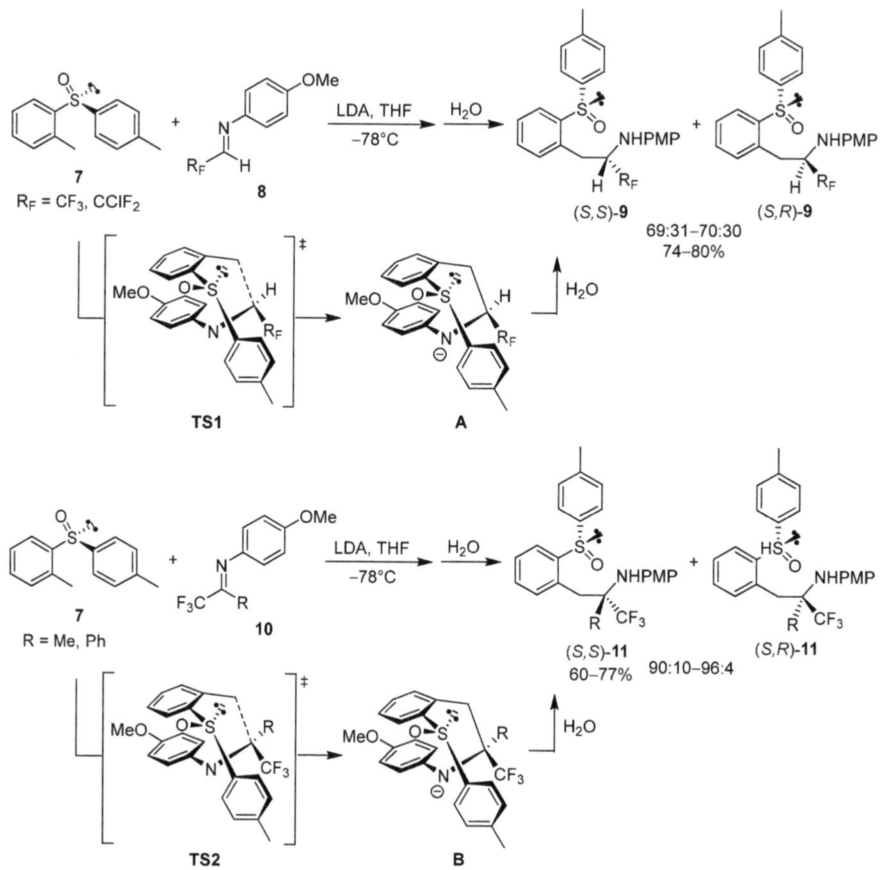

Scheme 2. Synthesis of optically active fluorinated sulfinyl-derived amines from aldimines (**A**) and ketimines (**B**).

The indoline skeleton is a ubiquitous moiety in the structures of many alkaloids and natural products. Indolines are generally considered to be key privileged structures for their diverse biological activities. To develop a step-economic and efficient asymmetric synthetic method of biologically important and optically active indoline derivatives, the above developed strategy was extended to the synthesis of non-fluorinated indoline derivatives via tandem nucleophilic addition and intramolecular aromatic nucleophilic substitution with the 4-methylphenylsulfinyl group as the leaving group. When fluorinated aldimines **8** and ketimines **10** were displaced with aromatic imines **17** generated from aromatic aldehydes and aromatic amines, the reaction of (*S*)-(2-ethylphenyl) 4-methylphenyl sulfoxide (**12a**) and aromatic imines **17** generated (2*S*,3*S*)-2,3-diaryl-4-methylindolines **18** in moderate to satisfactory yields of 25–62% with LDA as a base. When LHDMS or KHMDS was used as the base instead of LDA, the reaction was conducted at room temperature or at 70 °C to give the corresponding products **18** in higher yields than those with LDA as the base. The aromatic imines with electron-withdrawing substituents generally required longer reaction times and higher reaction temperatures than those with electron-donating groups. The reaction of electron-rich (*S*)-1-ethyl-4,5-dimethoxy-2-(4-tolylsulfinyl)benzene (**19**) and (*E*)-*N*-(4-methoxyphenyl)-1-phenylmethanimine (**17b**) stopped at the nucleophilic addition step in the presence of LDA as the base, generating the corresponding amine **20** as the final product after workup, rather than the desired indoline derivative because the electron-rich arylsulfinyl with two strong electron-donating methoxy groups could not undergo the intramolecular aromatic nucleophilic substitution. Upon further treatment of the amine **20** with LHMDS, no reaction occurred as well, indicating that the electron-rich substrate indeed hardly underwent the intramolecular aromatic nucleophilic substitution even in a step-wise fashion. For each of these cases, the π-stacking interaction between the *N*-aryl group of imines and the *S*-aryl group of sulfoxides in both the transition state **TS4** and intermediate state **D** played an important role in controlling the stereoselectivity (Scheme 4) [28].

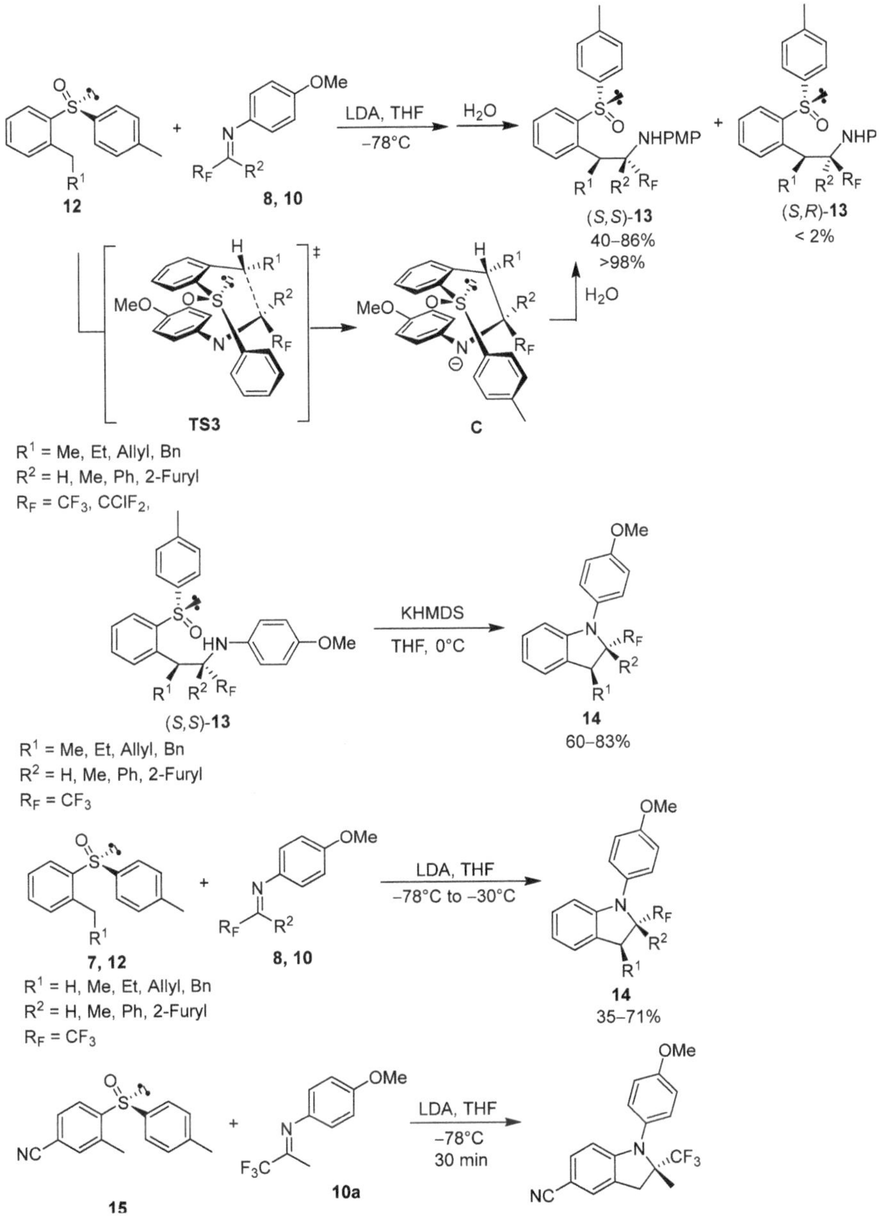

Scheme 3. Synthesis of optically active fluorinated sulfinyl-derived amines and indolines.

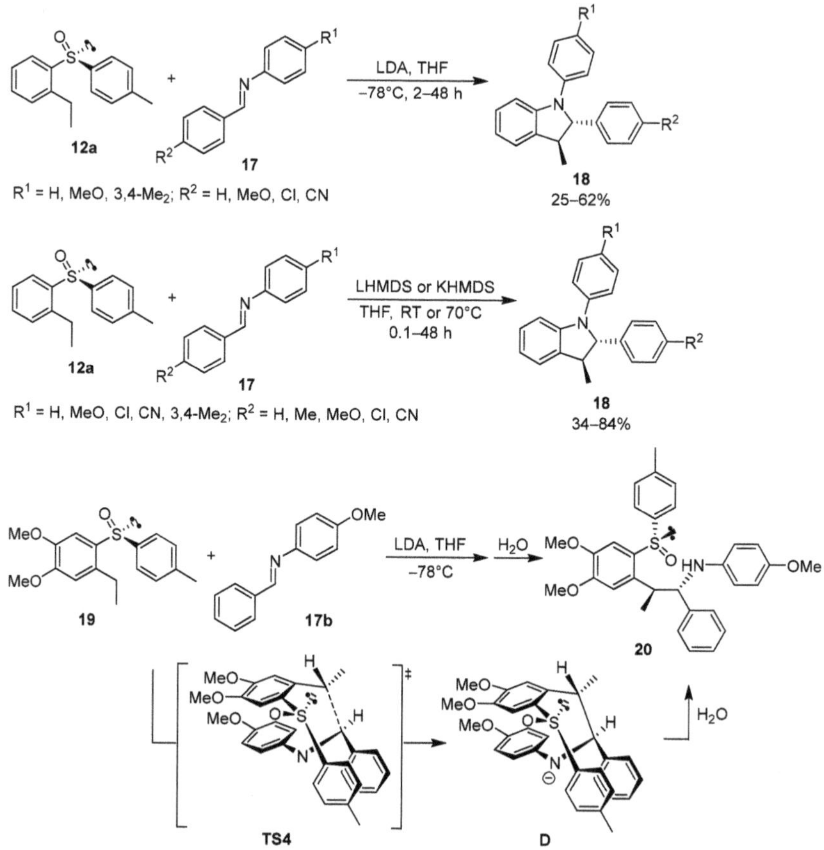

Scheme 4. Synthesis of optically active indolines.

2.2. *Adducts of Levoglucosenone and 9-(Aryloxymethyl)arthracenes as Chiral Auxiliaries in Asymmetric Synthesis*

The Diels–Alder cycloaddition of alkyl acrylates and cyclopentadiene can generate both *endo*-adducts and *exo*-adducts. If chiral alkyl acrylates were utilized, asymmetric induction would occur. Acrylate derivatives **23** bearing *para*-trifluoromethyl and methoxyphenoxymethyl substituents as the π-stacking templets and shelter were prepared via the Diels–Alder reaction of enantiomerically pure levoglucosenone (**20**) and 9-(*para*-trifluoromethyl and methoxyphenoxymethyl)arthracenes (**21**), and the subsequent reduction and acrylation. There is an intramolecular vinyl-aryl π-stacking interaction between the acrylate and aryloxy groups. When they were applied in the Diels–Alder reaction with cyclopentadiene, evident π-stacking-controlled asymmetric synthesis was observed, generating *endo*-(*S*)-bicyclo [2.2.1]hept-5-ene-2-carboxylates **24** in 65% and 59% yields as major products through more the stable transition state **TS5a**. Cyclopentadiene would approach the C=C bond in the acrylate moiety only from its top direction in all transition states because the aryloxy group was fixed below the C=C bond due to the existence of the vinyl-aryl π-stacking interaction between the acrylate and aryloxy groups (Scheme 5) [29].

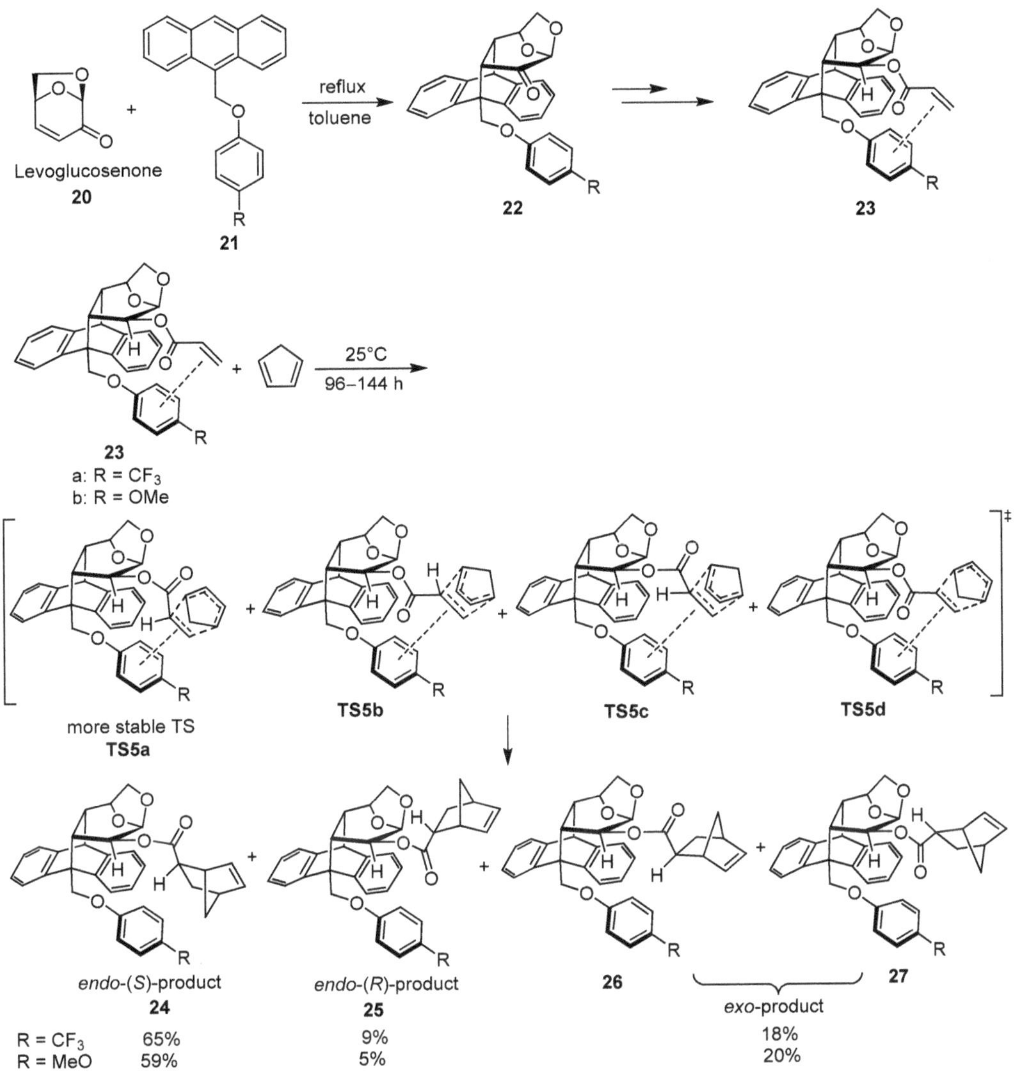

Scheme 5. π-stacking-controlled Diels–Alder reaction with *endo*-(*S*)-products as major products.

Both enantiomerically pure levoglucosenone (**20**) and its dihydro derivative **28** are readily available from biomass because they are products of cellulose pyrolysis. Enantiomerically pure dihydrolevoglucosenone (**28**) was also applied as a chiral auxiliary in the diastereoselective Diels–Alder reaction. It was first converted to dibenzylated dihydrolevoglucosenols (**29**) through double benzylation with benzyl halides under basic conditions followed by a reduction with sodium borohydride. Differently, dibenzylated dihydrolevoglucosenols (**29**) were further acrylated and then reacted with cyclopentadiene in the presence of Lewis acids in DCM, affording *endo*-(*R*)-bicyclo [2.2.1]hept-5-ene-2-carboxylates (*R*)-**31** as major products due to the existence of the vinyl-aryl π-stacking interaction (Scheme 6) [30]. Through the utilization of both levoglucosenone (**20**) and its dihydro derivative **28** as auxiliaries, both enantiomeric bicyclo [2.2.1]hept-5-ene-2-carboxylates were prepared in good to high yields. Both diastereomeric monobenzylated dihydro derivatives were also attempted as auxiliaries in the diastereoselective Diels–Alder reactions. They show excellent *endo/exo* selectivity, but their *R/S* stereoselectivity is generally lower than the corresponding dibenzylated dihydrolevoglucosenone (**28**).

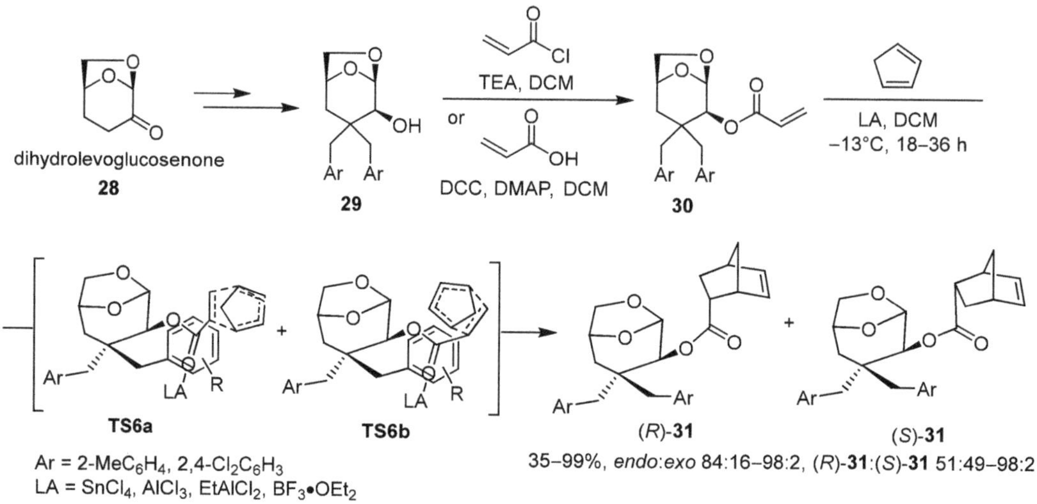

Scheme 6. π-stacking-controlled Diels–Alder reaction with *endo*-(*R*)-products as major products.

2.3. Chiral Oxazolidinone-Based Auxiliaries in Asymmetric Synthesis

Lignin is a class of natural plant-based polymers and ranks second in abundance only after cellulose, making it a potentially valuable raw material for biorefinery. However, it is a considerable challenge to use lignin as a feedstock for the production of biobased chemicals in either catalytic or enzymatic processes due to the structural heterogeneity of lignin. The heterogeneity is the result of the biosynthesis of lignin from the radical coupling of three primary monolignols. To improve lignin's utility as a renewable carbon feedstock, it is necessary to understand the assembly, stereostructure, and reactivity of the separation of lignin and the enzymatic lignin disassembly process. To realize these purposes, it is required to synthesize lignin models with different configurations in a stereospecific manner. To enantioselectively synthesize lignin models, Njiojob and coworkers selected the Evans auxiliary as the chiral source. They first prepared *N*-(2-methoxyphenyloxy)acetylated (*R*)-4-isopropyloxazolidin-2-ones **32** as starting materials. The reaction of aryloxyacetyl-derived (*R*)-4-isopropyloxazolidin-2-ones **32** and 4-benzyloxybenzaldehydes **33** stereospecifically generated lignin dimer models **34** in 60–68% yields in the presence of di-*n*-butylboron triflate and diisopropylethylamine (DIPEA) via condensation through six-membered Zimmerman-Traxler transition states **TS7**, in which the π-stacking interaction between benzaldehydes and aryloxy groups plays an important role in controlling the stereoselectivity. After subsequent transformations, including reduction, protection of the hydroxyl group, and oxidation, one of the lignin dimer models **34** was converted into aldehyde **35**. Following a similar strategy, the reaction of aldehyde **35** with (*S*)-4-isopropyl-*N*-(2-methoxyphenyloxy)acetyloxazolidin-2-ones **36** as the chiral starting material, a lignin trimer model **38** was synthesized in a 53% yield. (*R*)- and (*S*)-Evans auxiliaries **32** and **36** show completely opposite stereoselectivities (Scheme 7) [31]. The current synthetic strategy is an efficient way to prepare enantiopure lignin dimers and trimers with different stereochemical configurations from aryloxyacetylate oxazolidin-2-ones derivatives and appropriate aromatic aldehydes.

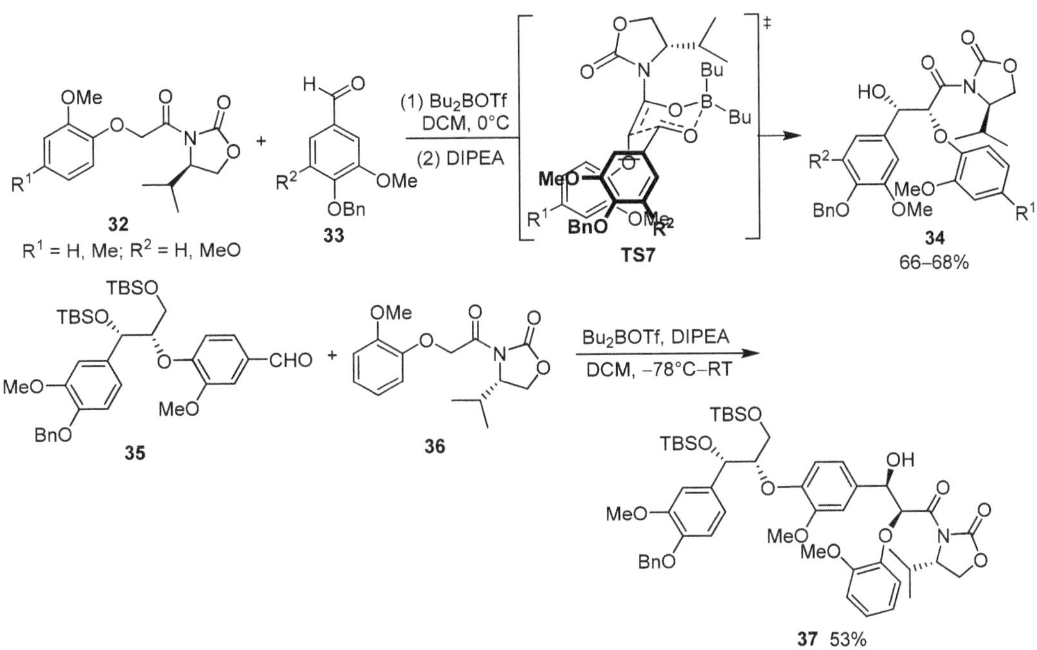

Scheme 7. Asymmetric synthesis of dimers and trimer of lignin models.

3. Acyl 2,3-Dihydroimidazo[1,2-*a*]pyridine and 1,2-Dihydroimidazo[1,2-*a*]quinolines in Kinetic Resolution

Although the asymmetric catalytic borane reduction of ketones is one of the most efficient methods for the preparation of enantiopure secondary alcohols [32–34], chemical resolution has been widely utilized in the industrial manufacturing of secondary alcohols as well [35,36]. Kinetic resolution is also an alternative choice for obtaining enantiopure secondary alcohols via selected reactions, for instance, acylation [37,38].

Birman's group first developed 6-substituted (*R*)-2-phenyl-2,3-dihydroimidazo[1,2-*a*]pyridines **38** as enantioselective acylation catalysts. In the resolution, the catalysts first were acylated with acetic or propanoic anhydrides in the presence of DIPEA. The acylated 2,3-dihydro-1*H*-imidazo[1,2-*a*]pyridin-4-iums **41** predominantly reacted with (*R*)-alcohols (*R*)-**39** due to less steric hindrance in the π-stacking attractive interaction. Generally, non-substituted and 6-bromo-2,3-dihydroimidazo[1,2-*a*]pyridines (**38a** and **38b**) exhibited lower selectivities than 6-nitro- and 6-trifloromethyl-2,3-dihydroimidazo[1,2-*a*]pyridines (**38c** and **38d**). Bulky alcohols (R = $CHMe_2$ and CMe_3) generally required long reaction times (30–52 h) to reach a similar conversion. However, aliphatic 1-cyclohexylethan-1-ol showed only less than 4% conversion for 50 h due to the absence of the π-stacking interaction. Although indan-1-ol was converted in 16% for 50 h, no selectivity was observed because of the nonexistence of the π-stacking interaction. These results support the function of the π-stacking attractive interaction in the kinetic resolution. The C=C double bond containing substrate 1-(cyclohexen-1-yl)ethan-1-ol (**42**) was resoluted in a moderate selectivity of 11 in 40% conversion (Scheme 8) [37].

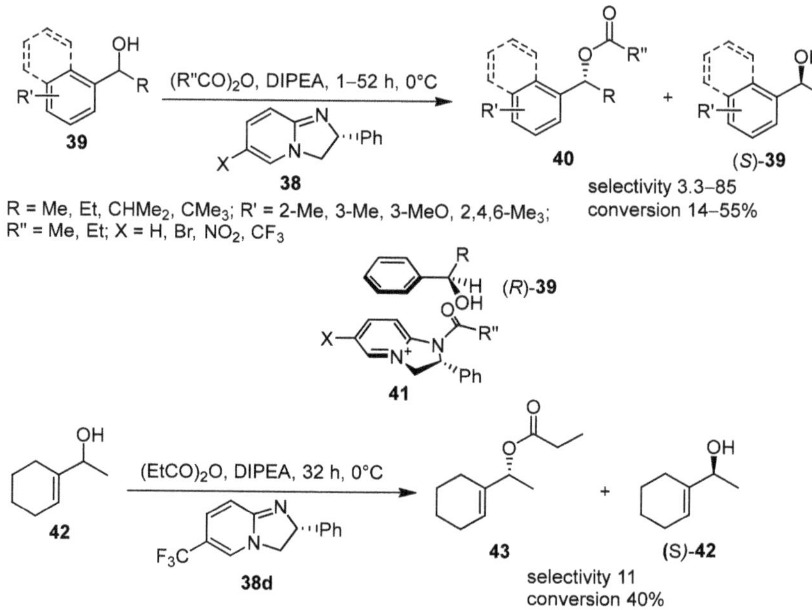

Scheme 8. Synthesis of phosphonopeptides with DCC as a coupling reagent.

To realize the kinetic resolution of conjugated cinnamyl alcohols and their derivatives **45**, they further developed new (*R*)-7-chloro-2-phenyl-1,2-dihydroimidazo[1,2-*a*]quinoline (**44**) with an additional fused benzene ring as the enantioselective acylation catalyst. In comparison with 2,3-dihydroimidazo[1,2-*a*]pyridines **38**, 1,2-dihydroimidazo[1,2-*a*]quinoline **44** extended the conjugative system and achieved efficient a kinetic resolution of both conjugated cinnamyl alcohols **45** with selectivities of 17–57 in conversions of 32–56% and non-conjugated 1-arylethan-1-ols **48** with selectivities of 33–117 in conversions of 42–56%; even the C=C double bond containing substrate 1-(cyclohexen-1-yl)ethan-1-ol (**42**) with a selectivity of 17 in a conversion of 47% with propanoyl anhydride was applied as the acylating reagent. In each of the cases, (*R*)-alcohols were acylated to the corresponding propanoates and (*S*)-alcohols remained (Scheme 9) [38]. All of these three classes of substrates were efficiently resoluted with (*R*)-7-chloro-2-phenyl-1,2-dihydroimidazo[1,2-*a*]quinoline (**44**) as the catalyst and propanoyl anhydride as the acylating reagent. However, bulky alcohols generally presented high selectivities. For example, 2,2-dimethyl-1-phenylpropan-1-ol was resoluted with the selectivity of 117. The results revealed that 1,2-dihydroimidazo[1,2-*a*]quinoline **44** with an extended conjugative system showed better behavior than 2,3-dihydroimidazo[1,2-*a*]pyridines **38** in the kinetic resolution of alcohols.

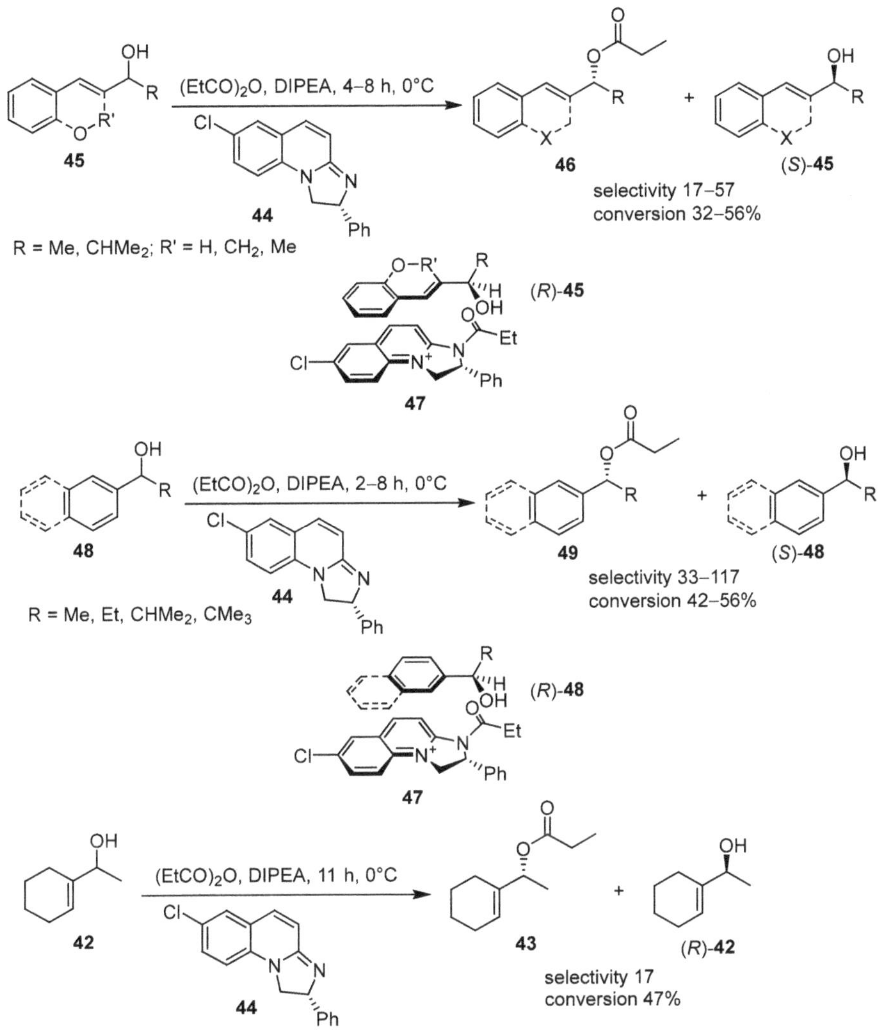

Scheme 9. Kinetic resolution of alcohols with 1,2-dihydroimidazo[1,2-*a*]quinolone as the catalyst.

4. Asymmetric Synthesis of β-Lactams and Deoxygenation of Oxiranecarbonitriles via Intramolecular π–π Stacking Interaction

β-lactam has been a key structural moiety of widely applied β-lactam antibiotics all over the world since the 1940s. β-lactam antibiotics have helped millions of people. Most β-lactam antibiotics are prepared from semisynthesis. The key structures of the β-lactam ring with defined stereostructures are generally constructed by organisms. The stereoselective synthesis of β-lactam derivatives is a crucial issue in constructing the β-lactam ring in both organic and medicinal chemistry [39]. The Staudinger cycloaddition is a versatile method of synthesizing β-lactams from imines and ketenes, generated from acyl chlorides or α-diazoketones [40,41]. The diastereoselectivity in the Staudinger cycloaddition is controlled by the competition between the direct ring closure and isomerization of zwitterionic intermediates **E** generated from imines **55** and ketenes **54** (Scheme 10) [42]. In contrast with other factors [43], the substituents of imines and ketenes and reaction temperature evidently impact diastereoselectivity [44]. On the other hand, torquoselectivity also plays an important role in the diastereocontrol in disubstituted ketene-participating Staudinger cycloaddition [45].

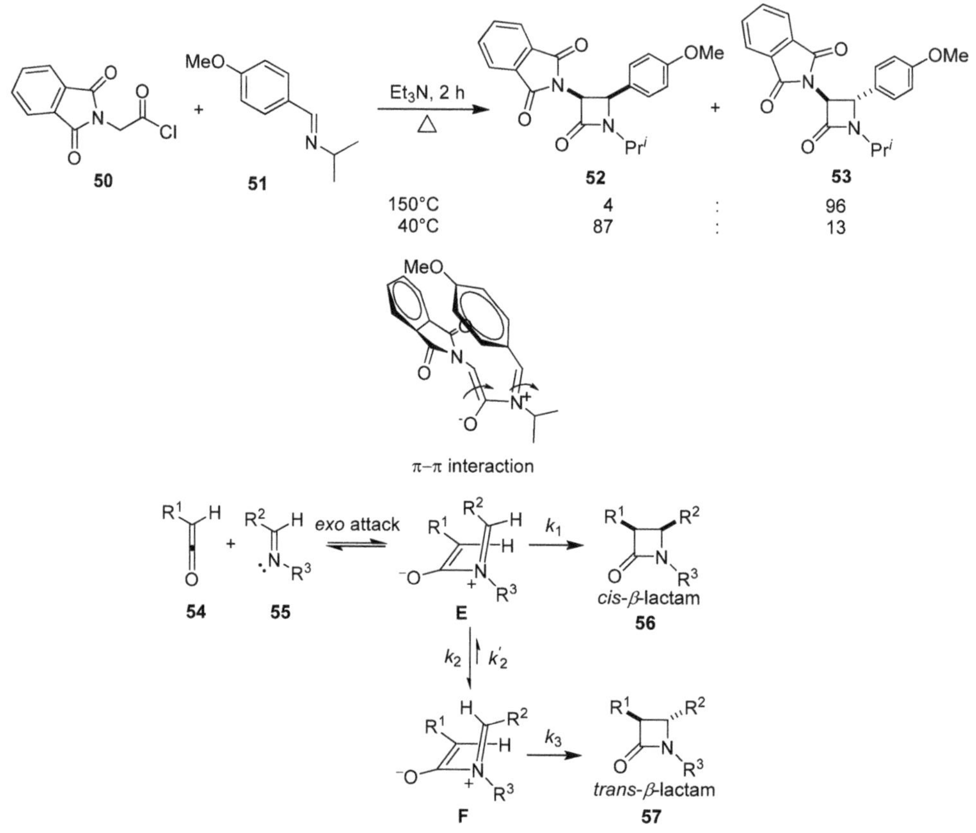

Scheme 10. Diastereoselective synthesis of β-lactams.

To illustrate the influence of different substituted ketenes and imines on diastereoselectivity in the formation of β-lactams at various reaction temperatures, the reaction of phthalimidoacetyl chloride (**50**) and *N*-(4-methoxybenzylidene)isopropylamine (**51**) was conducted and exhibited an evident increase in temperature on the diastereoselectivity. There was a favorable formation of *trans*-β-lactam product **53** at a higher temperature (150 °C) and a predominant generation of *cis*-β-lactam product **52** at a lower temperature (40 °C). It was rationalized that the strong π-stacking interaction existed between the electron-withdrawing phthalimido and electron-donating 4-methoxyphenyl groups in intermediate **E** at low temperatures and played an important role in stabilizing the intermediate **E**, leading to the formation of *cis*-β-lactam product **52**. However, the stability of the π-stacking interaction decreased along with the increase in the reaction temperature, resulting in the intermediate **E** converting into intermediate **F**, which generated *trans*-β-lactam product **53** (Scheme 10) [44]. The intramolecular π-stacking interaction played an important role in controlling the formation of *cis*-β-lactam product **54** diastereoselectively at low temperatures.

Oxiranecarbonnitriles are very important synthetic intermediates and have been applied in several transformations [46–50]. During the transformation of 3-substituted *trans*-oxiranecarbonnitriles **58** to 3-substituted (Z)-propenonitriles **60** through the thiourea-mediated stereospecific deoxygenation, *trans*-oxiranecarbonnitriles **58** generated (Z)-propenonitriles

60 stereospecifically due to the π-stacking interaction in intermediates **G** and **61**. *N,N,N′,N′*-Tetramethylthiourea (**59**) first nucleophilically attacked the cyano-attached ring carbon atom to form intermediates **G** via an S_N2 ring opening. Intermediates **G** rotated 180° to generate intermediates **H**, which further underwent an intramolecularly nucleophilic addition to produce 1,3-oxathiolane derivatives **61**. In an amine-induced fragmentation, they transformed into intermediates **I**, which rotated 180° to yield intermediates **J**. Intermediates **J** further underwent an intramolecularly nucleophilic substitution, resulting in the formation of *trans*-thiiranecarbonitriles **62** stereospecifically. *Trans*-Thiiranecarbonitriles **62** lost their sulfur atom to lead to (*Z*)-propenonitriles **60** stereospecifically (Scheme 11) [51].

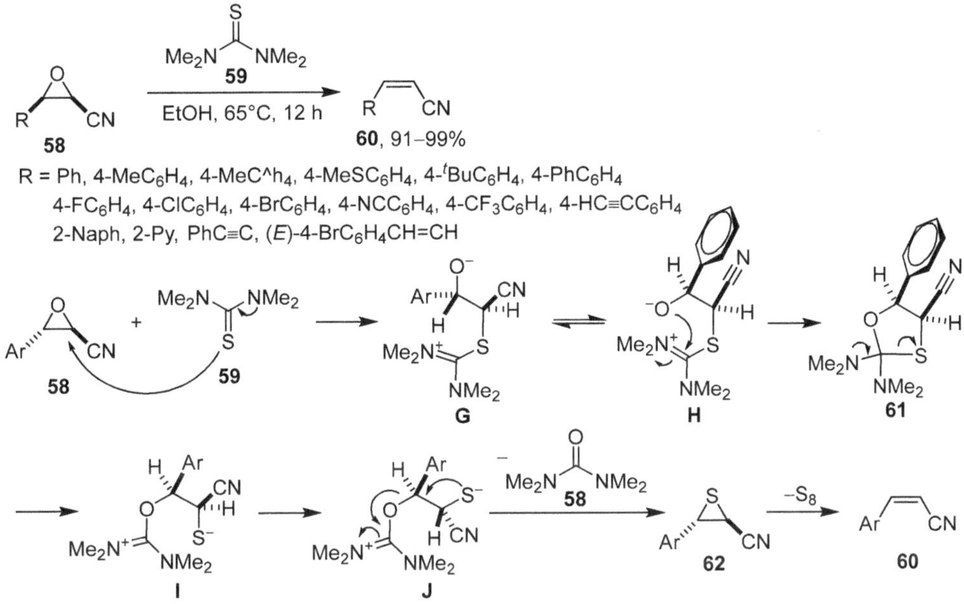

Scheme 11. Stereospecific synthesis of (*Z*)-3-arylpropenonitriles from *trans*-3-aryloxiranecarbonitriles.

The nucleophilic ring expansion of saturated three-membered heterocycles has been well investigated [52]. Recently, the electrophilic ring expansion of saturated three-membered heterocycles was also realized [18,53,54]. In contrast with the nucleophilic ring expansion of saturated three-membered heterocycles [52], their electrophilic ring expansion is a new avenue to construct new heterocyclic compounds [18,53,54]. The electrophilic ring expansion of polycyclic arylthiiranes **64** and ketenes **K** generated from aryloxyacetyl chlorides **63** in the presence of triethylamine is a new strategy for the synthesis of areno[*d*]-ε-thiolactones **66** directly without catalysts or additives. In the reaction, aryloxyacetyl chlorides **63** are first eliminated in the presence of TEA to form aryloxyketenes **K**.

Arylthiiranes **64** and aryloxyketenes **K** undergo a dearomatic sulfur-shifted ene reaction to directly generate intermediates **66** through the *endo* transition states **TS8*endo*** due to the existence of the π-stacking interaction, which was verified by theoretical calculation. After aromatization, intermediates **65** transform into the final products **66** in 22–94% yields. The current reaction features atom and step-economic characteristics via a tandem sequence of in situ ketene generation, π-stacking-controlled dearomatic sulfur-shifted ene, and aromatization and is a novel strategy for the electrophilic ring expansions of three-membered saturated heterocycles (Scheme 12) [55].

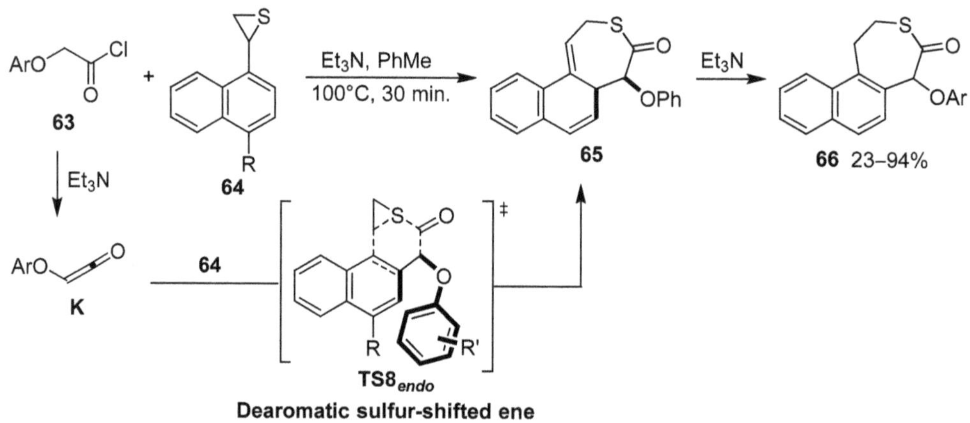

R = H, Me, cyclopropyl, Ph, F, Cl, Br, 2-Furyl
R' = H, 4-Me, 4-^{t}Bu, 4-MeO, 2-Cl, 3-Cl, 4-Cl, 4-Br, 4-I (for Ar)

Scheme 12. Synthesis of areno[*d*]-ε-thiolactones from arylthiiranes and aryloxyacetyl chlorides.

5. Asymmetric Synthesis of Optically Active Helicenes

Helicenes are a class of important carbon-rich materials. Helical supramolecular structures are important in chemistry and materials science. Helical structures that are derived from conjugated *ortho*-annelated arenes or heteroarenes are known as [*n*]helicenes. The n presents the number of the *ortho*-annelated arene or heteroarene ring. Helicenes are 3-dimensional polycyclic aromatic systems, which consist of all *ortho*-fused aromatic or heteroaromatic rings and are inherently chiral, presenting helical conformation. Different from fullerenes, carbon nanotubes, and graphene, helicenes are chiral organic compounds and exist in left-hand or right-hand helical structures. Different racemic [11]helicenes consisting of all *ortho*-fused benzene rings or *ortho*-fused benzene and conjugated cyclohexadiene rings were prepared successfully from 4,6-bis((2-(4-(triisopropylsilyl)but-3-yn-1-yl)naphthalen-1-yl)ethynyl)isophthalaldehyde (**67**), which was synthesized from 4,6-diformyl-1,3-phenylene bis(trifluoromethanesulfonate) and (4-(1-ethynylnaphthalen-2-yl)but-1-yn-1-yl)triisopropylsilane via the CuI-catalyzed coupling. The dialdehyde **67** reacted with *n*-butyl lithium-treated prop-1-yn-1-yltripropylsilane, generating a pair of enantiomeric double addition products and a *meso*-double addition product in an equivalent amount. After a series of transformations, they were converted into diacetates *rel*-(*R*,*R*)-**68** and *meso*-**68** as precursors for the next [2 + 2 + 2] annulation. The [2 + 2 + 2] annulations were conducted under three different conditions. For precursor *rel*-(*R*,*R*)-**68**, it asymmetrically produced double-annulated product [11]helicene-like derivatives *rel*-(*M*,*R*,*R*)-**69** and *rel*-(*P*,*R*,*R*)-**69** through intermediates *rel*-(*M*,*R*,*R*)-**69** and *rel*-(*P*,*R*,*R*)-**69**. The CpCo(CO)$_2$-catalyzed annulation preferred the formation of *rel*-(*P*,*R*,*R*)-**70** with a diastereoselectivity of 17:83 for *rel*-(*M*,*R*,*R*)-**70**:*rel*-(*P*,*R*,*R*)-**70**, while the Ni(COD)$_2$-catalyzed annulation predominately generated *rel*-(*M*,*R*,*R*)-**70** with a diastereoselectivity of 72:28 for *rel*-(*M*,*R*,*R*)-**70**:*rel*-(*P*,*R*,*R*)-**70**. However, only very low diastereoselectivity was observed in the CpCo(C$_2$H$_4$)$_2$-catalyzed annulation, with a slightly favorable formation of *rel*-(*M*,*R*,*R*)-**70** with a diastereoselectivity of 57:43 for *rel*-(*M*,*R*,*R*)-**70**:*rel*-(*P*,*R*,*R*)-**70**. For the double [2 + 2 + 2] annulation of *meso*-**68**, the Co-catalyzed double [2 + 2 + 2] annulations gave rise to the annulated product [11]helicene-like derivative *rel*-(*M*,*R*,*R*)-**70** in higher yields (40% and 30%) than those in the Ni-catalyzed annulation. The [11]helicene-like derivative *rel*-(*M*,*R*,*R*)-**63** was further converted to racemic [11]helicene *rac*-**70** in excellent yield via the elimination of acetate and dehydrogenation. In all [2 + 2 + 2] annulations, the edge-to-face π-stacking interaction plays a key role in controlling the diastereoselectivity (Scheme 13) [56]. Interestingly, the $CpCo(CO)_2$- and $Ni(COD)_2$-catalyzed double [2 + 2 + 2] annulations exhibit different diastereoinduction in the stereoselective synthesis of [11]helicene-like molecules.

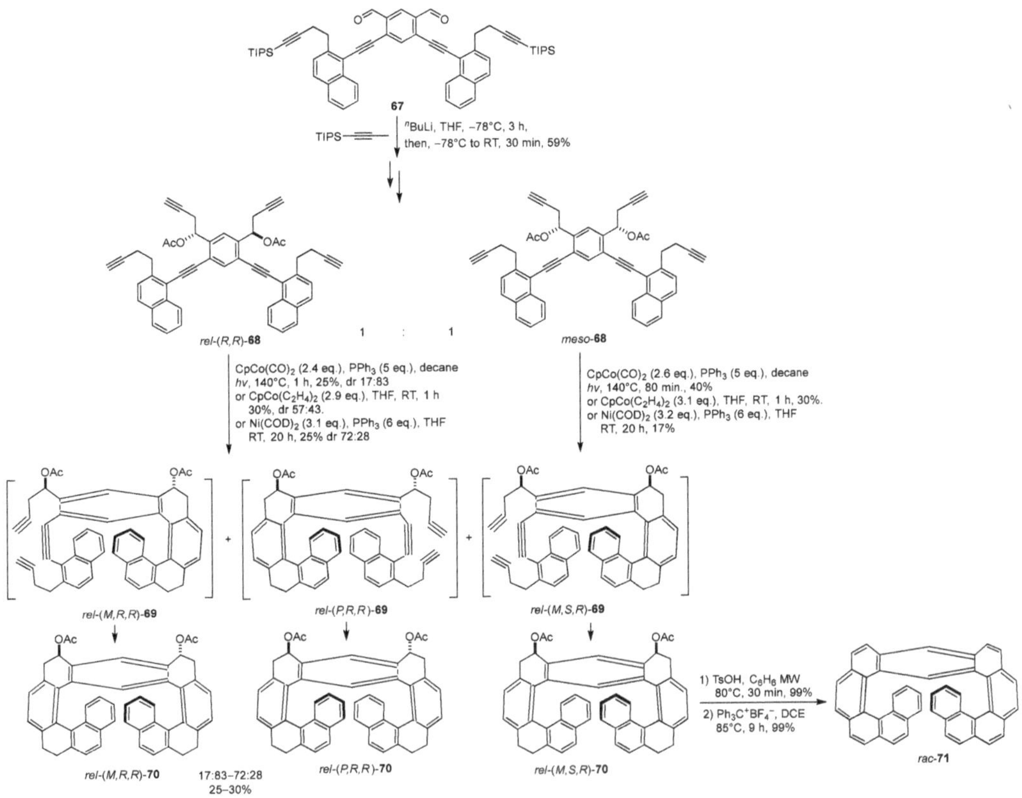

Scheme 13. Synthesis of optically active [11]helicene-like molecules.

Helicenes containing one or more nonaromatic heterocycles are also a class of important carbon-rich materials. Similar to helicenes derived from conjugated *ortho*-annelated arenes or heteroarenes, helicene-like molecules also present helical conformation. After the successful preparation of racemic [11]helicenes, the asymmetric synthesis of optically active (*P*,*S*,*S*)- and (*M*,*S*,*S*)-[11]helicene-like molecules (*P*,*S*,*S*)-**74** and (*M*,*S*,*S*)-**74** was realized from 1,1′-((4,6-bis((((S)-4-(*p*-tolyl)but-3-yn-2-yl)oxy)methyl)-1,3-phenylene)bis(ethyne-2,1-diyl))bis(2-(but-3-yn-1-yl)naphthalene) (−)-(*S*,*S*)-**72** through the Co-catalyzed double [2 + 2 + 2] annulation. In the [2 + 2 + 2] annulation, the edge-to-face π-stacking interaction plays a crucial role in controlling the stereoselectivity. Under all the used conditions, (*P*,S,S)-annulated product (*P*,*S*,*S*)-**74** was obtained as the major product. High stereoselectivity ((*M*,*S*,*S*)-**74**:(*P*,*S*,*S*)-**74** = 10:90) was observed under photo-irradiation (in 26% yield) and microwave irradiation (in 33% yield) conditions. However, a low stereoselectivity ((*M*,*S*,*S*)-**74**:(*P*,*S*,*S*)-**74** = 25:75) and yield (17%) were obtained at room temperature. The results indicated that a higher reaction temperature was favorable to improve both the yield and stereoselectivity (Scheme 14) [56]. The current Co-catalyzed double [2 + 2 + 2] annulation strategy is an efficient route to synthesize optically active [11]helicene-like molecules.

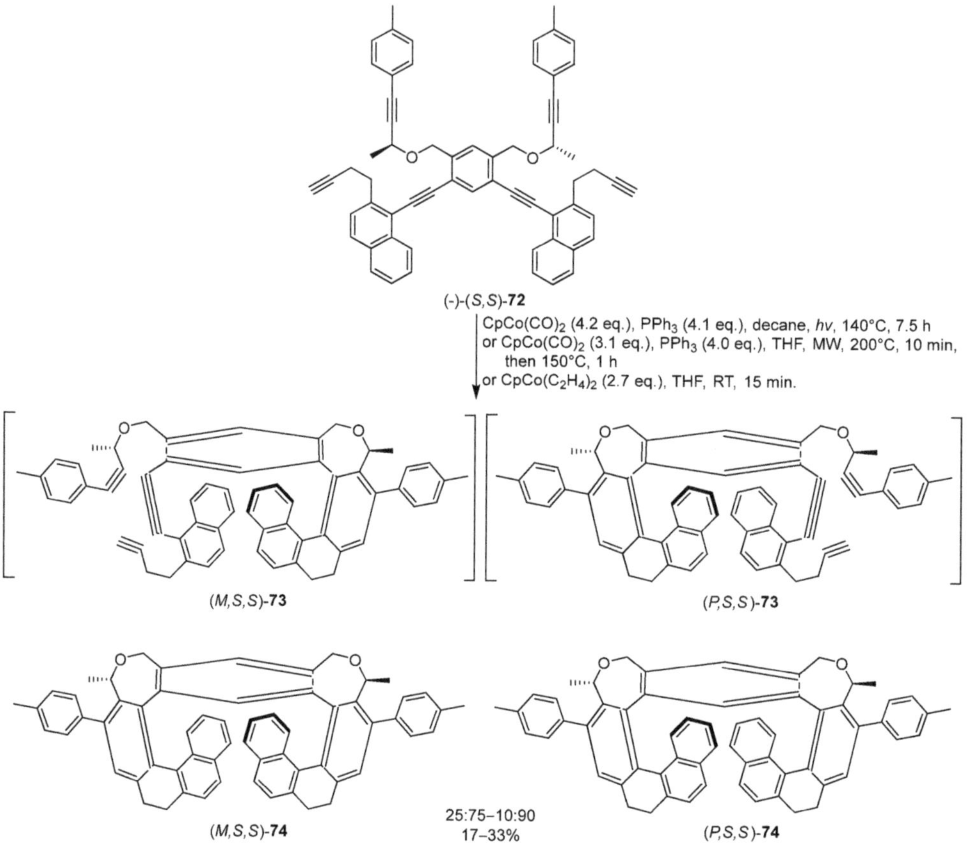

Scheme 14. Synthesis of optically active [11]helicene-like molecules.

Heterohelical supramolecular structures are important in chemistry and materials science as well. Oligothiophenes are the most studied π-systems due to their favorable electrical and optical properties. Much attention has been paid to their preparation. Several classical oligothiophenes, such as α-sexithiophene, quasi-linear oligomer pentathienoacene, and helical oligomer [7]helicene, have been synthesized until now. The asymmetric synthesis of optically active (*M*,*R*,*M*)-bis[7]helicene (*M*,*R*,*M*)-**79** was carried out from bis(β-trithiophene) **75** [57]. Bis(β-trithiophene) **75** was obtained from 3,4-dibromo-2-trimethylsilyldithieno[2,3-*b*:3′,2′-*d*]thiophene via lithiation with butyl lithium and the CuCl$_2$-catalyzed coupling followed with the treatment with TFA in dichloromethane. 3,4-Dibromo-2-trimethylsilyldithieno[2,3-*b*:3′,2′-*d*]thiophene was prepared from 3,4-dibromothiophene via lithiation with butyl lithium and the CuCl$_2$-catalyzed coupling followed by dilithiation with LDA and oxidation with bis(phenylsulfonyl)sulfide [58]. To circumvent solubility problems, bis(β′-bromo-β-trithiophene) **75** was protected with tripropylsilyl chloride with the treatment with LDA in the presence of a chiral diamine (-)sparteine, affording optically active mono-tripropylsilyl-protected (*R*)-tetrakis(β-trithiophene) (*R*)-**76** and ditripropylsilyl-protected tetrakis(β-trithiophene) **77**, which was hydrolyzed to starting material **75** with TFA in chloroform for reuse. The (*R*)-tetrakis(β-trithiophene) (*R*)-**76** was further coupled under the catalysis of a palladium catalyst in the presence of potassium phosphate to generate tetrakis[7]helicene (*R*,*R*,*R*)-**78** with (*R*,*R*,*R*) configuration. Finally, the (*R*,*R*,*R*)-tetrakis[7]helicene (*R*,*R*,*R*)-78 was transformed to optically active (*M*,*R*,*M*)-bis[7]helicene (*M*,*R*,*M*)-**79** via the formation of two new thiophene rings through the sequential treatments with LDA in the presence of (−)-sparteine and oxidation with bis(benzenesulfonyl) sulfide. During the transformation, the strong noncovalent π–π stacking interaction plays an important role in stereoinduction in the formation of (*M,R,M*)-bis[7]helicene (*M,R,M*)-**79**, which showed a helical structure. However, the transformation from (*R,R,R*)-tetrakis[7]helicene (*R,R,R*)-**78** into (*M,R,M*)-bis[7]helicene (*M,R,M*)-**79** exhibited low to moderate stereoselectivity, 62–64% ee for 3–4% conversion, and 22–47% ee for 12–23% conversion. The stereoselectivity decreased evidently along with the increase in conversion (Scheme 15) [57]. Although the synthetic method is not so efficient in conversion and stereoselectivity, it is still useful for the preparation of optically active bis[7]helicene and will show further application in the preparation of other [n]helicene derivatives in the future.

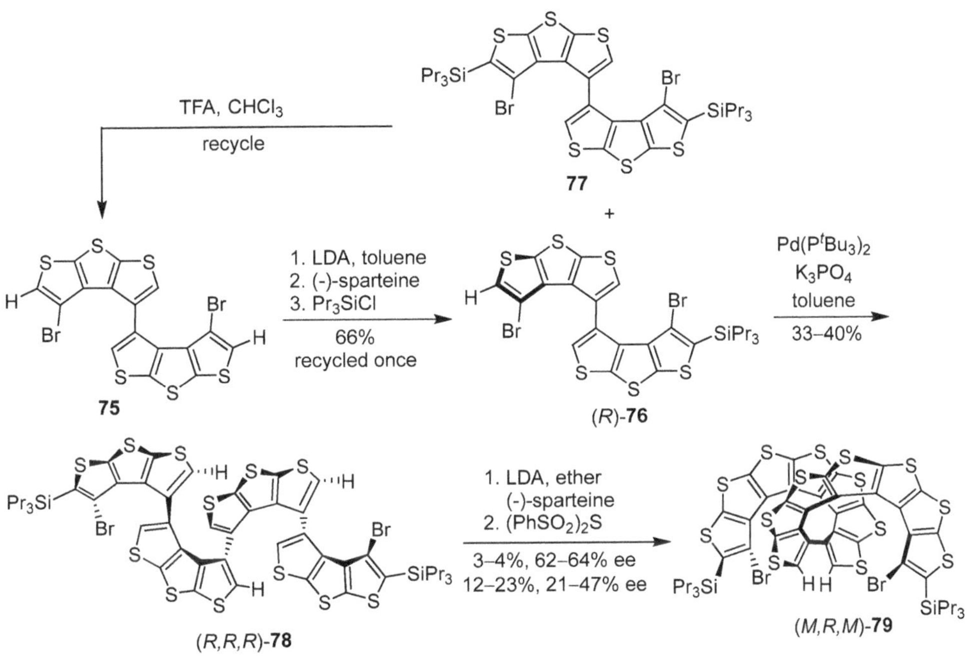

Scheme 15. Synthesis of helically locked bis[7]helicene from helically folded tetrakis(β-trithiophene).

6. Asymmetric Synthesis of Multilayer 3D Chiral Molecules

Multilayer 3D chiral molecules are a new class of macromolecular sandwich-shaped organic materials. They possess a new form of chirality which is different from traditional planar and helical counterparts. They are composed of both planar and axial chirality. Their middle part includes three parallel top, medium, and bottom layers of aromatic (hetero)arene systems, which fold together by an aromatic π-stacking interaction, while the medium aromatic (hetero)arene is linked generally with the top and bottom aromatic arenes, respectively, on its *para*-positions through two naphthalene derivatives, existing in axial chirality. They show a strong luminescence of different colors under UV irradiation and some of them display aggregation-induced emission (AIE) properties. Thus, they exhibit potential applications in chemical, medicinal, and material sciences including optoelectronic materials in the future [59,60].

Racemic multilayer 3D chiral molecules were first prepared from 8-arylnaphthalen-1-amines **80** and 1,2-dibromobenzene (**81**) via the palladium-catalyzed dual Buchwald–Hartwig couplings, generating vicinal *N,N'*-bis(8-arylnaphthalen-1-yl)benzene-1,2-diamines **82** and *ent*-**82**, which exist in an equilibrium of two enantiomeric conformers **82** and *ent*-**82**. However, after the treatment with phosphorus oxychloride in THF, they cyclized into 2-chloro-1,3-bis(8-arylnaphthalene-1-yl)-1,3-dihydrobenzo[*d*][1,3,2]diazaphosphole

2-oxides **83** and *ent*-**83**, existing in two stable enantiomers 83 and *ent*-83, which were resoluted. They were further converted to 2-amino-1,3-bis(8-arylnaphthalene-1-yl)-1,3-dihydrobenzo[*d*][1,3,2]diazaphosphole 2-oxides **84** and **85**, respectively, under sequential treatments with sodium azide and hydrogenolysis in the presence of Pd/C (Scheme 16).

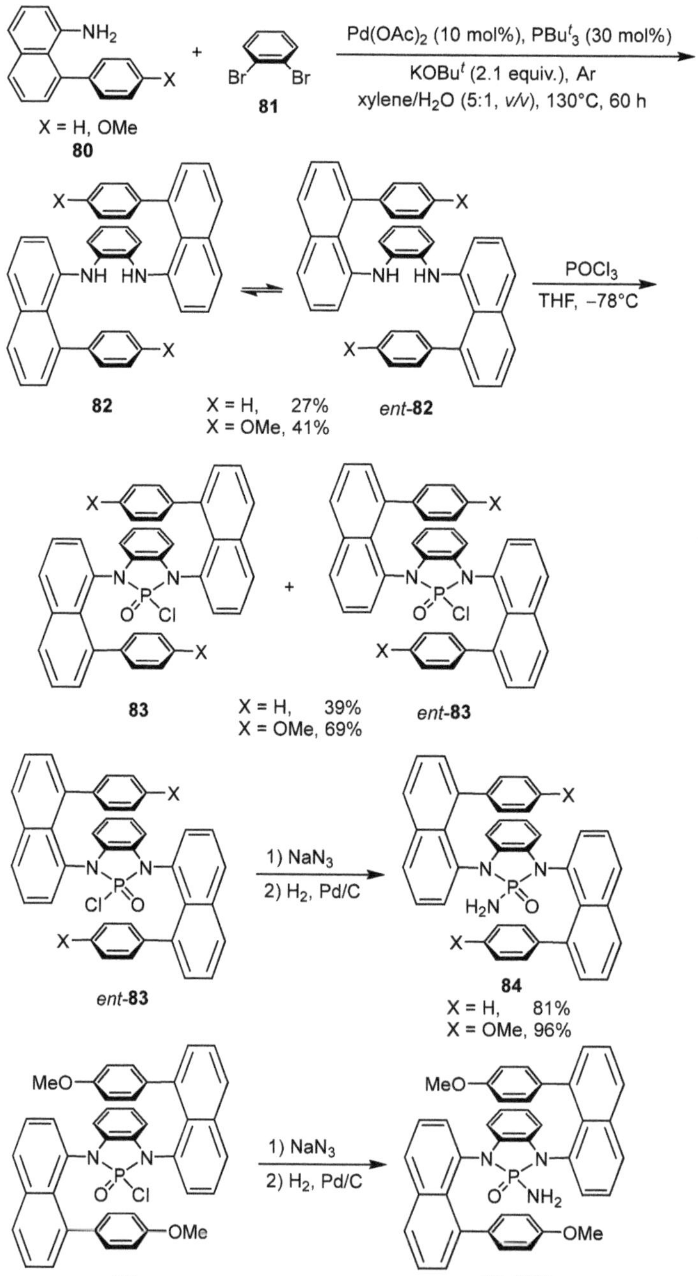

Scheme 16. Racemic synthesis of multilayer 3D chiral molecules.

2-Chloro-1,3-bis(8-(4-methoxyphenyl)naphthalene-1-yl)-1,3-dihydrobenzo[*d*][1,3,2] diazaphos-phole 2-oxide (**83b**) was also transformed into 2-benzylamino-1,3-bis(8-(4-methoxyphenyl)naphthalene-1-yl)-1,3-dihydrobenzo[*d*][1,3,2]diazaphosphole 2-oxide (**86**) and 2-methyl-1,3-bis(8-(4-methoxyphenyl)naphthalene-1-yl)-1,3-dihydrobenzo[*d*][1,3,2] diazaphosphole 2-oxide (**87**), respectively, through the reaction with lithium benzylamide and methyl lithium (Scheme 17) [59].

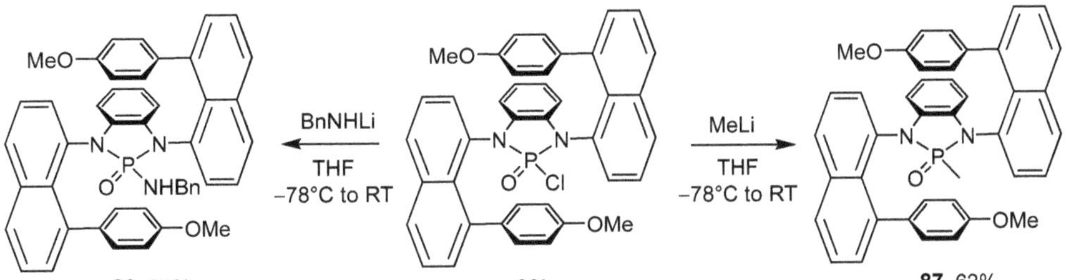

86 81% **83b** **87** 62%

Scheme 17. Derivatization of 2-chloro-1,3-bis(8-(4-methoxyphenyl)naphthalene-1-yl)-1,3-dihydrobenzo[*d*][1,3,2]diazaphosphole 2-oxide.

In the same year, two different strategies for the enantioselective assembly of multilayer 3D chiral compounds were exploited as well. In the first one, 4,7-bis(8-bromonaphthalen-1-yl)benzo[*c*][1,2,5]thiadiazole (**88**) as a core molecule was reacted with various (*R*)-4-carbamoylphenylboronic acids **89**, respectively, via the palladium-catalyzed dual Suzuki–Miyaura coupling to give rise to the desired products **90** in low 16–48% yields in diastereomeric ratios of 1.49:1 to 3.00:1. To improve the yields, the second strategy was attempted. 4,7-Dibromobenzo[*c*][1,2,5]thiadiazole (**91**) as the central molecule was reacted with (*R*)-(8-(4-((1-phenylethyl)carbamoyl)phenyl)naphthalene-1-yl)boronic acid (**92**) to generate the target product **90a** in 53% yield with a diastereomeric ratio of 2.09:1. Alternatively, the reaction of benzo[*c*][1,2,5]thiadiazole-4,7-diyldiboronic acid (**93**) and (*R*)-4-(8-bromonaphthalen-1-yl)-*N*-(1-phenylethyl)benzamide (**94**) gave rise to the expected product **90a** in 60% yield with a diastereomeric ratio of 1.91:1. The yield was improved slightly, but diastereoselectivity decreased (Scheme 18) [60].

To increase the steric bulkiness, different substituted (*R*)-4-(1-phenylethyl) carbamoylphenylboronic acids **95** were applied in the Suzuki–Miyaura coupling with 4,7-bis(8-bromonaphthalen-1-yl)benzo[*c*][1,2,5]thiadiazole (**88**) as a core molecule, giving the desired products **96** in 20–73% yields with diastereomeric ratios of 1.64:1 to 2.17:1. The reaction of 4,9-dibromonaphtho [2,3-*c*][1,2,5]thiadiazole (**97**) and (8-phenylnaphthalen-1-yl)boronic acid (**98**) led to the target product **99** in 70% yield (Scheme 19) [60]. The results indicate that the synthetic strategies involving difunctionalized planar aromatic (hetero)arenes as the central molecules generally result in the formation of the desired products in high yields.

Asymmetric catalytic Suzuki–Miyaura coupling was also developed with diaryl(1-bromonaphthalene-2-yl)phosphine oxides **100** and (7-(8-arylnaphthalen-1-yl)benzo[*c*][1,2,5] thiadiazol-4-yl)boronic acid pinacol esters **101** as starting materials, generating the target products **103** in 40–57% yields with diastereomeric ratios of 58:42 to 87:13 with enantiopure phosphine **102** as chiral ligand (Scheme 20) [61]. Similar asymmetric catalytic Suzuki–Miyaura coupling was also investigated for the synthesis of desired products **96** with a naphthyl group in the middle part instead of the aryl group by displacement of the 8-aryl group with the 8-(naphthalen-1-yl) group in substrates **101** [62].

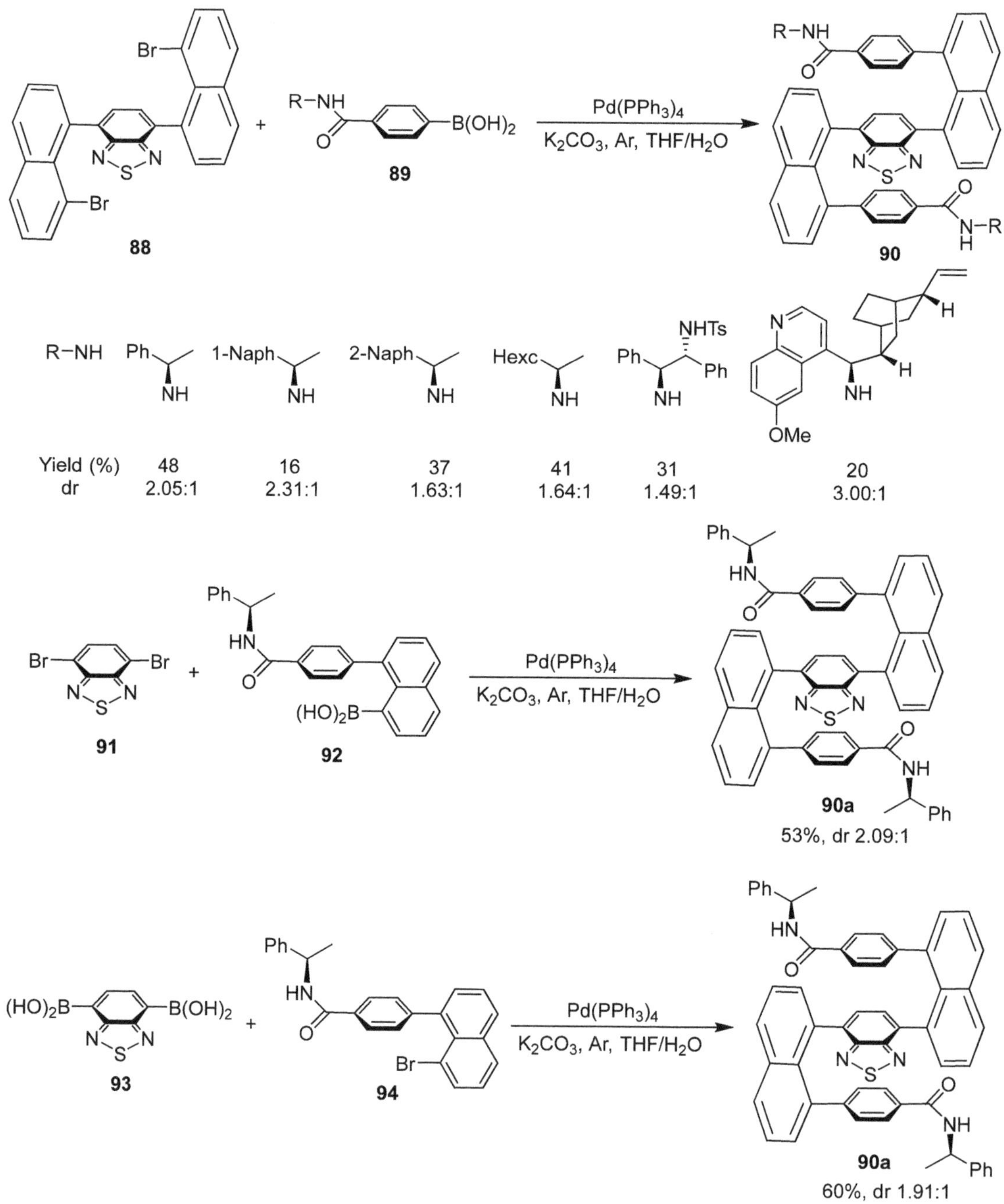

Scheme 18. Enantioselective synthesis of multilayer 3D chiral molecules.

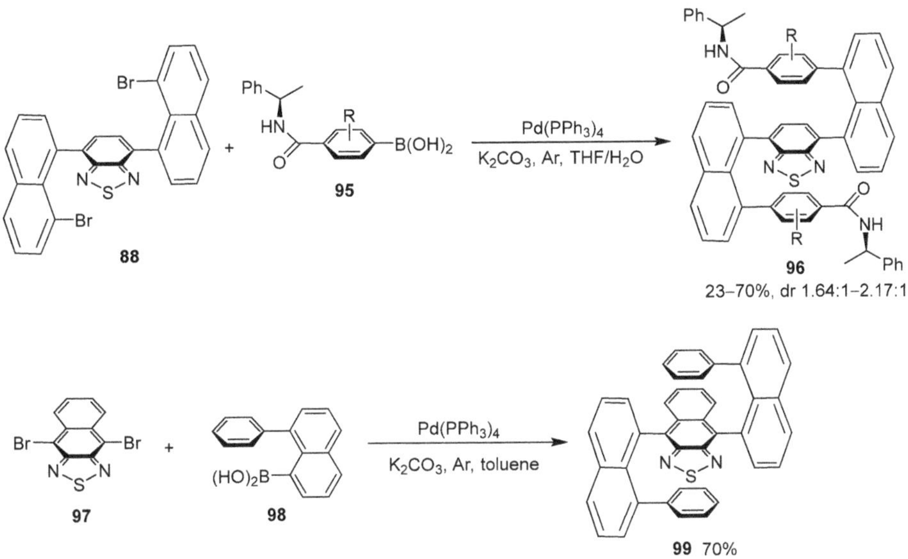

Scheme 19. Stereoselective synthesis of multilayer 3D chiral molecules.

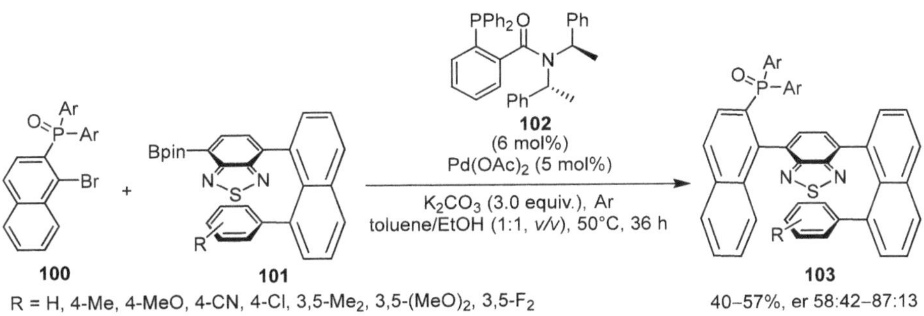

R = H, 4-Me, 4-MeO, 4-CN, 4-Cl, 3,5-Me_2, 3,5-$(MeO)_2$, 3,5-F_2
Ar = Ph, 4-MeC_6H_4, 4-MeC_6H_4, 3,5-C_6H_3

Scheme 20. Asymmetric catalytic synthesis of multilayer 3D chiral molecules.

Chiral amide (*R*)-4-(8-bromonaphthalen-1-yl)-*N*-(1-phenylethyl)benzamide **94** was prepared in a 58% yield from 1,8-dibromonaphthalene (**104**) and (*R*)-(4-((1-phenylethyl)carbamoyl) phenyl)boronic acid (**89a**) via the Suzuki–Miyaura coupling. It further dually coupled with pinacol 2,3-dihydronaphtho[2,3-*b*][1,4]dioxine-5,10-diyldiboronate (**97**) to generate a pair of diastereomeric products **106** and **107** in 2:1 in a 53% total yield. The XRD single crystal diffraction analysis reveals that two naphthalenes are located in opposing directions at a dihedral angle of approximately 60°. For the middle part, the top and bottom layers are nearly parallel to the central layer due to the π-stacking interaction. The phenyl group of one of the amide groups is oriented nearly parallel to the naphthalene ring. In a unit cell, the intermolecular distances between proximate aromatic rings are very similar to those of intramolecular distances due to the π-stacking interaction. The functional group amide was further transferred to nitrile, amide, carbamate, amino, and hydroxyl groups, sequentially (Scheme 21) [63].

Similarly, chiral amide (*R*)-4-(8-bromonaphthalen-1-yl)-*N*-(1-phenylethyl)benzamide (**94**) was dually coupled with pinacol naphtho[2,3-*c*][1,2,5]thiadiazole-4,9-diyldiboronate (**108**) to generate a pair of diastereomeric products **109** and **110** in 1.6:1 in a 45% total yield. The XRD single crystal diffraction analysis indicates that two naphthalenes are located in opposing directions nearly perpendicular to the central anphthothiadiazole ring, different from the previous 2,3-dihydronaphtho[2,3-*b*][1,4]dioxine central one, which exists at a dihedral angle of approximate 60°. Other structural features are very similar to the previous 2,3-dihydronaphtho[2,3-*b*][1,4]dioxine central one. The functional group amide was further transferred to nitrile, amide, and carbamate groups, sequentially (Scheme 22) [64].

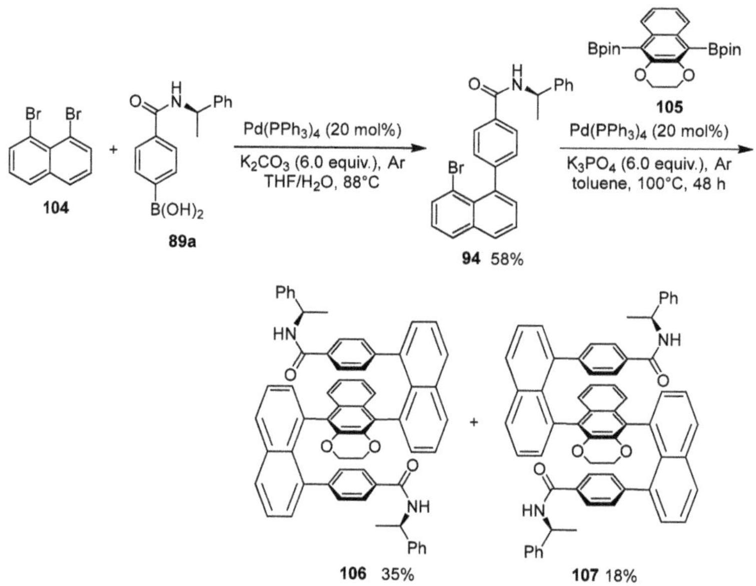

Scheme 21. Asymmetric synthesis of multilayer 3D chiral molecules with arenediboronate as the central moiety.

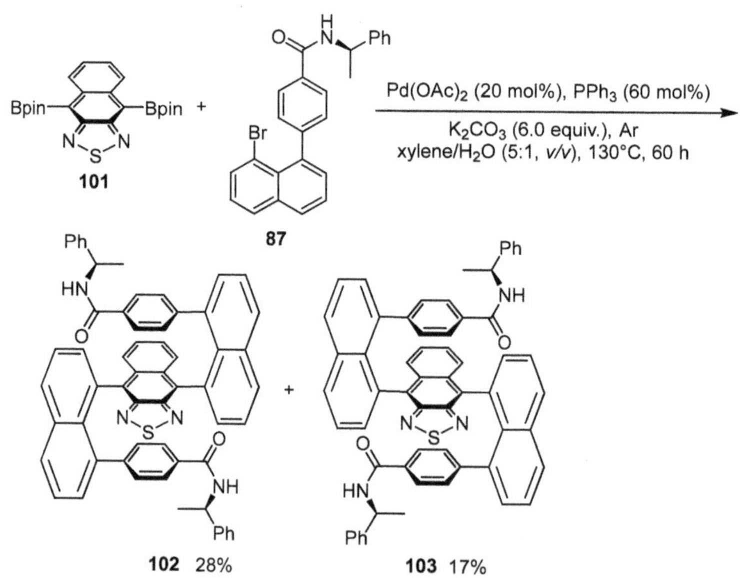

Scheme 22. Asymmetric synthesis of multilayer 3D chiral molecules with heteroarenediboronate as the central moiety.

Recently, a similar synthetic strategy was utilized in the synthesis of polymeric multilayer 3D chiral compounds **115–117** with pinacol (naphtho[2,3-*c*][1,2,5]thiadiazole-4,9-diyl)bis(8-naphthalen-1-ylboronate) and (naphtho[2,3-*c*][1,2,5]selenodiazole-4,9-diyl)bis(8-naphthalen-1-ylboronate) **104** with three different 1,8-dibromonaphthalene derivatives **104**, **112**, and **113** as monomers (Scheme 23) [65]. They are very interesting polymers and oligomers of multilayer 3D chiral compounds and show potential application in materials science.

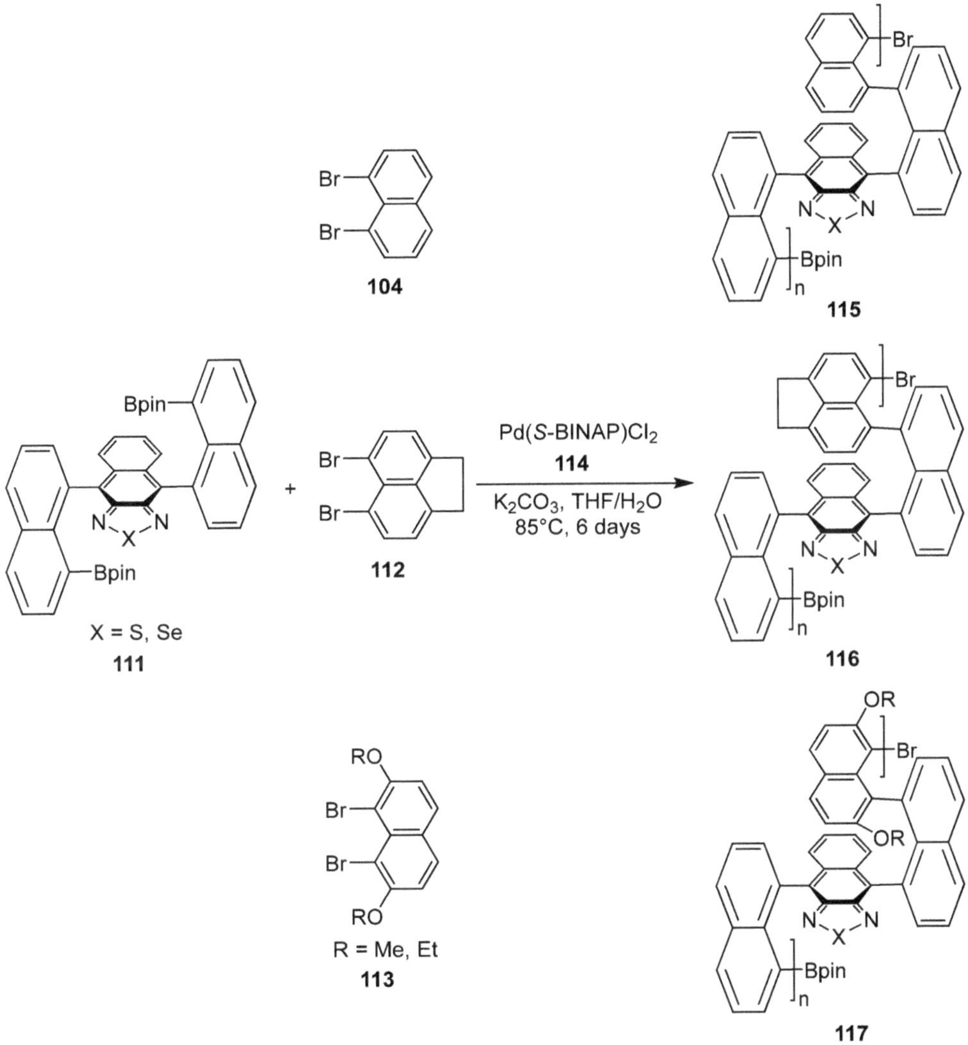

Scheme 23. Asymmetric catalytic synthesis of polymeric multilayer 3D chiral molecules.

7. Conclusions

As one of the most important intramolecular and intermolecular noncovalent interactions in organic chemistry, the π-stacking interaction can exist in different fashions, for instance, in face-to-face, edge-to-face, and even T-shape interactions. It exists widely, such as in biological macromolecules, organic materials, and organic reactions. It plays an important role not only in the stabilization of biological macromolecules and complexations of biomolecules with small organic compounds but also in the stabilization of conformations and transition states in organic reactions via intramolecular and intermolecular attractive interactions. This minireview summarized the recently developed examples of π-stacking interaction-controlled asymmetric synthesis, including auxiliary-induced asymmetric synthesis, kinetic resolution for asymmetric synthesis, diastereoselective synthesis, the asymmetric synthesis of helicenes and heterohelicenes, and the synthesis of multilayer 3D chiral molecules. The π-stacking interaction has been applied in the stabilizations of biomacromolecules, complexations of biomacromolecules and small organic compounds, design of organic materials, organocatalysts, and chiral ligands for asymmetric catalysis. It will show wide applications in understanding the biological function of biomacromolecules and the development of medicines in the future. Before, steric hindrance was considered to be one of the most crucial issues in the design of chiral auxiliaries and catalysts. Recently, steric hindrance, electronic effect, and noncovalent interaction have been recognized as important factors in realizing high stereoselectivity in the design of chiral auxiliaries and catalysts. Recently, several highly efficient asymmetric catalytic reactions have been achieved through the π-stacking interaction between substrates with organocatalysts or chiral ligands under the catalysis of organocatalysts [66–70] and transition metal–chiral ligand complexes [71–73], respectively. The enantiomerization of [5]helicene was also successful under the catalysis of a perylene bisimide cyclophane through the π-stacking interaction [74]. As one of the most important intramolecular and intermolecular noncovalent interactions in organic chemistry, the π-stacking interaction will be paid much attention to in the design of novel chiral auxiliaries for stereocontrol in asymmetric synthesis and in the design of new organocatalysts and chiral ligands for asymmetric catalysis in the preparation of biologically important organic compounds, medicines, and their intermediates in the future.

Funding: This research was funded by the National Key R&D Program of China (No. 2022YFF0709803) and the National Natural Science Foundation of China (Nos. 21772010 and 21911530099).

Institutional Review Board Statement: Not applicable.

Informed Consent Statement: Not applicable.

Data Availability Statement: The data underlying this study are available in the published article.

Conflicts of Interest: The author declares no conflict of interest.

References

1. Hunter, C.J.; Sanders, J.K.M. The nature of π-π interactions. *J. Am. Chem. Soc.* **1990**, *112*, 5525–5534. [CrossRef]
2. Sherrill, C.D. Energy component analysis of π interactions. *Acc. Chem. Res.* **2013**, *46*, 1020–1028. [CrossRef]
3. Rahman, M.M.; Muhseen, Z.T.; Junaid, M.; Zhang, H.J. The aromatic stacking interactions between proteins and their macromolecular ligands. *Curr. Protein Pept. Sci.* **2015**, *16*, 502–512. [CrossRef]
4. Davis, J.T.; Spada, G.P. Supramolecular architectures generated by self-assembly of guanosine derivatives. *Chem. Soc. Rev.* **2007**, *36*, 296–313. [CrossRef]
5. Paliwal, S.; Geib, S.; Wilcox, C.S. Molecular torsion balance for weak molecular recognition forces. Effects of "tilted-T" edge-to-face aromatic interactions on conformational selection and solid-state structure. *J. Am. Chem. Soc.* **1994**, *116*, 4497–4498. [CrossRef]
6. Mo, S.Y.; Xu, J.X. Chemospecific intramolecular Büchner reaction catalyzed by copper(II) acetylacetonate. *ChemCatChem* **2014**, *6*, 1679–1683. [CrossRef]
7. Hein, S.J.; Lehnherr, D.; Arslan, H.; Uribe-Romo, F.J.; Dichtel, W.R. Alkyne benzannulation reactions for the synthesis of novel aromatic architectures. *Acc. Chem. Res.* **2017**, *50*, 2776–2788. [CrossRef]
8. Ma, L.G.; Jiao, P.; Zhang, Q.H.; Du, D.-M.; Xu, J.X. Ligand and substrate pi-stacking interaction controlled enantioselectivity in the asymmetric aziridination. *Tetrahedron Asymmetry* **2007**, *18*, 878–884. [CrossRef]
9. Ma, L.G.; Jiao, P.; Zhang, Q.H.; Xu, J.X. Rigid backbone 1,8-anthracene-linked bis-oxazolines (AnBOXes): Design, synthesis, application and characteristics in catalytic asymmetric aziridination. *Tetrahedron Asymmetry* **2005**, *16*, 3718–3734. [CrossRef]
10. Rezazgui, O.; Boëns, B.; Teste, K.; Vergnaud, J.; Trouillas, P.; Zerrouki, R. One-pot and catalyst-free amidation of ester: A matter of non-bonding interactions. *Tetrahedron Lett.* **2011**, *52*, 6796–6799. [CrossRef]
11. Mo, S.Y.; Li, X.H.; Xu, J.X. In situ generated iodonium ylides as safe carbene precursors for the chemoselective intramolecular Buchner reaction. *J. Org. Chem.* **2014**, *79*, 9186–9195. [CrossRef]
12. Langdon, S.M.; Legault, C.Y.; Gravel, M. Origin of chemoselectivity in N-heterocyclic carbene catalyzed cross-benzoin reactions: DFT and experimental insights. *J. Org. Chem.* **2015**, *80*, 3597–3610. [CrossRef]

13. Liu, T.T.; Tang, S.Y.; Hu, B.; Liu, P.; Bi, S.W.; Jiang, Y.Y. Mechanism and origin of Chemoselectivity of Ru-catalyzed cross-coupling of secondary alcohols to β-disubstituted ketones. *J. Org. Chem.* **2020**, *85*, 12444–12455. [CrossRef]
14. Qi, T.; Fu, S.; Zhang, X.; Liu, T.H.; Li, Q.Z.; Gou, C.A.; Li, J.L. Theoretical insight into the origins of chemo- and diastereo-selectivity in the palladium-catalysed (3+2) cyclisation of 5-alkenyl thiazolones. *Org. Chem. Front.* **2021**, *8*, 6203–6214. [CrossRef]
15. Chen, G.; Zhang, J.; Wu, Y. Synthesis and regiochemistry of spiro indane-1,3-dione compounds. *Res. Chem. Intermediat.* **2012**, *38*, 413–420. [CrossRef]
16. Chen, G.; Yang, J.; Gao, S.; Zhang, Y.; Hao, X.J. Theoretical study of the regioselectivity of the Huisgen reaction. *Res. Chem. Intermediat.* **2013**, *39*, 1245–1250. [CrossRef]
17. Yu, H.; Cao, S.L.; Zhang, L.L.; Liu, G.; Xu, J.X. Synthesis of α-aliphatic and β-aromatic substituted taurines via the regioselective ammonia ring-opening of thiiranes. *Synthesis* **2009**, *13*, 2205–2209. [CrossRef]
18. Wu, Q.Y.; Xu, J.X. Regio- and stereoselective synthesis of thiazoline derivatives via the thioketene-induced ring expansion of aziridines. *Chem. Commun.* **2022**, *58*, 2714–2717. [CrossRef]
19. Shen, B.M.; Chen, Y.; Yu, P.Y. Mechanism and origin of regioselectivity in Rh-catalyzed desymmetric [2+2+2] cycloaddition: Charge versus π-π stacking interaction. *Org. Chem. Front.* **2022**, *9*, 4625–4632. [CrossRef]
20. Jones, G.B.; Chapman, B.J. π–Stacking effect in asymmetric synthesis. *Synthesis* **1995**, *1995*, 475–497. [CrossRef]
21. Xu, J.X. Influence of the electronic effect of catalysts on the enantioselectivity: Applicability and complexity. *Curr. Org. Synth.* **2010**, *7*, 650–676. [CrossRef]
22. Yamada, S. Intramolecular cation-π interaction in organic synthesis. *Org. Biomol. Chem.* **2007**, *5*, 2903–2912. [CrossRef]
23. Fanourakis, A.; Docherty, P.J.; Chuentragool, P.; Phipps, R.J. Recent developments in enantioselective transition metal catalysis featuring attractive noncovalent interactions between ligand and substrate. *ACS Catal.* **2020**, *10*, 10672–10714. [CrossRef]
24. Fustero, S.; Navarro, A.; Pina, B.; Soler, J.G.; Bartolome, A.; Asensio, A.; Simon, A.; Bravo, P.; Fronza, G.; Volonterio, A.; et al. Enantioselective synthesis of fluorinated α-amino acids and derivatives in combination with ring-closing methathesis: Intramolecular π–stacking interactions as a source of stereocontrol. *Org. Lett.* **2001**, *3*, 2621–2624. [CrossRef]
25. García Ruano, J.L.; Alemán, J.; Catalán, S.; Marcos, V.; Monteagudo, S.; Parra, A.; del Pozo, C.; Fustero, S. Anionic–anionic asymmetric tandem reactions: One-pot synthesis of optically pure fluorinated indolines from 2-*p*-tolylsulfinyl alkylbenzenes. *Angew. Chem. Int. Ed.* **2008**, *47*, 7941–7944. [CrossRef]
26. Garcia Ruano, J.L.; Alemán, J.; Soriano, J.F. Facile synthesis of optically pure 1,2-diaryl (and 1-alkyl-2-aryl) ethyl and propylamines. *Org. Lett.* **2003**, *5*, 677–680. [CrossRef]
27. Garcia Ruano, J.L.; Alemán, J.; Parra, A. Highly stereoselective benzylation of *N*-sulfinylketimines. *J. Am. Chem. Soc.* **2005**, *127*, 13048–13054. [CrossRef]
28. Garcia Ruano, J.L.; Parra, A.; Marcos, V.; del Pozo, C.; Catalan, S.; Monteagudo, S.; Fustero, S.; Poveda, A. Asymmetric synthesis of indolines through intramolecular shifting of aromatic sulfinyl groups. Role of the π-stacking interactions in these unusual S_NAr processes. *J. Am. Chem. Soc.* **2009**, *131*, 9432–9441. [CrossRef]
29. Corne, V.; Sarotti, A.M.; de Arellano, C.R.; Spanevello, R.A.; Suarez, A.G. Experimental and theoretical insights in the alkene-arene intramolecular π-stacking interaction. *Beilstein J. Org. Chem.* **2016**, *12*, 1616–1623. [CrossRef]
30. Klepp, J.; Sumby, C.J.; Greatrex, B.W. Synthesis of a chiral auxiliary family from levoglucosenone and evaluation in the Diels–Alder reaction. *Synlett* **2018**, *29*, 1441–1446.
31. Njiojob, C.N.; Bozell, J.J.; Long, B.K.; Elder, T.; Key, R.E.; Hartwig, W.T. Enantioselective synthesis of lignin models: An efficient synthesis of β-O-4 dimers and trimers by using the Evans chiral auxiliary. *Chem. Eur. J.* **2016**, *22*, 12506–12517. [CrossRef]
32. Xu, J.X.; Wei, T.Z.; Zhang, Q.H. Effect of temperature on the enantioselectivity in the oxazaborolidine-catalyzed asymmetric reduction of ketones. non-catalytic borane reduction, a non-neglectable factor in reduction system. *J. Org. Chem.* **2003**, *68*, 10146–10151. [CrossRef]
33. Xu, J.X.; Wei, T.Z.; Zhang, Q.H. Influences of electronic effects and anions on the enantioselectivity in the oxazaborolidine-catalyzed asymmetric borane reduction of ketones. *J. Org. Chem.* **2004**, *69*, 6860–6866. [CrossRef]
34. Xu, J.X.; Wei, T.Z.; Xia, J.K.; Zhang, Q.H.; Wu, H.S. Preparation of highly optically pure homochiral sulfide-containing alcohols via oxazaborolidine-catalyzed asymmetric borane reduction of ketones. *Chirality* **2004**, *16*, 341–346. [CrossRef]
35. Su, X.B.; Zhang, Q.H.; Wu, Y.Q.; Xu, J.X. Preparation of linear secondary alcohols of high optical purity through chemical resolution with chiral α-phenylethylamines. *Chin. J. Org. Chem.* **2002**, *22*, 496–500.
36. Zhang, Q.H.; Lin, C.X.; Zhang, S.S.; Xu, J.X. Preparation of optically active *trans*-1-(4-alkylcyclohexyl) alcohols. *Chin. J. Org. Chem.* **2004**, *24*, 1069–1074.
37. Birman, V.B.; Uffman, E.W.; Jiang, H.; Li, X.; Kilbane, C.J. 2,3-Dihydroimidazo[1,2-*a*]pyridines: A new class of enantioselective acyl transfer catalysts and their use in kinetic resolution of alcohols. *J. Am. Chem. Soc.* **2004**, *126*, 12226–12227. [CrossRef] [PubMed]
38. Birman, V.B.; Jiang, H. Kinetic resolution of alcohols using a 1,2-dihydroimidazo[1,2-*a*]quinolone enantioselective acylation catalyst. *Org. Lett.* **2005**, *7*, 3445–3447. [CrossRef] [PubMed]
39. Xu, J.X. Stereoselectivity in the synthesis of 2-azetidinones from ketenes and imines via the Staudinger reaction. *ARKIVOC* **2009**, *9*, 21–44. [CrossRef]
40. Jiao, L.; Zhang, Q.F.; Liang, Y.; Zhang, S.W.; Xu, J.X. A versatile method for the synthesis of 3-alkoxycarbonyl β-lactam derivatives. *J. Org. Chem.* **2006**, *71*, 815–818. [CrossRef] [PubMed]

41. Liang, Y.; Jiao, L.; Zhang, S.W.; Xu, J.X. Microwave and photoirradiation-induced Staudinger reactions of cyclic imines and ketenes generated from α-diazoketones. A further investigation into the Stereochemical process. *J. Org. Chem.* **2005**, *70*, 334–337. [CrossRef] [PubMed]
42. Jiao, L.; Liang, Y.; Xu, J.X. Origin of the relative stereoselectivity of the β-lactam formation in the Staudinger reaction. *J. Am. Chem. Soc.* **2006**, *128*, 6060–6069. [CrossRef]
43. Wang, Y.K.; Liang, Y.; Jiao, L.; Du, D.-M.; Xu, J.X. Do reaction conditions affect the stereoselectivity in the Staudinger reaction? *J. Org. Chem.* **2006**, *71*, 6983–6990. [CrossRef]
44. Li, B.N.; Wang, Y.K.; Du, D.-M.; Xu, J.X. Notable and obvious ketene substituent-dependent effect of temperature on the stereoselectivity in the Staudinger reaction. *J. Org. Chem.* **2007**, *72*, 990–997. [CrossRef]
45. Liang, Y.; Jiao, L.; Zhang, S.W.; Yu, Z.X.; Xu, J.X. New insights into the torquoselectivity of the Staudinger reaction. *J. Am. Chem. Soc.* **2009**, *131*, 1542–1549. [CrossRef]
46. Xu, C.C.; Xu, J.X. Synthetic applications of oxiranecarbonitriles. *Chem. Heterocycl. Compd.* **2021**, *57*, 731–733. [CrossRef]
47. Xu, C.C.; Xu, J.X. Oxygenophilic Lewis acid-promoted synthesis of 2-arylindoles from anilines and cyanoepoxides in alcohol. *J. Org. Chem.* **2018**, *83*, 14733–14742. [CrossRef]
48. Xu, C.C.; Xu, J.X. $BF_3{\cdot}OEt_2$-promoted tandem Meinwald rearrangement and nucleophilic substitution of oxiranecarbonitriles. *Org. Biomol. Chem.* **2020**, *18*, 127–134. [CrossRef]
49. Xu, C.X.; Lu, Y.; Xu, K.N.; Xu, J.X. $BF_3{\cdot}OEt_2$-Catalyzed synthesis of *anti*-β-(*N*-arylamino)-α-hydroxynitriles by regio- and diastereospecific ring-opening of 3-aryloxirane-2-carbonitriles with anilines. *Synthesis* **2020**, *52*, 602–608. [CrossRef]
50. Xu, C.C.; Xie, W.L.; Xu, J.X. Metal-free and regiospecific synthesis of 3-arylindoles. *Org. Biomol. Chem.* **2020**, *18*, 2661–2671. [CrossRef]
51. Zhang, Y.J.; Shi, S.J.; Yang, Z.H. Thiourea-mediated stereospecific deoxygenantion of cyanoepoxides to access highly diastereopure alkenyl nitriles. *J. Org. Chem.* **2024**, *89*, 2748–2758. [CrossRef] [PubMed]
52. D'hooghe, M.; Ha, H.-J. *Synthesis of 4-to 7-Membered Heterocycles by Ring Expansion*; Springer: Cham, Switzerland, 2016.
53. Chen, X.P.; Huang, Z.S.; Xu, J.X. Catalyst-free electrophilic ring expansion of *N*-unprotected aziridines with α-oxoketenes to efficient access 2-alkylidene-1,3-oxazolidines. *Adv. Synth. Catal.* **2021**, *363*, 3098–3108. [CrossRef]
54. Lei, Y.L.; Xu, J.X. Efficient synthesis of ethyl 2-(oxazolin-2-yl)alkanoates via ethoxycarbonylketene-Induced electrophilic ring expansion of aziridines. *Beilstein J. Org. Chem.* **2022**, *18*, 70–76. [CrossRef] [PubMed]
55. Wang, Y.Q.; Chen, Y.X.; Xu, J.X. π-Stacking-controlled dearomatic sulfur-shifted ene reaction of ketenes and polycyclic arylthiiranes: Access to areno[*d*]-ε-thiolactones. *J. Org. Chem.* **2024**, *89*. [CrossRef]
56. Sehnal, P.; Stará, I.G.; Šaman, D.; Tichý, M.; Mišek, J.; Cvačka, J.; Rulišek, L.; Chocholoušová, J.; Vacek, J.; Goryl, G.; et al. An organometallic route to long helicenes. *Proc. Nat. Acad. Sci. USA* **2009**, *106*, 13169–13174. [CrossRef]
57. Miyasaka, M.; Pink, M.; Rajca, S.; Rajca, A. Noncovalent interactions in the asymmetric synthesis of rigid, conjugated helical structures. *Angew. Chem. Int. Ed.* **2009**, *48*, 5954–5957. [CrossRef]
58. Rajca, A.; Miyasaka, M.; Pink, M.; Wang, H.; Rajca, S. Helically annelated and cross-conjugated oligothiophenes: Asymmetric synthesis, resolution, and characterization of a carbon-sulfur [7]helicene. *J. Am. Chem. Soc.* **2004**, *126*, 15211–15222. [CrossRef]
59. Liu, Y.X.; Wu, G.Z.; Yang, Z.; Rouh, H.; Katakam, N.; Ahmed, S.; Unruh, D.; Cui, Z.H.; Lischka, H.; Li, G.G. Multi-layer 3D chirality: New synthesis, AIE and computional studies. *Sci. China Chem.* **2020**, *63*, 692–698. [CrossRef]
60. Wu, G.Z.; Liu, Y.X.; Yang, Z.; Jiang, T.; Katakam, N.; Rouh, H.; Ma, L.L.; Tang, Y.; Ahmed, S.; Rahman, A.; et al. Enantioselective assembly of multi-layer 3D chirality. *Nat. Sci. Rev.* **2020**, *7*, 588–599. [CrossRef]
61. Wu, G.Z.; Liu, Y.X.; Rouh, H.; Ma, L.L.; Tang, Y.; Zhang, S.; Zhou, P.; Wang, J.-Y.; Jin, S.Z.; Unruh, D.; et al. Asymmetric catalytic approach to multilayer 3D chirality. *Chem. Eur. J.* **2021**, *27*, 8013–8020. [CrossRef]
62. Liu, H.X.; Rouh, H.; Tang, Y.; Wu, G.Z.; Yuan, Q.K.; Zhang, S.; Wang, J.Y.; Jin, S.Z.; Xu, T.; Wang, Y.; et al. Enantio- and diastereoselective assembly of multi-layer folding chiral targets via asymmetric catalytic single C–C bond formation. *Synlett* **2023**, *34*, 153–158. [CrossRef]
63. Tang, Y.; Wu, G.Z.; Jin, S.W.; Liu, Y.X.; Ma, L.L.; Zhang, S.; Rouh, H.; Ali, A.I.M.; Wang, J.-Y.; Xu, T.; et al. From center-to-multilayer chirality: Asymmetric synthesis of multilayer targets with electron-rich bridges. *J. Org. Chem.* **2022**, *87*, 5976–5986. [CrossRef]
64. Jin, S.W.; Wang, J.-Y.; Tang, Y.; Rouh, H.; Zhang, S.; Xu, T.; Wang, Y.; Yuan, Q.K.; Chen, D.X.; Unruh, D.; et al. Central-to-folding chirality control: Asymmetric synthesis of multilayer 3D targets with electron-deficient bridges. *Front. Chem.* **2022**, *10*, 860398. [CrossRef]
65. Tang, Y.; Jin, S.Z.; Zhang, S.; Wu, G.-Z.; Wang, J.-Y.; Xu, T.; Wang, Y.; Unruh, D.; Surowiec, K.; Ma, Y.Y.; et al. Multilayer 3D chiral folding polymers and their asymmetric catalytic assembly. *Research* **2022**, *2022*, 9847949. [CrossRef]
66. Kanomata, K.; Toda, Y.; Shibata, Y.; Yamanaka, M.; Tsuzuki, S.; Gridneva, I.D.; Terada, M. Secondary stereocontrolling interactions in chiral Brønsted acid catalysis: Study of a Petasis–Ferrier-type rearrangement catalyzed by chiral phosphoric acids. *Chem. Sci.* **2014**, *5*, 3515–3523. [CrossRef]
67. Yang, H.B.; Zhao, Y.Z.; Sang, R.; Wei, Y.; Shi, M. Asymmetric synthesis of bioxindole-substituted hexahydrofuro[2,3-*b*]furans via hydroquinine anthraquinone-1,4-diyl diether-catalyzed domino annulation of acylidenoxindoles/isatins, acylidenoxindoles and allenoates. *Adv. Synth. Catal.* **2014**, *356*, 3799–3808. [CrossRef]

68. Machuca, E.; Juaristi, E. Organocatalytic activity of α,α-dipeptide derivatives of (*S*)-proline in the asymmetric aldol reaction in absence of solvent. Evidence for non-covalent π–π interactions in the transition state. *Tetrahedron Lett.* **2015**, *56*, 1144–1148. [CrossRef]
69. Bai, H.-Y.; Tan, F.-X.; Liu, T.-Q.; Zhu, G.-D.; Tian, J.-M.; Ding, T.-M.; Chen, Z.-M.; Zhang, S.-Y. Highly atroposelective synthesis of nonbiaryl naphthalene-1,2-diamine N-C atropisomers through direct enantioselective C-H amination. *Nat. Commun.* **2019**, *10*, 3063. [CrossRef]
70. Zhou, Q.-Y.; Xin Li, X. Atroposelective construction of axially chiral enamides via *N*-allylic alkylation. *Chem. Commun.* **2022**, *58*, 4727–4730. [CrossRef] [PubMed]
71. Zhou, Y.-Y.; Li, J.; Ling, L.; Liao, S.-H.; Sun, X.-L.; Li, Y.-X.; Wang, L.-J.; Tang, Y. Highly enantioselective [3+3] cycloaddition of aromatic azomethine imines with cyclopropanes directed by π–π stacking interactions. *Angew. Chem. Int. Ed.* **2013**, *52*, 1452–1456. [CrossRef]
72. Jin, M.Y.; Zhen, Q.Q.; Xiao, D.M.F.; Tao, G.Y.; Xing, X.Y.; Yu, P.Y.; Chen Xu, C. Engineered non-covalent π interactions as key elements for chiral recognition. *Nat. Commun.* **2022**, *13*, 3276. [CrossRef] [PubMed]
73. Baek, D.; Ryu, H.; Hahm, H.; Lee, J.; Hong, S. Palladium catalysis featuring attractive noncovalent interactions enabled highly enantioselective access to β-quaternary δ-lactams. *ACS Catal.* **2022**, *12*, 5559–5564. [CrossRef]
74. Weh, M.; Kroeger, A.A.; Shoyama, K.; Grüne, M.; Karton, A.; Würthner, F. π-π Catalysis made asymmetric—Enantiomerization catalysis mediated by the chiral π-system of a perylene bisimide cyclophane. *Angew. Chem. Int. Ed.* **2023**, *62*, e202301301. [CrossRef] [PubMed]

molecules

MDPI

Review

Recent Advances in the Utilization of Chiral Covalent Organic Frameworks for Asymmetric Photocatalysis

Peng Liu [1], Weijun Dai [2], Xianfu Shen [1], Xiang Shen [1] Yuxiang Zhao [1,*] and Jian-Jun Liu [1,*]

[1] College of Chemistry and Environmental Science, Qujing Normal University, Qujing 655011, China; liupengxjlp@163.com (P.L.); xianfu_shen@163.com (X.S.); sx008100@163.com (X.S.)

[2] School of Ethnic Medicine, Yunnan Minzu University, Kunming 650504, China; dwj1015@163.com

* Correspondence: zhaoyuxiang@mail.qjnu.edu.cn (Y.Z.); jjliu302@163.com (J.-J.L.)

Abstract: The use of light energy to drive asymmetric organic transformations to produce high-value-added organic compounds is attracting increasing interest as a sustainable strategy for solving environmental problems and addressing the energy crisis. Chiral covalent organic frameworks (COFs), as porous crystalline chiral materials, have become an important platform on which to explore new chiral photocatalytic materials due to their precise tunability, chiral structure, and function. This review highlights recent research progress on chiral COFs and their crystalline composites, evaluating their application as catalysts in asymmetric photocatalytic organic transformations in terms of their structure. Finally, the limitations and challenges of chiral COFs in asymmetric photocatalysis are discussed, with future opportunities for research being identified.

Keywords: covalent organic frameworks; asymmetric photocatalysis; chirality; synthetic method

1. Introduction

Photocatalysis is widely accepted as a convenient method in the field of asymmetric catalysis and has shown promise for bulk production [1–6]. Since MacMillan's group reported in 2008 that photoredox catalysis combined with organocatalysis promotes asymmetric α-alkylation of aldehydes through a process catalyzed by ruthenium complexes [7], efforts have been made to develop more environmentally benign asymmetric photocatalysts and to extend their application to more meaningful reactions [8–15]. In recent years, hybrid photocatalysts consisting of inorganic semiconductor photocatalysts such as $PbBiO_2Br$ and $[W_{10}O_{32}]^{4-}$ and chiral organic catalysts have shown excellent photocatalytic performance in stereoselective carbon–carbon bond formation [16,17]. These materials combine the advantages of the high stereoselectivity of organic catalysis with the stability and easy separation of heterogeneous catalysts [18,19]. However, related reports are still very limited, and designing new photocatalyst systems for asymmetric photochemical reactions is challenging.

In the past few years, an increasing number of studies based on reticular framework materials and their composites have been published in the field of photocatalysis, and their extraordinary features have attracted great attention [20–22]. Metal–organic frameworks (MOFs) are a promising reticular framework material for photocatalysis because they are composed of metal ions or clusters and organic ligands, which can act as active sites separately or cooperatively, affecting the light-harvesting, charge generation and separation, and surface reactions for remarkable photocatalytic performances [23–25]. At present, a large number of reviews focused on the application of MOFs in the field of photocatalysis have been reported [26–28]. Unlike MOFs constructed by coordination bonds, covalent organic frameworks (COFs) are a novel class of organic reticular framework materials, which are composed of organic building units linked only through covalent bonds, and they exhibit well-defined structures and permanent porosity. Crystalline COF materials hold potential for applications in energy storage, optoelectronics, catalysis, gas storage and separation, and drug delivery [29–36]. It is well known that the chirality of molecules can be amplified by a reticular structure [37,38]. In addition, the enantioselectivity of asymmetric transformations can be improved by fixing chirality into constrained framework spaces. Therefore, introducing chiral functional groups and photoactive units into COFs through a rational selection of organic monomers can pave the way for chiral functionalities to have viable applications, such as in asymmetric photocatalysis.

In 2014, Jiang and co-workers reported the first chiral COF and its application in asymmetric catalysis [39]. Since then, numerous chiral COFs have been created, and their diverse applications in asymmetric catalysis, chiral separation, and enantioselective sensing have been investigated [40–44]. Recently, Cui et al. published two seminal reviews on chiral COFs [45,46], concentrating on their synthesis, structural features, and applications. Although a large number of examples of chiral COFs and some reviews of their applications have been reported, the construction of chiral COFs with chiral photocatalysts is still a challenge due to the need to consider both chiral catalysis and photocatalysis and their synergistic relationship. To date, a comprehensive review outlining the synthesis and structures of chiral COFs and their applications in asymmetric photocatalytic organic transformations has not been published.

This review mainly focuses on summarizing the recent research progress on chiral COFs and their crystalline composites for asymmetric photocatalytic organic transformations evaluated in terms of their structure. The photochemical properties and asymmetric photocatalytic performance of these materials are illustrated to provide a useful guide for the preparation and application of chiral COFs in the future.

2. Synthesis and Structure of Chiral COFs

Since Yaghi's group first reported COF powders in 2005, many research groups have employed a variety of synthetic methods to prepare COFs [47]. To date, COFs are mainly synthesized by solvothermal [48–50], microwave-assisted [51,52], ionothermal [53,54], flux synthesis [55], and room temperature methods [56–58]. Despite the success of these approaches, the first chiral COF was not synthesized until 2014 [39]. A major challenge in chiral COFs is balancing the asymmetry of chiral monomers and the crystallinity of materials because of the low-symmetry nature of chiral monomers, which contradicts with the observations that most COFs are crystallized in high-symmetry space groups, with multiple crystallographic symmetry elements residing on each monomer. Moreover, in the preparation of chiral COFs with specific functions, the solvent and temperature are crucial factors in obtaining crystalline products.

To date, all chiral COFs have been synthesized by the solvothermal method. However, the solvothermal synthesis of chiral COFs can also be categorized into the following three different synthesis approaches [59–64]. First, in de novo synthesis, homochiral COFs are prepared using either enantiopure monomers or achiral monomers with chiral crosslinking units. Second, in a post-synthesis approach, achiral COFs are modified at the level of organic monomers or by inserting chiral substances into the pores. Third, in a chiral induction approach, achiral monomers are used as precursors in the chiral catalytic or chirality-induced synthesis of homochiral COFs. In this section, we briefly introduce the various strategies and synthetic conditions for producing chiral COFs and describe the relationship between their structure and properties. Figure 1 shows the structures of chiral monomers used to construct chiral COFs via the de novo synthesis method.

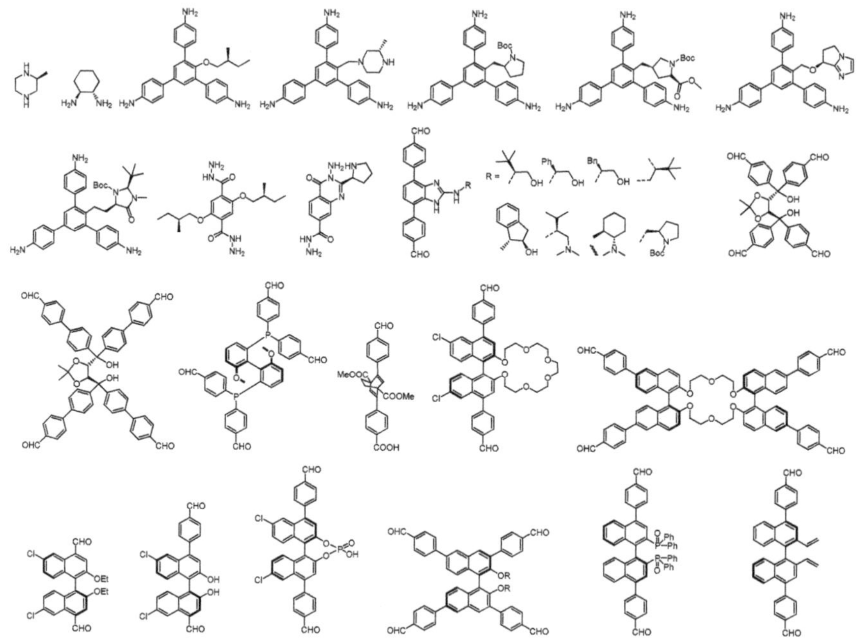

Figure 1. The chemical structures of chiral monomers used in the construction of chiral COFs.

2.1. *De Novo Synthesis*

Constructing predesigned chiral COFs through the polymerization of rationally selected enantiopure monomers, guided by framework chemistry design rules, is a promising yet challenging approach. The inherent chirality of the building blocks is preserved in the overall framework, maintaining the absolute configuration of the resulting homochiral structures [65]. The inherent chirality of the building blocks is conserved in the absolute configuration of the resulting homochiral framework, and chiral COFs synthesized by de novo synthesis exhibit a well-defined structure and uniformly distributed chiral sites. Chiral COFs can be categorized into two structural types: (1) a structure in which the framework itself is chiral and (2) a structure in which the framework is achiral but the substituents on the framework are chiral. However, both of them crystallize in chiral space groups.

In 2016, Wang and colleagues reported the first de novo synthesis process [66], preparing two chiral COFs, LZU-72 and LZU-76, which crystallized in the *P*3 space group and exhibited a 2D layered hexagonal network with high crystallinity. The frameworks of these COFs are achiral but the substituents on the framework are chiral. The chiral monomer was synthesized from rigid 4,4′-(1*H*-benzo[*d*]imidazole-4,7-diyl)dianiline with chiral pyrrolidine substituents. The resulting chiral COFs are structurally robust and highly active as heterogeneous catalysts in asymmetric aldol reactions. Subsequently, Wang's group explored a general method for introducing different chiral functional groups into an achiral COF [67]. They used the achiral molecule DBCBI (4,7-dibromo-2-chloro-1H-benzo[*d*]imidazole) as a platform on which to immobilize eight chiral functional groups via nucleophilic substitutions and Suzuki coupling to obtain eight chiral monomers. These chiral monomers were condensed with 1,3,5-tri(4-aminophenyl) benzene (TAPB) to produce eight 2D chiral COFs, which crystallize in the *P*3 space group with a hexagonal network. The high-throughput synthesis of chiral COFs from a platform molecule not only provided eight catalysts for the asymmetric reaction of β-ketone esters, but also offered a general method for structure–activity studies of COFs.

In 2016, Yan et al. reported a bottom-up strategy for the synthesis of chiral COFs [68]. First, they prepared a C_3-symmetrical 1,3,5-triformylphloroglucinol (Tp) to bind linear chiral carboxylic acids. Then, this chiral building block was directly condensed with benzidine (BD), 1,4-phenylenediamine (Pa-1), and 2,5-dimethylp-phenylenediamine (Pa-2) affording the 2D hexagonal chiral COFs CTpBD, CTpPa-1, and CTpPa-2, respectively. On this basis, they used an in situ growth approach to prepare chiral COF-bound capillary columns for chiral separation.

Another report by Zhang's group described a hydrazone-linked chiral COF based on 1,3,5-benzene-tricarboxaldehyde with chiral (*S*)-2-methylbutoxy groups [69] that showed good chemical stability and high performance for separating enantiomers. At the same time, Liu, Cui, and co-workers explored a multivariate method to prepare several multi-component 2D homochiral COFs based on 1,3,5-tris(4-aminophenyl)benzene with chiral L-imidazolidine and L-imidazolidine side chain modifications on the central aromatic ring [70]. These chiral COFs were successfully obtained by the multi-component crystallization of two dialdehydes and triamines with or without chiral organic catalysts.

Using a similar strategy, Dong's group synthesized a chiral COF in 2017 by directly condensing the chiral monomer of S-(+)-2-methylpiperazine (S-MP) and cyanuric chloride [71]. The cyanuric chloride is connected by S-MP and extends on the *ab* plane to form a 2D layer network (CCOF-MPC) with hexagonal cavities (Figure 2). The chirality of this chiral COF is derived from the chiral S-MP affixed to the inner wall of the COF channel. The crystalline CCOF-MPC is then subjected to $Pd(NO_3)_2$ solution impregnation and metal reduction steps to obtain the final metal nanoparticle-loaded chiral composite (Pd@CCOF-MPC). Pd@CCOF-MPC can effectively promote the Henry reaction and reductive Heck reaction in a high yield with excellent stereoselectivity.

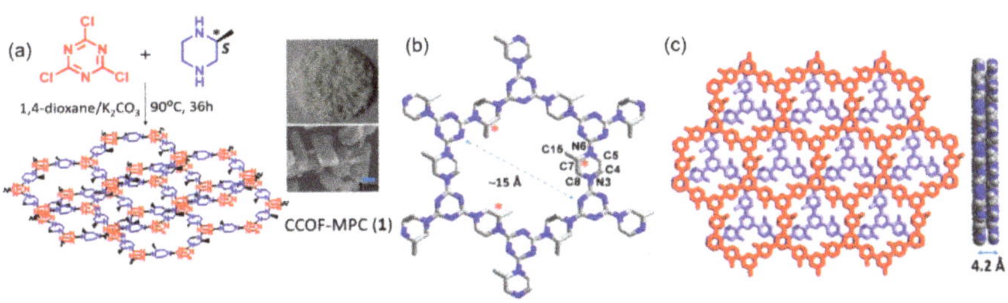

Figure 2. (**a**) A schematic diagram of the preparation of CCOF-MPC; (**b**) a single-layer hexagonal building unit of CCOF-MPC; (**c**) the stacking patterns of CCOF-MPC. Adapted from ref. [71] with permission, copyright 2017, the American Chemical Society.

Chiral COFs are prepared using achiral skeleton molecules. By nucleophilic substitution or condensation, the chiral centers are modified or attached to the surface or inner wall of the COFs, but the COF framework remains achiral. To further release the chiral expression potential of the materials, in 2018, Cui's group reported two chiral COFs with chiral scaffolds using enantiopure monomers TADDOL (tetraaryl-1,3-dioxolane-4,5-dimethanol) and tetra(4-anilyl)methane (TAM) [72]. The TADDOL-based *tetra*-aldehydes react with a tetrahedral TAM to form 3D imine-based chiral COFs with permanent porosity and high crystallinity. Powder X-ray diffraction, computer simulation, and an aperture distribution analysis indicated that both chiral COFs crystallized in the *P*2 space group and can be used as chiral stationary phases for high-performance liquid chromatographic enantioseparation.

BINOL is a crucial chiral compound in organic chemistry and material science. However, it was not until Cui's group synthesized two BINOL-based chiral COFs that the application of chiral COFs was realized. They achieved this by designing enantiopure BINOL-based dialdehydes with 1,3,5-tris(3,5-diisopropyl-4-aminophenyl)benzene (CCOF 7) and tetrakis(4-aminophenyl)ethane (CCOF 8) [62]. In this work, CCOF 7 crystallized in the *C*2 space group with 2D layered tetragonal networks. CCOF 8 crystallized in the *R*3 space group, exhibiting a staggered stacking mode of 2D hexagonal layered frameworks. Notably, due to weak interlayer interactions, CCOF 7 can be easily exfoliated into two-dimensional sheets and exhibits superior capabilities for the detection of chiral odor vapors in both the solution and membrane. In 2021, the same group reported two 3D chiral COFs that have 9-fold or 11-fold interpenetrated diamondoid frameworks by condensing the tetrahedral tetramine and enantiopure BINOL [73]. The chiral COFs demonstrated exceptional asymmetric catalytic activities for the cyclocondensation of aldehydes and anthranilamides into 2,3-dihydroquinazolinones. This is attributed to chiral BINOL units that are uniformly distributed within the tubular channels, which facilitate preferential secondary interactions between the framework and substrate, thereby inducing enantioselectivity.

In addition, Dong et al. synthesized a chiral COF using a gold *N*-heterocyclic carbene (NHC-Au) building block and chiral secondary amine through a modified de novo synthesis approach [74]. The chiral COF (*S*)-NHC-Au-SA-COF is a heterogeneous catalyst for the asymmetric aryl methanol oxidation-aldol relay reaction. Moreover, this chiral COF-based setup was developed using a facile templating freeze-drying method based on an eco-friendly chitosan material, promoting the highly asymmetric aerobic alcohol oxidation-aldol relay reaction on a gram scale.

Recently, Cui's group designed two 3D chiral COFs via the polycondensation of a chiral 1,1′-binaphthyl-20-crown-6-derived dialdehyde and tetraamines with diisopropyl substituents (Figure 3a) [75]. Structural characterization showed that both COFs had 11-fold interpenetrating diamond frameworks characterized by tubular open channels decorated with chiral crown ethers as enantioselective recognition and binding sites (Figure 3b). These chiral COFs are particularly useful for the chiral separation of epoxides, ketones, and drugs when applied as coatings for chiral columns (Figure 3c).

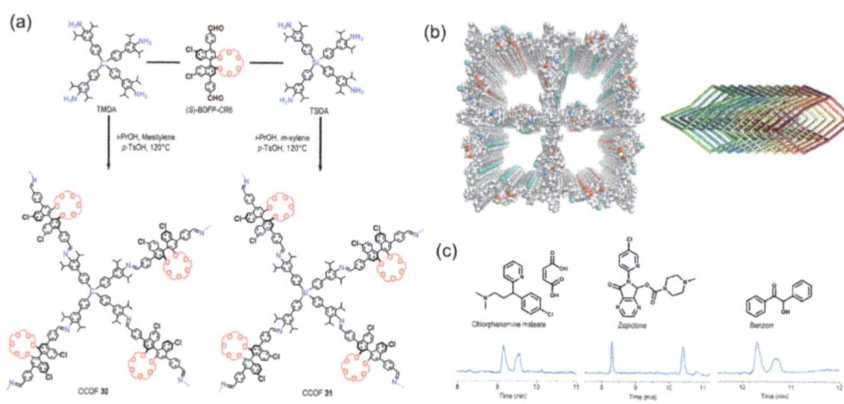

Figure 3. (**a**) Construction of 3D chiral COFs; (**b**) 3D structure and 11-fold interpenetration of diamond framework; (**c**) chiral crown ether-decorated COFs are used for separation of racemates. Adapted from ref. [75] with permission, copyright 2024, Oxford University Press.

In the past year, to further explore the properties of chiral COF materials, Cui and co-workers synthesized a series of chiral COFs with different topological structure types by a de novo synthesis method [38,76–80]. For example, they reported a mixed-linker strategy to design multicomponent 2D chiral COFs by the condensation of a mixture of triamines (with or without the MacMillan imidazolidinone catalyst or aryl substituent (ethyl and isopropyl)) and a dialdehyde derivative of thieno-[3,2-*b*]thiophene [76]. Structural characterization showed that these three-component homochiral 2D COFs feature either AA or ABC stacking. The stacking modes that can be synthetically controlled through steric tuning using different aryl substituents affect their chemical stability and electrochemical performance. With the MacMillan catalyst periodically appended on their channels, all of these COFs can be highly enantioselective and recyclable electrocatalysts for the asymmetric α-arylation of aldehydes, affording alkylated anilines with up to 97% enantiomeric excess by an anodic oxidation/organocatalytic protocol. Soon afterwards, they created a range of two- and three-component 3D chiral COFs by utilizing enantiopure BINOL-derived tetraaldehydes with varying alkyl substituents [77]. By imine condensations of BINOL-derived tetraaldehydes bearing different alkyl substituents with the monomer *tetra*(*p*-aminophenyl)-methane, a series of two-component and three-component chiral COFs with different interpenetrated dia networks were obtained (Figure 4a). Structural characterization showed that alkyl groups are appended on the walls of the channels, and their types/contents that can be synthetically tuned control the interpenetration degree of COFs by minimizing repulsive interactions between the alkyl groups (Figure 4b–d). This work provides a potential way to use steric hindrance to regulate and control the interpenetration, stability, porosity, and functionalities of chiral COFs.

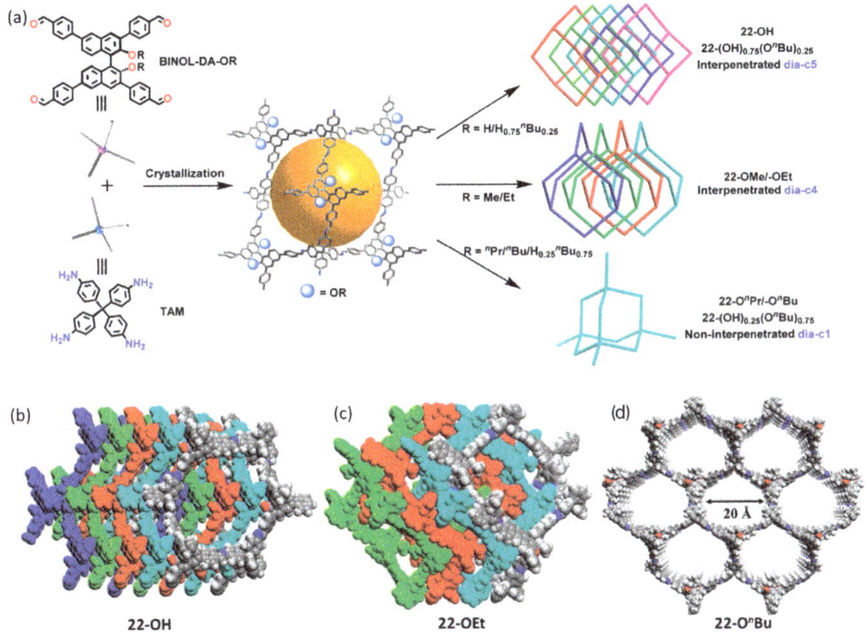

Figure 4. (**a**) Synthesis diagram of 3D chiral COFs with different alkyl substituents; (**b**) five-fold interpenetrated dia network of 22-OH; (**c**) four-fold interpenetrated dia network of 22-OEt; (**d**) noninterpenetrated dia network of 22-O^{n}Bu. Adapted from ref. [77] with permission, copyright 2024, American Chemical Society.

The chirality-induced spin selectivity (CISS) effect describes a phenomenon wherein the charge transport through certain chiral structures or molecules exhibits a distinct chirality-dependent electronic spin polarization. The use of chiral COFs to achieve CISS remains a largely untapped area of research. More recently, Cui's group synthesized four 2D Zn(salen)-based chiral COFs, namely CCOFs 9–12, by the imine condensation of chiral 1,2-diaminocyclohexane and *tri*- or *tetra*-(salicylaldehyde) derivatives (Figure 5a) [79]. CCOF-9 has a unique C2-symmetric "armchair" tetra-substituent pyrene conformation and exhibits the most pronounced chirality among these chiral COF materials, which means it can act as a solid state host capable of achieving the enantioselective adsorption of racemic drugs with up to 97% ee (Figure 5b). After paramagnetic Co(II) replaces diamagnetic Zn(II) ions, the obtained CCOF-9-Co not only maintains its high crystallinity and porosity and excellent chirality, but also exhibits enhanced conductivity (Figure 5c). Magnetic conductive atomic force microscopy shows that CMOF-9-Co has an impressive spin polarization ratio of 88–94% (Figure 5d). This work shows the great potential of chiral COFs for controlling spin selectivity and will motivate the development of novel crystalline polymers with the CISS effect. Using a similar strategy, they reported a one-step, template-free approach for synthesizing higher-order superhelical nanofibrous chiral COFs with preferred helicity by controlling alkyl chain lengths in organic linkers [80]. This approach yields chiral 3D COFs (13-OR, R = H, Me, Et, ^{n}Pr, ^{n}Bu) with a 10-fold interpenetrated diamondoid structure from enantiopure 1,1′-bi-2-naphthol (BINOL)-based tetraaldehydes and tetraamine. The superhelical nature of these chiral COFs is evident in their chiral recognition and spin-filter properties, displaying significantly enhanced enantiodiscrimination in carbohydrate binding and a remarkable CISS effect with a 48–51% spin polarization ratio. This work provides a robust and effective method for manipulating COF superhelical nanostructures, offering new insights into CISS in chiral materials.

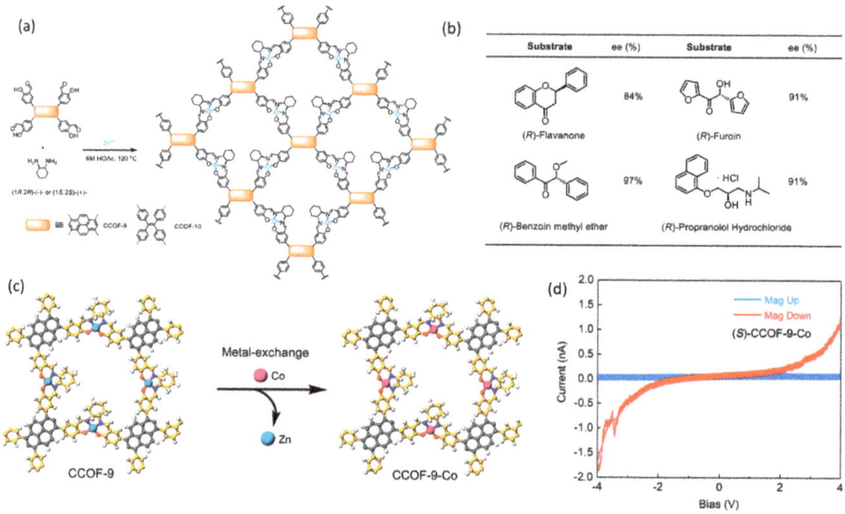

Figure 5. (**a**) Synthesis diagram of Zn(salen)-based chiral COFs; (**b**) enantiosorption of CCOF-9 toward racemic drugs; (**c**) synthesis diagram of CCOF-9-Co; (**d**) *I*–*V* curves of (S)-CCOF-9-Co obtained from mc-AFM. Adapted from ref. [79] with permission, copyright 2024, American Chemical Society.

According to the above research progress, the chiral COFs synthesized via the de novo synthesis method exhibit uniformly distributed chiral sites throughout the framework. Chiral scaffold COFs generally exhibit higher porosity and larger pore channels, whereas in chiral COFs with non-chiral scaffolds, a material porosity is typically reduced to a certain extent due to the chiral groups on the side chains of the framework. Therefore, the de novo synthesis of chiral COFs with chiral scaffolds has garnered significant attention, accelerating the rapid development of chiral COFs.

2.2. Post-Synthesis

The post-synthesis strategy is more commonly employed than de novo synthesis in the field of chiral porous solid materials. Post-synthetic modification can easily introduce chiral groups, accurately locate the chiral functions at anchor points of the achiral framework, and maintain the overall structure of the achiral framework [81,82]. Therefore, post-synthetic modification is an efficient and practical method for regulating the structure and function of COFs. This approach is ideal for achiral COFs with suitable functional groups, enabling the creation of functionalized chiral COFs by attaching the chiral component to the achiral COF framework through chemical bonds or non-covalent interactions.

The first chiral COF synthesized by post-synthetic modification, reported by Jiang's group, was based on imine-linked porphyrins [39]. In 2014, they used the acetylene-azide click reaction to anchor optically pure pyrrolidine to the achiral COF framework, allowing the rational construction of the chiral COFs $[Pyr]_x$-H_2P-COFs (x = 0, 25, 75, and 100). The experimental results indicate that the modified COFs exhibit reduced BET surface areas and pore sizes compared to the parent COF, while the crystallinity remains unchanged. By regulating pyrrolidine loading on the pore walls, the modified COFs demonstrate superior catalytic activity for asymmetric Michael additions. Subsequently, the same group reported three chiral COFs using the same click reaction [83]. First, they synthesized an achiral 2D COF that was stable and resistant to strong bases and acids by the condensation of methoxy-containing terephthalaldehyde with 1,3,5-*tri*(4-aminophenyl)benzene (Figure 6a). The (*S*)-pyrrolidine was then anchored to the COF channel wall through the azide-ethyl click reaction, converting the achiral COF into a catalytically active chiral COF while maintaining the porosity and crystallinity (Figure 6b). As organocatalysts, the obtained chiral COFs show excellent activity toward Michael additions under ambient conditions, with over 90% enantioselectivity (Figure 6c).

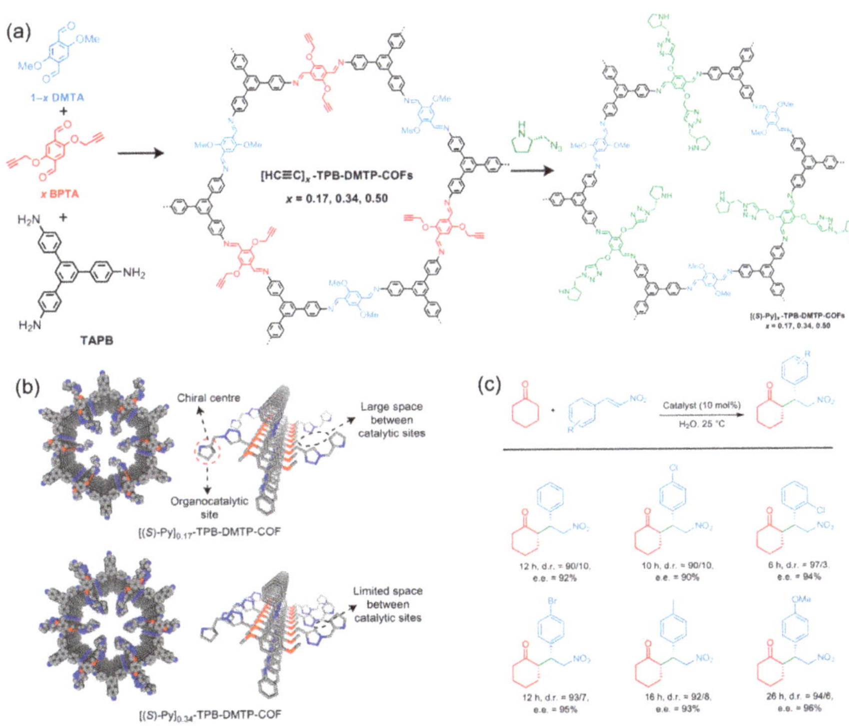

Figure 6. (**a**) Construction of chiral COFs through post-synthetic modification. (**b**) Channel wall structure of chiral COFs. (**c**) Substrate scope of Michael addition reactions catalyzed with chiral COFs. Adapted from ref. [83] with permission, copyright 2015, Nature Publishing Group.

In addition to the covalent binding of small organic molecules as chiral attachments, increasing numbers of examples have been obtained by grafting optically pure natural biomolecules such as sugars, enzymes, and peptides onto COF channel walls through immobilization and stabilization [44,84–86]. For example, Chen and Ma et al. proposed a general strategy for enriching chirality by covalently anchoring a range of biomolecules (lysine, tripeptides, and lysozyme) into achiral polyimide COFs [85]. The resulting biomolecule⊂COFs retain the framework structure of the parent COF but exhibit reduced pore sizes and surface areas. Interestingly, the biomolecule⊂COFs showed high chiral separation efficiency when used as chiral stationary phases due to the specific interactions and strong chirality inherited from the immobilized biomolecules. In the same year, Ma's group incorporated optically pure lipase into achiral 2D COFs, creating chiral enzyme⊂COFs in phosphate buffer solution [42]. Compared to amorphous analogues, the porous channels of enzyme⊂COFs not only make the enzyme more accessible to the reagents, but also act as strong shields to protect the enzymes from inactivation, demonstrating superior activity and tolerance to harsh environments.

Chen et al. have also made significant advancements in constructing chiral COFs by encapsulating bioactive molecules into achiral COF channel walls through post-synthetic modifications [87]. For example, in 2022, they introduced a scalable and green method for creating high-performance biocatalysts by assembling enzymes (lipase) with COFs under ambient conditions (Figure 7a) [88]. The obtained lipase@COFs demonstrated remarkable reusability and stability, serving as more effective catalysts for key industrial reactions than free enzymes or those immobilized by traditional methods (Figure 7b–d).

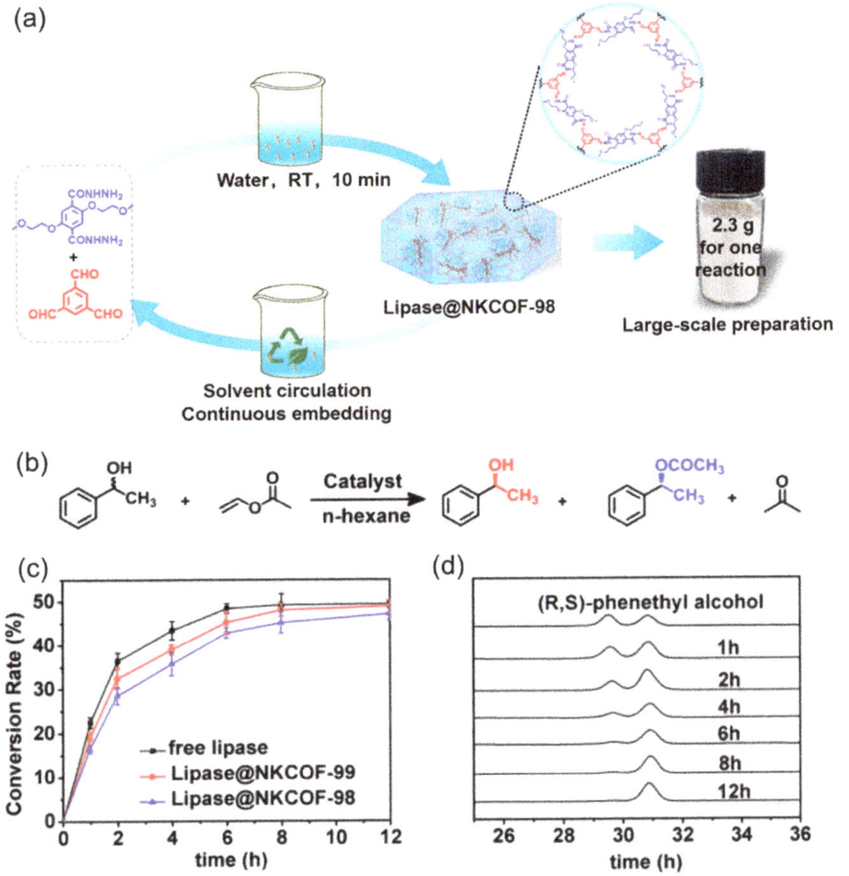

Figure 7. (**a**) The synthesis process of lipase@NKCOF-98 and a photograph of a lipase@NKCOF-98 sample from a scaled-up synthesis process. (**b**) The chiral catalytic resolution of 1-phenylethanol. (**c**) The conversion rate curves of different catalysts over 12 h. (**d**) The catalytic reaction results of NKCOF-99 at different reaction times (HPLC). Adapted from ref. [88] with permission, copyright 2022, Wiley-VCH.

Cyclodextrins are a class of important biomolecules with intrinsic chiral cavities. In 2019, Cui et al. constructed chiral COFs by anchoring chiral β-cyclodextrin (β-CD) into channels of a 2D COF through a thiol-ene click reaction [89]. The modified COFs exhibited reduced pore sizes and surface areas compared to the parent COF while retaining the same crystal structure. Nevertheless, modified chiral COFs with low β-CD loading showed high enantiomeric discrimination of amino acids. Recently, Yi et al. constructed a carboxyl-functionalized achiral COF (TpBD-3COOH) with hexagonal channels and then integrated a chiral molecule, heptakis(6-amino-6-deoxy)-β-CD (Am7CD), to obtain a chiral COF (TpBD-Am7CD) [90]. Compared with TpBD-3COOH, the chiral selectivity of TpBD-Am7CD is significantly improved in adsorption experiments. At the same time, TpBD-Am7CD exhibited chiral selectivity to adsorb amino acid enantiomers similar to the previously reported β-CD COF, indicating that the chiral Am7CD functionality provides a chiral microenvironment.

The research described above indicates post-synthetic modification to be an effective and simple method for synthesizing chiral COFs. However, several shortcomings of the approach remain to be resolved. First, a uniform distribution of the chiral reagents introduced within the COF cannot be ensured. Second, the introduction of chiral functional groups inevitably leads to a decrease in the porosity of the COF. Third, the chiral appendants must be smaller than the COF channel diameter, limiting the scope of the approach.

2.3. *Chiral Induction Synthesis*

Chirality-induced synthesis is a highly appealing method employing chiral small-molecule catalysts to convert achiral reagents into optically active polymers, supramolecular systems, and other small molecules. Chiral induction has produced homochiral polymers with diverse application potential and has been applied in the construction of chiral COFs from achiral monomers.

In 2018, inspired by metal–organic framework chiral induction studies, Cui's group first used chiral induction synthesis to construct nine 2D chiral COFs [91]. These chiral COFs were solvothermally condensed from achiral C_3-symmetric 1,3,5-triformylphloroglucinol (Tp) and triamine or diamine linkers in the presence of catalytic amounts of (*S*)- or (*R*)-1-phenylethylamine (Figure 8a). The chirality of these COFs is derived from the chiral catalyst-induced immobilization of tris(*N*-salicylideneamine) cores by intramolecular hydrogen bonding during crystallization. These chiral COFs showed high enantioselectivity toward chiral carbohydrates. Moreover, the post-synthetic modification of enaminone groups with Cu^{2+} enables metalated chiral COFs to serve as highly stereoselective heterogeneous catalysts for the asymmetric Henry reaction. Subsequently, Dong's group prepared a chiral COF with high crystallinity via organocatalytic asymmetric Schiff base condensation under ambient conditions [92]. In this case, they employed chiral 2-methylpyrrolidine as a catalyst for the asymmetric condensation of tricarbonyl phloroglucinol with various hydrazides or diamines to prepare a series of tris(*N*-salicylideneamine)-derived β-ketoenamine chiral COFs (Figure 8b). Furthermore, the obtained chiral COFs can be metalated using a solid-state coordination method, providing heterogeneous catalysts for asymmetric A^3-coupling reactions. Similarly, using chiral amines as chiral inducers, Gu et al. synthesized ultra-thin 2D chiral COF nanosheets from achiral monomers which exhibit tunable circularly polarized luminescence [93].

In 2020, Dong et al. prepared, for the first time, two homochiral COFs through the polymerization of achiral triamine dialdehydes and terminal aryl alkynes with the assistance of catalytic inducer CuOTf·toluene/(*S*,*S*)-pydox (pydox = 2,6-*bis*(4-phenyl-2-oxazolinyl)pyridine) [94]. The resulting propargylamine-linked chiral COFs can be highly reusable chiral catalysts of asymmetric Michael addition reactions with moderate activity and enantioselectivity.

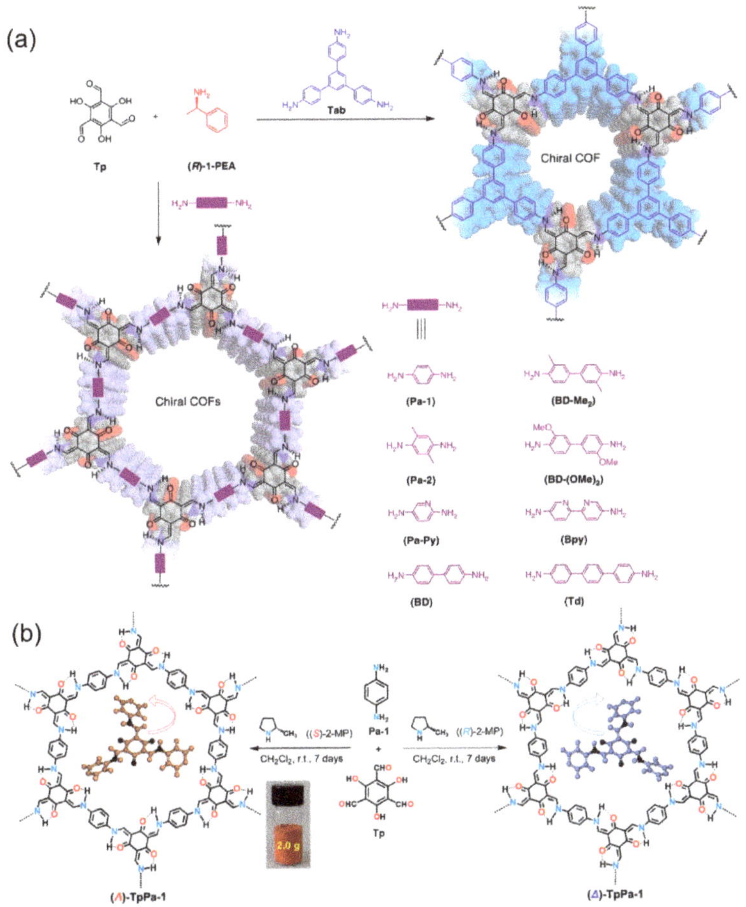

Figure 8. (**a**) Synthesis of chiral COFs by chiral 1-phenylethylamine asymmetric condensation. Adapted from ref. [91] with permission, copyright 2018, Nature Publishing Group. (**b**) Synthesis of (Λ)- and (Δ)-TpPa-1 by chiral 2-methylpyrrolidine asymmetric polymerization. Adapted from ref. [92] with permission, copyright 2022, Wiley-VCH.

Although achiral monomers can be assembled into chiral COFs by introducing chiral inducers or chiral catalysts, the chirality-induced synthesis of COFs is still in its infancy. To date, only a limited scope of chiral small molecules or chiral catalysts has been investigated. In addition, the development of alternative chiral inducers like circularly polarized light or chiral solvents is essential for advancing new chiral COFs, and the mechanism of chiral induction should be further studied.

3. Application of Chiral COFs in Asymmetric Photocatalysis

Asymmetric catalysis is one of the most efficient methods to obtain homochiral compounds. Photocatalytic organic synthesis utilizing sunlight is a green organic synthesis protocol offering the advantage of low energy consumption. The combination of these methods in asymmetric photocatalysis has attracted much attention in recent years, not least due to its application in drug synthesis [95]. Chiral COFs, as emerging porous materials, show significant potential as robust heterogeneous catalysts for asymmetric photocatalysis as they are readily functionalized and structurally regulated at the molecular level.

The work of Dong's group has been crucial in advancing chiral COFs and asymmetric photocatalytic organic synthesis. Copper porphyrin is a common organic photothermal conversion material that can enhance the non-radiative transition of copper (II) [96,97]. Dong's group reported several Cu(II) porphyrinyl-based chiral COFs for photo-assisted thermocatalysis [98–101]. In 2019, the group reported two metal nanoparticle-loaded chiral COFs (denoted Au@CCOF-CuTPP and Pd@CCOF-CuTPP) comprising a porphyrin-derived chiral COF [98]. The chiral COF was prepared by assembling copper tetrabromophenolphthalein (Cu-TBrPP) and the chiral organic monomer (*S*)-(+)-2-methylpiperazine (*S*-MP) in anhydrous 1,4-dioxane using $Pd(PPh_4)_3$ as a catalyst (Figure 9a). The crystalline COF was then subjected to successive solution impregnation and metal reduction steps to obtain the final metal nanoparticle-loaded chiral composite. Interestingly, both nanoparticle-loaded composites show effective photothermal conversion with a maximum temperature rise of 25–31 °C within 1080 s (Figure 9b,c). Meanwhile, Pd@CCOF-CuTPP can catalyze the asymmetric A^3 coupling reaction by photothermal conversion, providing the desired products with yields of 68–98% and ee values of 90–98%. Under visible light, Au@CCOF-CuTPP can catalyze the asymmetric Henry reaction of benzyl alcohols and nitromethane with excellent yields (93–99%) and enantioselectivity (94–98% ee) (Figure 9d,e).

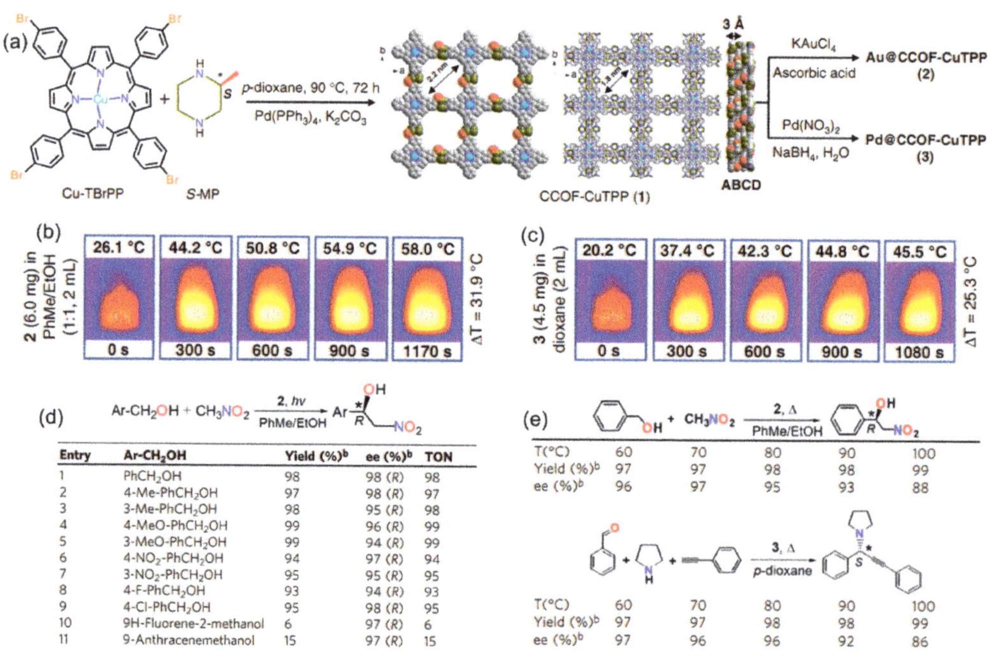

Entry	Ar-CH₂OH	Yield (%)ᵇ	ee (%)ᵇ	TON
1	$PhCH_2OH$	98	98 (*R*)	98
2	4-Me-$PhCH_2OH$	97	98 (*R*)	97
3	3-Me-$PhCH_2OH$	98	95 (*R*)	98
4	4-MeO-$PhCH_2OH$	99	96 (*R*)	99
5	3-MeO-$PhCH_2OH$	99	94 (*R*)	99
6	4-NO_2-$PhCH_2OH$	94	97 (*R*)	94
7	3-NO_2-$PhCH_2OH$	95	95 (*R*)	95
8	4-F-$PhCH_2OH$	93	94 (*R*)	93
9	4-Cl-$PhCH_2OH$	95	98 (*R*)	95
10	9H-Fluorene-2-methanol	6	97 (*R*)	6
11	9-Anthracenemethanol	15	97 (*R*)	15

T(°C)	60	70	80	90	100
Yield (%)ᵇ	97	97	98	98	99
ee (%)ᵇ	96	97	95	93	88

T(°C)	60	70	80	90	100
Yield (%)ᵇ	97	97	98	98	99
ee (%)ᵇ	97	96	96	92	86

Figure 9. (**a**) Synthesis of Pd@CCOF-CuTPP and Au@CCOF-CuTPP. (**b**) Photothermal effect of Au@CCOF-CuTPP. (**c**) Photothermal effect of Pd@CCOF-CuTPP. (**d**) Scope of Au@CCOF-CuTPP-catalyzed one-pot asymmetric Henry reaction. (**e**) Effect of temperature on asymmetric reactions. Adapted from ref. [98] with permission, copyright 2019, Nature Publishing Group.

Later, Dong's group designed and synthesized the multifunctional chiral (*R*)-CuTAPBN-COF through the condensation reaction between 5,10,15,20-tetrakis-(4-aminophenyl)-porphyrin-Cu(II) (Cu-TAPP) and 6,6′-dichloro-2,2′-diethoxy-1,1′-binaphthyl-4,4′-dialdehyde ((*R*)-BINOL-DA) (Figure 10) [101]. It is noteworthy that the chiral control and catalytic centers of the (*R*)-CuTAPBN-COF are not on the same molecular body, a structural motif not dissimilar to the structure of enzymes. As an ADP receptor blocker, (*S*)-clopidogrel ((*S*)-CIK) is currently one of the best-selling antiplatelet and antithrombotic drugs in the world.

Under visible light irradiation, (*R*)-CuTAPBN-COF can catalyze the one-step synthesis of (*S*)-CIK by a photothermally triggered three-component one-pot asymmetric Strecker reaction (Figure 10). Interestingly, the reaction can be carried out effectively under sunlight irradiation (the reaction system temperature is ~47 °C), providing (*S*)-CIK in up to a 70% yield and 92% ee within 3 h. Furthermore, the catalytic reaction has demonstrated excellent substrate scope. This work provides an experimental basis for the development of green energy-saving "windowsill" reactions based on chiral COF photocatalysis.

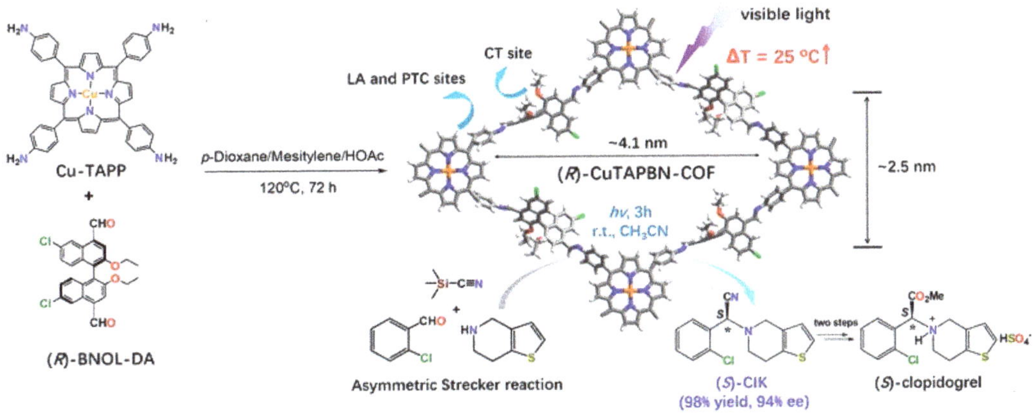

Figure 10. Synthesis of multifunctional chiral (*R*)-CuTAPBN-COF and its application in catalytic synthesis of drug intermediates. Adapted from ref. [101] with permission, copyright 2020, American Chemical Society.

In 2022, Dong and Chen et al. reported another Cu(II)-metalated chiral COF using a similar method. The obtained (*R*)-CuTAPBP-COF contains both Lewis acid (copper porphyrin) and Brønsted acid (phosphoric acid) catalytic sites [100]. Under visible light irradiation, (*R*)-CuTAPBP-COF can catalyze the intermolecular asymmetric α-benzylation of aldehydes through photothermal conversion induced by visible light. For example, 4-(bromoethyl)pyridine reacts with propanaldehyde in methanol to produce (*R*)-2-methyl-3-(pyridine-4-yl)propane with a yield of 98% and with 95% ee. Moreover, this kind of photothermally catalyzed asymmetric reaction can also be effectively carried out under natural light.

Although chiral COFs are increasingly used in photothermal asymmetric catalysis, their application in asymmetric photocatalysis remains scarcely documented. Recently, Dong's group developed a quaternary ammonium bromide-decorated chiral photocatalytic system by integrating the functional photosensitizer 4-(10,15,20-triphenylporphyrin-5-yl)-aniline (TAPP), the phase transfer species quaternary ammonium bromide-decorated phenylacetylene (PA-QA), and chiral controller propargylamine into a chiral COF (denoted (*R*)-DTP-COF-QA) [102]. Under ambient conditions, (*R*)-DTP-COF-QA was synthesized from achiral 2,5-dimethoxyterephthaldehyde (DMTP), TAPP, and PA-QA through asymmetric A^3-coupling polymerization in the presence of the chiral catalyst CuOTf-pybox (pybox = (*R*,*R*)-2,6-bis(4-phenyl-2-oxazolinyl)pyridine) (Figure 11a). (*R*)-DTP-COF-QA crystallizes in the chiral space group P4 and shows a broad absorption band covering the entire visible spectral range. Due to the synergistic effect between porphyrin, chiral propargylamine, and amphiphilic quaternary ammonium bromide, (*R*)-DTP-COF-QA can promote visible light-driven enantioselective photooxidation of sulfides to sulfoxides in water and in air with high activity and enantioselectivity (Figure 11b,c).

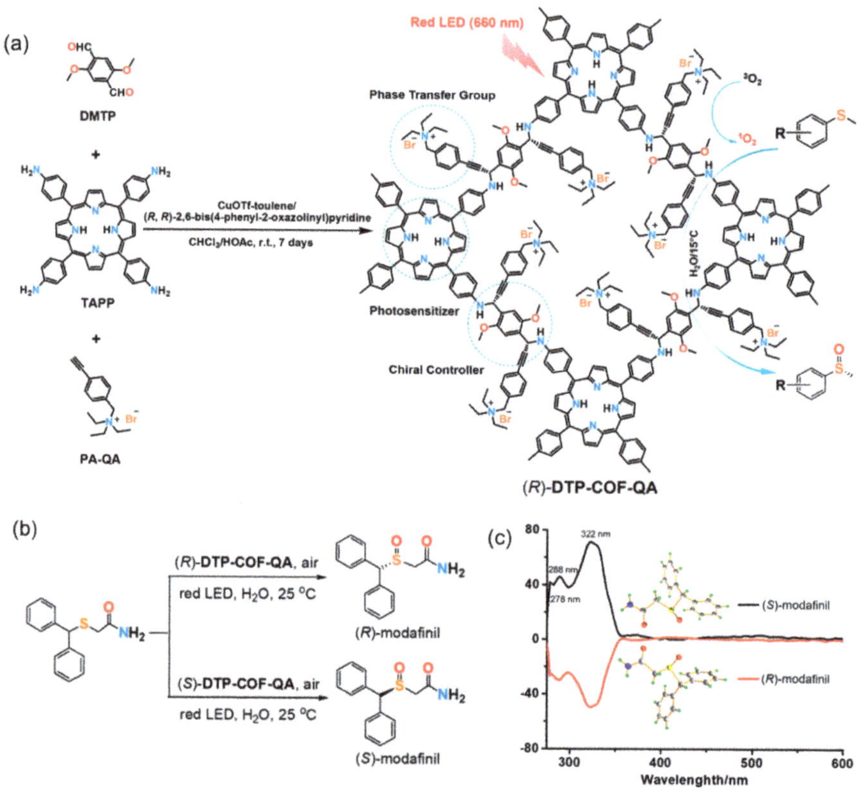

Figure 11. (**a**) Synthesis of quaternary ammonium bromide-decorated chiral photocatalyst (*R*)-DTP-COF-QA. (**b**) Photocatalytic synthesis of (*R*)- and (*S*)-modafinil by chiral DTP-COF-QA. (**c**) Circular dichroism spectra of (*R*)- and (*S*)-modafinil. Adapted from ref. [102] with permission, copyright 2022, American Chemical Society.

More recently, Zhao et al. reported a common bottom-up strategy for the successful synthesis of several photoactive chiral COFs [103]. They chose photoactive porphyrins as building blocks, immobilizing various secondary amine chiral catalytic centers on the COF pore walls by rationally designing benzimidazole linkers. In these chiral COFs, the porphyrin units act as antennas, capturing UV and visible light in the range of 250–1000 nm. At the same time, various secondary amine chiral functional groups were fixed on the channel wall of the chiral COFs through 1*H*-benzo[*d*]-imidazole building blocks, which weakened the influence of the flexible chiral functional groups on the crystallinity of the chiral COFs. Moreover, the reactants can easily reach the chiral catalytic site (the chiral secondary amine) through 1D nanotubes. Thus, well-designed chiral COFs can incorporate photoactive building blocks (porphyrin units) and chiral catalytic sites, resulting in high yields (97%) and excellent enantioselectivities (93%) in the photocatalytic asymmetric alkylation of aldehydes (Figure 12).

Besides merging COFs with chiral catalytic sites like binaphthol derivatives and small organic molecules, integrating optically pure natural biomolecules such as enzymes and peptides into COFs with open channels also shows promise for the generation of photocatalysts of asymmetric organic transformations. For example, in 2022, Chen's group designed and synthesized a photoenzymatic platform (WGL@COF) using mesoporous porphyrin-based COFs (NKCOF-118(M), M = H, Zn, Cu, and Ni) as solid carriers to immobilize wheat germ lipase (WGL) (Figure 13a) [104]. PXRD patterns, N_2 adsorption and desorption isotherms, and confocal microscopy images demonstrate that WGL is successfully immobilized on the channel wall of porphyrin-based COFs (Figure 13b–d). Due to the proximity effect between the enzymes and porphyrin-based COF in one system, the resulting WGL@COFs exhibit good chemical stability and excellent photocatalytic activity in the asymmetric Mannich reaction under visible light irradiation (Figure 13e). Moreover, control experiments show that this asymmetric Mannich reaction cannot be achieved by porphyrin-based COFs or WGL independently. This work provides a simple and efficient method for producing highly efficient biomolecule-containing asymmetric photocatalysts by enzyme immobilization in photoactive COFs.

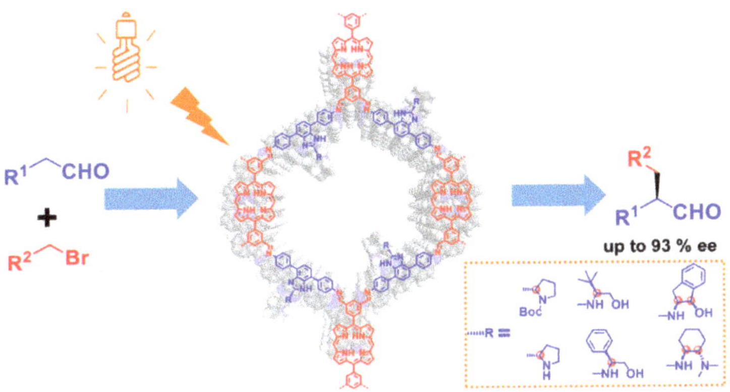

Figure 12. Chiral porphyrin-based COF-photocatalyzed asymmetric alkylation of aldehydes. Adapted from ref. [103] with permission, copyright 2023, American Chemical Society.

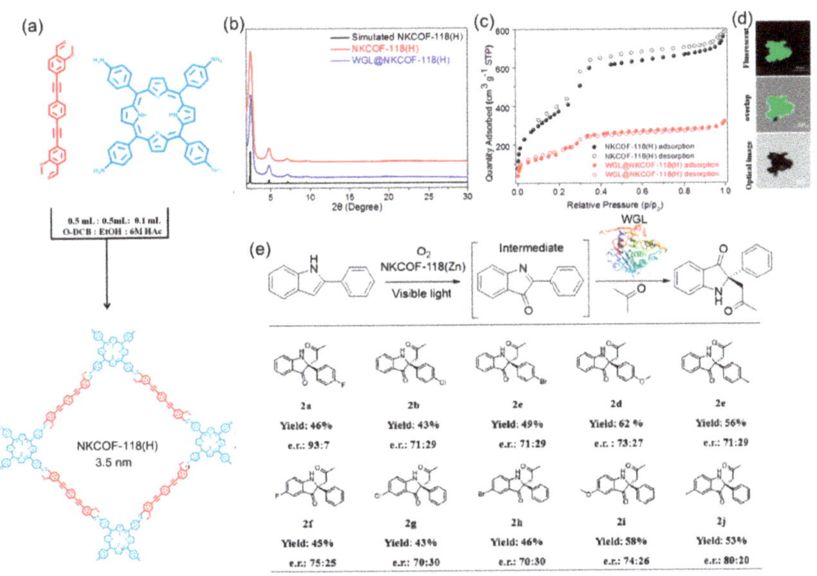

Figure 13. (**a**) Synthesis of mesoporous porphyrin-based COFs. (**b**) PXRD patterns of WGL@COFs. (**c**) N_2 adsorption and desorption isotherms of NKCOF-118(H). (**d**) Confocal microscopy images of WGL@COFs. (**e**) Substrate scope of asymmetric Mannich reaction catalyzed by WGL@COFs. Adapted from ref. [104] with permission, copyright 2022, American Chemical Society.

4. Conclusions

Homochiral COF materials possess diverse characteristics and applications and have been extensively studied in recent years. This review highlights the synthesis strategies, structural characteristics, and research progress related to chiral COFs in asymmetric photocatalysis. By applying three different synthesis strategies, many chiral COFs have been prepared. Post-modified synthesis can easily introduce chiral sites while preserving the framework of COFs. De novo synthesis is beneficial to obtain chiral COFs with permanent pores and defined chiral sites, but the protocol still presents significant challenges. Other methods besides solvothermal synthesis, like room temperature synthesis, microwave assisted reactions, and ionic thermal synthesis, should also be explored.

Due to their unique structural characteristics and ready tunability of their channels, chiral COFs show great potential in asymmetric photocatalysis. Typical asymmetric photocatalytic reactions like the Henry reaction, the Mannich reaction, and the alkylation of aldehydes have been demonstrated with excellent yields and enantioselectivities. However, compared with other porous materials, COFs have not yet been utilized in more complex and advanced asymmetric catalytic reactions. Their potential in asymmetric photocatalysis remains underdeveloped. Moreover, the fundamental mechanism of asymmetric photocatalysis using chiral COFs remains unclear. Although several COFs have recently been characterized by X-ray single-crystal diffraction, the structures of reported COFs are most usually determined by an analysis of powder X-ray diffraction and by taking into account the geometry principles developed in network chemistry. More detailed structural information is very useful for the study of the properties of COFs, so studying the crystallization process to prepare single crystals of chiral COFs and determining their atomic-level structure through a single-crystal X-ray diffraction analysis will provide more evidence for the elucidation of the catalytic mechanism of these materials. More 2D and 3D chiral COFs should be designed and synthesized to study their differences for asymmetric photocatalysis. When chiral COFs are used as photocatalysts, the dimensionality and pore size may particularly influence the asymmetric photocatalytic activity and catalyst sites of chiral COFs. In addition, new chiral COFs with achiral pore/channels need to be further investigated, and examples of their application to asymmetric photocatalysis are still unexplored.

Author Contributions: Conceptualization, P.L. and W.D.; methodology, P.L.; validation, X.S. (Xianfu Shen) and X.S. (Xiang Shen); writing—original draft preparation, P.L. and Y.Z.; writing—review and editing, J.-J.L.; supervision, J.-J.L.; project administration, J.-J.L. and Y.Z.; funding acquisition, J.-J.L. All authors have read and agreed to the published version of the manuscript.

Funding: This work is supported by Application Basis Research Project of the Yunnan Province Science and Technology Department (202201AT070016), Yunnan Local Colleges Research Projects (202301BA070001-101 and 202301BA070001-083), the Key Laboratory of the Yunnan Provincial Department of Education, the Scientific and Technological Innovation Team for Function-oriented Crystalline Porous Materials in Yunnan Institutions of Higher Learning, and the Program for Innovative Research Team in Qujing Normal University.

Institutional Review Board Statement: Not applicable.

Informed Consent Statement: Not applicable.

Data Availability Statement: The research data are available by contacting the corresponding author.

Conflicts of Interest: The authors declare no conflicts of interest.

References

1. Foubelo, F.; Nájera, C.; Retamosa, M.G.; Sansano, J.M.; Yus, M. Catalytic asymmetric synthesis of 1,2-diamines. *Chem. Soc. Rev.* **2024**, *53*, 7983–8085. [CrossRef] [PubMed]
2. Huang, X.; Meggers, E. Asymmetric Photocatalysis with Bis-cyclometalated Rhodium Complexes. *Acc. Chem. Res.* **2019**, *52*, 833–847. [CrossRef]

3. Genzink, M.J.; Kidd, J.B.; Swords, W.B.; Yoon, T.P. Chiral Photocatalyst Structures in Asymmetric Photochemical Synthesis. *Chem. Rev.* **2022**, *122*, 1654–1716. [CrossRef]
4. Prier, C.K.; Rankic, D.A.; MacMillan, D.W.C. Visible Light Photoredox Catalysis with Transition Metal Complexes: Applications in Organic Synthesis. *Chem. Rev.* **2013**, *113*, 5322–5363. [CrossRef] [PubMed]
5. Zou, Y.-Q.; Hörmanna, F.M.; Bach, T. Iminium and enamine catalysis in enantioselective photochemical reactions. *Chem. Soc. Rev.* **2018**, *47*, 278–290. [CrossRef]
6. Yoon, T.P.; Ischay, M.A.; Du, J. Visible light photocatalysis as a greener approach to photochemical synthesis. *Nat. Chem.* **2010**, *2*, 527–532. [CrossRef] [PubMed]
7. Nicewicz, D.A.; MacMillan, D.W.C. Merging Photoredox Catalysis with Organocatalysis: The Direct Asymmetric Alkylation of Aldehydes. *Science* **2008**, *322*, 77–80. [CrossRef]
8. Shih, H.-W.; Vander Wal, M.N.; Grange, R.L.; MacMillan, D.W.C. Enantioselective α-Benzylation of Aldehydes via Photoredox Organocatalysis. *J. Am. Chem. Soc.* **2010**, *132*, 13600–13603. [CrossRef]
9. Capacci, A.G.; Malinowski, J.T.; McAlpine, N.J.; Kuhne, J.; MacMillan, W.C.D. Direct, enantioselective α-alkylation of aldehydes using simple olefins. *Nat. Chem.* **2017**, *9*, 1073–1077. [CrossRef]
10. Yin, Y.; Dai, Y.; Jia, H.; Li, J.; Bu, L.; Qiao, B.; Zhao, X.; Jiang, Z. Conjugate Addition–Enantioselective Protonation of N-Aryl Glycines to α-Branched 2-Vinylazaarenes via Cooperative Photoredox and Asymmetric Catalysis. *J. Am. Chem. Soc.* **2018**, *140*, 6083–6087. [CrossRef]
11. Cao, K.; Tan, S.M.; Lee, R.; Yang, S.; Jia, H.; Zhao, X.; Qiao, B.; Jiang, Z. Catalytic Enantioselective Addition of Prochiral Radicals to Vinylpyridines. *J. Am. Chem. Soc.* **2019**, *141*, 5437–5443. [CrossRef]
12. Miller, Z.D.; Lee, B.J.; Yoon, T.P. Enantioselective Crossed Photocycloadditions of Styrenic Olefins by Lewis Acid Catalyzed Triplet Sensitization. *Angew. Chem. Int. Ed.* **2017**, *56*, 11891–11895. [CrossRef] [PubMed]
13. Ding, W.; Lu, L.; Zhou, Q.; Wei, Y.; Chen, J.; Xiao, W. Bifunctional Photocatalysts for Enantioselective Aerobic Oxidation of β-Ketoesters. *J. Am. Chem. Soc.* **2017**, *139*, 63–66. [CrossRef] [PubMed]
14. Zhang, J.; Rueping, M. Metallaphotoredox catalysis for sp^3 C–H functionalizations through single-electron transfer. *Nat. Catal.* **2024**, *7*, 963–976. [CrossRef]
15. Wu, P.; He, C.; Wang, J.; Peng, X.; Li, X.; An, Y.; Duan, C. Photoactive Chiral Metal–Organic Frameworks for Light-Driven Asymmetric α-Alkylation of Aldehydes. *J. Am. Chem. Soc.* **2012**, *134*, 14991–14999. [CrossRef] [PubMed]
16. Cherevatskaya, M.; Neumann, M.; Füldner, S.; Harlander, C.; Kümmel, S.; Dankesreiter, S.; Pfitzner, A.; Zeitler, K.; König, B. Visible-Light-Promoted Stereoselective Alkylation by Combining Heterogeneous Photocatalysis with Organocatalysis. *Angew. Chem. Int. Ed.* **2012**, *51*, 4062–4066. [CrossRef]
17. Perry, I.B.; Brewer, T.F.; Sarver, P.J.; Schultz, D.M.; Dirocco, D.A.; MacMillan, D.W.C. Direct arylation of strong aliphatic C–H bonds. *Nature* **2018**, *560*, 70–75. [CrossRef]
18. Liang, J.; Liang, Z.B.; Zou, R.Q.; Zhao, Y.L. Heterogeneous Catalysis in Zeolites, Mesoporous Silica, and Metal-Organic Frameworks. *Adv. Mater.* **2017**, *29*, 1701139–1701159. [CrossRef]
19. Liu, Y.; Xuan, W.; Cui, Y. Engineering Homochiral Metal-Organic Frameworks for Heterogeneous Asymmetric Catalysis and Enantioselective Separation. *Adv. Mater.* **2010**, *22*, 4112–4135. [CrossRef]
20. Yaghi, O.M.; Kalmutzki, M.J.; Diercks, C.S. *Introduce to Reticular Chemistry: Metal–Organic Frameworks and Covalent Organic Frameworks*; Wiely-VCH: Weinheim, Germany, 2019.
21. Huang, N.-Y.; Zheng, Y.-T.; Chen, D.; Chen, Z.-Y.; Huang, C.-Z.; Xu, Q. Reticular framework materials for photocatalytic organic reactions. *Chem. Soc. Rev.* **2023**, *52*, 7949–8004. [CrossRef]
22. Chen, Z.; Kirlikovali, K.O.; Li, P.; Farha, O.K. Reticular Chemistry for Highly Porous Metal–Organic Frameworks: The Chemistry and Applications. *Acc. Chem. Res.* **2022**, *55*, 579–591. [CrossRef] [PubMed]
23. Yang, Q.; Xu, Q.; Jiang, H.-L. Metal–organic frameworks meet metal nanoparticles: Synergistic effect for enhanced catalysis. *Chem. Soc. Rev.* **2017**, *46*, 4774–4808. [CrossRef]
24. Rassu, P.; Ma, X.; Wang, B. Engineering of catalytically active sites in photoactive metal–organic frameworks. *Coord. Chem. Rev.* **2022**, *465*, 214561. [CrossRef]
25. Nordin, N.A.; Mohamed, M.A.; Salehmin, M.N.I.; Yusoff, S.F.M. Photocatalytic active metal–organic framework and its derivatives for solar-driven environmental remediation and renewable energy. *Coord. Chem. Rev.* **2022**, *468*, 214639. [CrossRef]
26. Luo, T.; Gilmanova, L.; Kaskel, S. Advances of MOFs and COFs for photocatalytic CO_2 reduction, H_2 evolution and organic redox transformations. *Coord. Chem. Rev.* **2023**, *490*, 215210. [CrossRef]
27. Luo, Y.-H.; Dong, L.-Z.; Liu, J.; Li, S.-L.; Lan, Y.-Q. From molecular metal complex to metal-organic framework: The CO_2 reduction photocatalysts with clear and tunable structure. *Coord. Chem. Rev.* **2019**, *390*, 86–126. [CrossRef]
28. Ding, M.; Flaig, R.W.; Jiang, H.-L.; Yaghi, O.M. Carbon capture and conversion using metal–organic frameworks and MOF-based materials. *Chem. Soc. Rev.* **2019**, *48*, 2783–2828. [CrossRef]
29. Wang, D.-G.; Qiu, T.; Guo, W.; Liang, Z.; Tabassum, H.; Xia, D.; Zou, R. Covalent organic framework-based materials for energy applications. *Energy Environ. Sci.* **2021**, *14*, 688–728. [CrossRef]
30. Guan, Q.; Zhou, L.-L.; Dong, Y.-B. Metalated covalent organic frameworks: From synthetic strategies to diverse applications. *Chem. Soc. Rev.* **2022**, *51*, 6307–6416. [CrossRef]

31. Qian, C.; Feng, L.; Teo, W.L.; Liu, J.; Zhou, W.; Wang, D.; Zhao, Y. Imine and imine-derived linkages in two-dimensional covalent organic frameworks. *Nat. Rev. Chem.* **2022**, *6*, 881–898. [CrossRef]
32. Geng, K.; He, T.; Liu, R.; Dalapati, S.; Tan, K.T.; Li, Z.; Tao, S.; Gong, Y.; Jiang, Q.; Jiang, D. Covalent Organic Frameworks: Design, Synthesis, and Functions. *Chem. Rev.* **2020**, *120*, 8814–8933. [CrossRef] [PubMed]
33. Liu, S.; Wang, M.; He, Y.; Qian, T.; Yan, C. Covalent organic frameworks towards photocatalytic applications: Design principles, achievements, and opportunities. *Coord. Chem. Rev.* **2023**, *475*, 214882. [CrossRef]
34. Guan, X.; Fang, Q.; Yan, Y.; Qiu, S. Functional Regulation and Stability Engineering of Three-Dimensional Covalent Organic Frameworks. *Acc. Chem. Res.* **2022**, *55*, 1912–1927. [CrossRef] [PubMed]
35. López-Magano, A.; Daliran, S.; Oveisi, A.R.; Mas-Ballesté, R.; Dhakshinamoorthy, A.; Alemán, J.; Garcia, H.; Luque, R. Recent Advances in the Use of Covalent Organic Frameworks as Heterogenous Photocatalysts in Organic Synthesis. *Adv. Mater.* **2023**, *35*, 2209475. [CrossRef]
36. Bai, L.; Phua, S.Z.F.; Lim, W.Q.; Jana, A.; Luo, Z.; Tham, H.P.; Zhao, L.; Gao, Q.; Zhao, Y. Nanoscale covalent organic frameworks as smart carriers for drug delivery. *Chem. Commun.* **2016**, *52*, 4128–4131. [CrossRef]
37. Du, C.; Zhu, X.; Yang, C.; Liu, M. Stacked Reticular Frame Boosted Circularly Polarized Luminescence of Chiral Covalent Organic Frameworks. *Angew. Chem. Int. Ed.* **2022**, *61*, e202113979. [CrossRef]
38. Zhang, X.; Chen, X.; Fu, S.; Cao, Z.; Gong, W.; Liu, Y.; Cui, Y. Homochiral π-Rich Covalent Organic Frameworks Enabled Chirality Imprinting in Conjugated Polymers: Confined Polymerization and Chiral Memory from Scratch. *Angew. Chem. Int. Ed.* **2024**, *63*, e202403878. [CrossRef]
39. Xu, H.; Chen, X.; Gao, J.; Lin, J.; Addicoat, M.; Irleb, S.; Jiang, D. Catalytic covalent organic frameworks via pore surface engineering. *Chem. Commun.* **2014**, *50*, 1292–1294. [CrossRef]
40. Dong, J.; Liu, Y.; Cui, Y. Emerging chiral two-dimensional materials. *Nat. Chem.* **2024**, *16*, 1398–1407. [CrossRef]
41. Weng, W.; Guo, J. The effect of enantioselective chiral covalent organic frameworks and cysteine sacrificial donors on photocatalytic hydrogen evolution. *Nat. Commun.* **2022**, *13*, 5768. [CrossRef]
42. Sun, Q.; Fu, C.-W.; Aguila, B.; Perman, J.; Wang, S.; Huang, H.-Y.; Xiao, F.-S.; Ma, S. Pore Environment Control and Enhanced Performance of Enzymes Infiltrated in Covalent Organic Frameworks. *J. Am. Chem. Soc.* **2018**, *140*, 984–992. [CrossRef] [PubMed]
43. Yuan, C.; Jia, W.; Yu, Z.; Li, Y.; Zi, M.; Yuan, L.; Cui, Y. Are Highly Stable Covalent Organic Frameworks the Key to Universal Chiral Stationary Phases for Liquid and Gas Chromatographic Separations? *J. Am. Chem. Soc.* **2022**, *144*, 891–900. [CrossRef]
44. Yuan, C.; Wang, Z.; Xiong, W.; Huang, Z.; Lai, Y.; Fu, S.; Dong, J.; Duan, A.; Hou, X.; Yuan, L.; et al. Cyclodextrin Incorporation into Covalent Organic Frameworks Enables Extensive Liquid and Gas Chromatographic Enantioseparations. *J. Am. Chem. Soc.* **2023**, *145*, 18956–18967. [CrossRef]
45. Kang, X.; Stephens, E.R.; Spector-Watts, B.M.; Li, Z.; Liu, Y.; Liu, L.; Cui, Y. Challenges and opportunities for chiral covalent organic frameworks. *Chem. Sci.* **2022**, *13*, 9811–9832. [CrossRef]
46. Han, X.; Yuan, C.; Hou, B.; Liu, L.; Li, H.; Liu, Y.; Cui, Y. Chiral covalent organic frameworks: Design, synthesis and property. *Chem. Soc. Rev.* **2020**, *49*, 6248–6272. [CrossRef] [PubMed]
47. Cote, A.P.; Benin, A.I.; Ockwig, N.W.; O'keeffe, M.; Matzger, A.J.; Yaghi, O.M. Porous, Crystalline, Covalent Organic Frameworks. *Science* **2005**, *310*, 1166–1170. [CrossRef] [PubMed]
48. Colson, J.W.; Woll, A.R.; Mukherjee, A.; Levendorf, M.P.; Spitler, E.L.; Shields, V.B.; Spencer, M.G.; Park, J.; Dichtel, W.R. Oriented 2D Covalent Organic Framework Thin Films on Single-Layer Graphene. *Science* **2011**, *332*, 228–231. [CrossRef]
49. Auras, F.; Ascherl, L.; Bon, V.; Vornholt, S.M.; Krause, S.; Döblinger, M.; Bessinger, D.; Reuter, S.; Chapman, K.W.; Kaskel, S.; et al. Dynamic two-dimensional covalent organic frameworks. *Nat. Chem.* **2024**, *16*, 1373–1380. [CrossRef]
50. Li, S.; Ma, R.; Xu, S.; Zheng, T.; Fu, G.; Wu, Y.; Liao, Z.; Kuang, Y.; Hou, Y.; Wang, D.; et al. Direct Construction of Isomeric Benzobisoxazol-Vinylene-Linked Covalent Organic Frameworks with Distinct Photocatalytic Properties. *J. Am. Chem. Soc.* **2022**, *144*, 13953–13960. [CrossRef]
51. Lavillunière, H.; Pham-Truong, T.-N.; Nguyen, T.-K.-L.; Vancaeyzeele, C.; Aubert, P.-H. Controlled Microwave-Assisted Synthesis of Covalent Organic Frameworks Opens the Way Toward More Suitable Porous Supercapacitor Electrodes. *ACS Appl. Energy Mater.* **2024**, *7*, 1723–1734. [CrossRef]
52. Dey, A.; Chakraborty, S.; Singh, A.; Rahimi, F.A.; Biswas, S.; Mandal, T.; Maji, T.K. Microwave Assisted Fast Synthesis of a Donor-Acceptor COF Towards Photooxidative Amidation Catalysis. *Angew. Chem. Int. Ed.* **2024**, *63*, e202403093. [CrossRef] [PubMed]
53. Song, K.S.; Fritz, P.W.; Abbott, D.F.; Poon, L.N.; Caridade, C.M.; Gándara, F.; Mougel, V.; Coskun, A. Mixed-metal Ionothermal Synthesis of Metallophthalocyanine Covalent Organic Frameworks for CO2 Capture and Conversion. *Angew. Chem. Int. Ed.* **2023**, *62*, e202309775. [CrossRef] [PubMed]
54. Wei, H.; Ning, J.; Cao, X.; Li, X.; Hao, L. Benzotrithiophene-Based Covalent Organic Frameworks: Construction and Structure Transformation under Ionothermal Condition. *J. Am. Chem. Soc.* **2018**, *140*, 11618–11622. [CrossRef] [PubMed]
55. Wang, Z.; Zhang, Y.; Liu, J.; Chen, Y.; Cheng, P.; Zhang, Z. Flux synthesis of two-dimensional covalent organic frameworks. *Nat. Protoc.* **2024**. [CrossRef] [PubMed]
56. Ruigómez, A.P.; Rodríguez-San-Miguel, D.; Stylianou, K.C.; Cavallini, M.; Gentili, D.; Liscio, F.; Milita, S.; Roscioni, O.M.; Ruiz-González, M.L.; Carbonell, C. Direct On-Surface Patterning of a Crystalline Laminar Covalent Organic Framework Synthesized at Room Temperature. *Chem.-Eur. J.* **2015**, *21*, 10666–10670. [CrossRef] [PubMed]

57. Biswal, B.P.; Chandra, S.; Kandambeth, S.; Lukose, B.; Heine, T.; Banerjee, R. Mechanochemical Synthesis of Chemically Stable Isoreticular Covalent Organic Frameworks. *J. Am. Chem. Soc.* **2013**, *135*, 5328–5331. [CrossRef]
58. Peng, Y.; Wong, W.K.; Hu, Z.; Cheng, Y.; Yuan, D.; Khan, S.A.; Zhao, D. Room Temperature Batch and Continuous Flow Synthesis of Water-Stable Covalent Organic Frameworks (COFs). *Chem. Mater.* **2016**, *28*, 5095–5101. [CrossRef]
59. Wang, Y.; Zhang, Y.; Yu, C.; Yin, B.H.; Chen, Q.; Sun, S.-P.; Wang, X. Chiral Covalent Organic Frameworks as Promising Materials for Racemate Resolution. *ACS Appl. Polym. Mater.* **2024**, *6*, 8706–8720. [CrossRef]
60. Wang, X.; Han, X.; Zhang, J.; Wu, X.; Liu, Y.; Cui, Y. Homochiral 2D Porous Covalent Organic Frameworks for Heterogeneous Asymmetric Catalysis. *J. Am. Chem. Soc.* **2016**, *138*, 12332–12335. [CrossRef]
61. Han, X.; Xia, Q.; Huang, J.; Liu, Y.; Tan, C.; Cui, Y. Chiral Covalent Organic Frameworks with High Chemical Stability for Heterogeneous Asymmetric Catalysis. *J. Am. Chem. Soc.* **2017**, *139*, 8693–8697. [CrossRef]
62. Wu, X.; Han, X.; Xu, Q.; Liu, Y.; Yuan, C.; Yang, S.; Liu, Y.; Jiang, J.; Cui, Y. Chiral BINOL-Based Covalent Organic Frameworks for Enantioselective Sensing. *J. Am. Chem. Soc.* **2019**, *141*, 7081–7089. [CrossRef] [PubMed]
63. Wang, R.; Wei, X.; Feng, Y. β-Cyclodextrin Covalent Organic Framework for Selective Molecular Adsorption. *Chem.-Eur. J.* **2018**, *24*, 10979–10983. [CrossRef] [PubMed]
64. Zhang, Y.; Duan, J.; Ma, D.; Li, P.; Li, S.; Li, H.; Zhou, J.; Ma, X.; Feng, X.; Wang, B. Three-Dimensional Anionic Cyclodextrin-Based Covalent Organic Frameworks. *Angew. Chem. Int. Ed.* **2017**, *56*, 16313–16317. [CrossRef] [PubMed]
65. Sánchez-Fuente, M.; López-Magano, A.; Moya, A.; Mas-Ballesté, R. Stabilized Chiral Organic Material Containing BINAP Oxide Units as a Heterogeneous Asymmetric Organocatalyst for Allylation of Aldehydes. *ACS Appl. Mater. Interfaces* **2023**, *15*, 30212–30219. [CrossRef]
66. Xu, H.-S.; Ding, S.-Y.; An, W.-K.; Wu, H.; Wang, W. Constructing Crystalline Covalent Organic Frameworks from Chiral Building Blocks. *J. Am. Chem. Soc.* **2016**, *138*, 11489–11492. [CrossRef]
67. Wang, L.-K.; Zhou, J.-J.; Lan, Y.-B.; Ding, S.-Y.; Yu, W.; Wang, W. Divergent Synthesis of Chiral Covalent Organic Frameworks. *Angew. Chem. Int. Ed.* **2019**, *58*, 9443–9447. [CrossRef]
68. Qian, H.-L.; Yang, C.-X.; Yan, X.-P. Bottom-up synthesis of chiral covalent organic frameworks and their bound capillaries for chiral separation. *Nat. Commun.* **2016**, *7*, 12104–12110. [CrossRef]
69. Zhang, K.; Cai, S.-L.; Yan, Y.-L.; He, Z.-H.; Lin, H.-M.; Huang, X.-L.; Zheng, S.-R.; Fan, J.; Zhang, W.-G. Construction of a hydrazone-linked chiral covalent organic framework–silica composite as the stationary phase for high performance liquid chromatography. *J. Chromatogr. A* **2017**, *1519*, 100–109. [CrossRef]
70. Zhang, J.; Han, X.; Wu, X.; Liu, Y.; Cui, Y. Multivariate Chiral Covalent Organic Frameworks with Controlled Crystallinity and Stability for Asymmetric Catalysis. *J. Am. Chem. Soc.* **2017**, *139*, 8277–8285. [CrossRef]
71. Ma, H.-C.; Kan, J.-L.; Chen, G.-J.; Chen, C.-X.; Dong, Y.-B. Pd NPs-Loaded Homochiral Covalent Organic Framework for Heterogeneous Asymmetric Catalysis. *Chem. Mater.* **2017**, *29*, 6518–6524. [CrossRef]
72. Han, X.; Huang, J.; Yuan, C.; Liu, Y.; Cui, Y. Chiral 3D Covalent Organic Frameworks for High Performance Liquid Chromatographic Enantioseparation. *J. Am. Chem. Soc.* **2018**, *140*, 892–895. [CrossRef] [PubMed]
73. Hou, B.; Yang, S.; Yang, K.; Han, X.; Tang, X.; Liu, Y.; Jiang, J.; Cui, Y. Confinement-Driven Enantioselectivity in 3D Porous Chiral Covalent Organic Frameworks. *Angew. Chem. Int. Ed.* **2021**, *60*, 6086–6093. [CrossRef] [PubMed]
74. Li, Y.; Wang, J.-M.; Kan, J.-L.; Li, F.; Dong, Y.; Dong, Y.-B. Combination of a Metal-N-Heterocyclic-Carbene Catalyst and a Chiral Aminocatalyst within a Covalent Organic Framework: A Powerful Cooperative Approach for Relay Asymmetric Catalysis. *Inorg. Chem.* **2022**, *61*, 2455–2462. [CrossRef] [PubMed]
75. Fu, S.; Qin, G.; Dong, J.; Yuan, C.; Liu, Y.; Yuan, L.; Cui, Y. Construction of chiral crown ethers into robust covalent organic frameworks for electrochromatographic enantioseparation. *Nat. Sci. Rev.* **2024**, *11*, nwae256. [CrossRef]
76. Yuan, C.; Fu, S.; Kang, X.; Cheng, C.; Jiang, C.; Liu, Y.; Cui, Y. Mixed-Linker Chiral 2D Covalent Organic Frameworks with Controlled Layer Stacking for Electrochemical Asymmetric Catalysis. *J. Am. Chem. Soc.* **2024**, *146*, 635–645. [CrossRef]
77. Wang, K.; Hou, B.; Dong, J.; Niu, H.; Liu, Y.; Cui, Y. Controlling the Degree of Interpenetration in Chiral Three-Dimensional Covalent Organic Frameworks via Steric Tuning. *J. Am. Chem. Soc.* **2024**, *146*, 21466–21475. [CrossRef] [PubMed]
78. Kang, X.; Cheng, C.; Chen, X.; Dong, J.; Liu, Y.; Cui, Y. Three-Dimensional Homochiral Covalent Organic Frameworks with Intrinsic Chiral qzd Topology. *J. Am. Chem. Soc.* **2024**, *146*, 8407–8416. [CrossRef]
79. Han, X.; Jiang, C.; Hou, B.; Liu, Y.; Cui, Y. Covalent Organic Frameworks with Tunable Chirality for Chiral-Induced Spin Selectivity. *J. Am. Chem. Soc.* **2024**, *146*, 6733–6743. [CrossRef]
80. Hou, B.; Wang, K.; Jiang, C.; Guo, Y.; Zhang, X.; Liu, Y.; Cui, Y. Homochiral Covalent Organic Frameworks with Superhelical Nanostructures Enable Efficient Chirality-Induced Spin Selectivity. *Angew. Chem. Int. Ed.* **2024**, *63*, e202412380.
81. Xi, W.; Liu, Y.; Xia, Q.; Li, Z.; Cui, Y. Direct and Post-Synthesis Incorporation of Chiral Metallosalen Catalysts into Metal-Organic Frameworks for Asymmetric Organic Transformations. *Chem.-Eur. J.* **2015**, *21*, 12581–12585. [CrossRef]
82. Cohen, S.M. The Postsynthetic Renaissance in Porous Solids. *J. Am. Chem. Soc.* **2017**, *139*, 2855–2863. [CrossRef] [PubMed]
83. Xu, H.; Gao, J.; Jiang, D. Stable, crystalline, porous, covalent organic frameworks as a platform for chiral organocatalysts. *Nat. Chem.* **2015**, *7*, 905–912. [CrossRef] [PubMed]
84. Zhou, Y.; Wei, Y.; Ren, J.; Qu, X. A chiral covalent organic framework (COF) nanozyme with ultrahigh enzymatic activity. *Mater. Horiz.* **2020**, *7*, 3291–3297. [CrossRef]

85. Zhang, S.; Zheng, Y.; An, H.; Aguila, B.; Yang, C.-X.; Dong, Y.; Xie, W.; Cheng, P.; Zhang, Z.; Chen, Y.; et al. Covalent Organic Frameworks with Chirality Enriched by Biomolecules for Efficient Chiral Separation. *Angew. Chem. Int. Ed.* **2018**, *57*, 16754–16759. [CrossRef]
86. Yuan, C.; Wu, X.; Gao, R.; Han, X.; Liu, Y.; Long, Y.; Cui, Y. Nanochannels of Covalent Organic Frameworks for Chiral Selective Transmembrane Transport of Amino Acids. *J. Am. Chem. Soc.* **2019**, *141*, 20187–20197. [CrossRef]
87. Qiao, S.; Jin, H.; Zuo, A.; Chen, Y. Integration of Enzyme and Covalent Organic Frameworks: From Rational Design to Applications. *Acc. Chem. Res.* **2024**, *57*, 93–105. [CrossRef]
88. Zheng, Y.; Zhang, S.; Guo, J.; Shi, R.; Yu, J.; Li, K.; Li, N.; Zhang, Z.; Chen, Y. Green and Scalable Fabrication of High-Performance Biocatalysts Using Covalent Organic Frameworks as Enzyme Carriers. *Angew. Chem. Int. Ed.* **2022**, *134*, e202208744. [CrossRef]
89. Zhang, J.; Han, X.; Wu, X.; Liu, Y.; Cui, Y. Chiral DHIP- and Pyrrolidine-Based Covalent Organic Frameworks for Asymmetric Catalysis. *ACS Sustainable Chem. Eng.* **2019**, *7*, 5065–5071. [CrossRef]
90. Zhuo, S.; Wang, X.; Li, L.; Yang, S.; Ji, Y. Chiral Carboxyl-Functionalized Covalent Organic Framework for Enantioselective Adsorption of Amino Acids. *ACS Appl. Mater. Interfaces* **2021**, *13*, 31059–31065. [CrossRef]
91. Han, X.; Zhang, J.; Huang, J.; Wu, X.; Yuan, D.; Liu, Y.; Cui, Y. Chiral induction in covalent organic frameworks. *Nat. Commun.* **2018**, *9*, 1294. [CrossRef]
92. Li, F.; Kan, J.-L.; Yao, B.-J.; Dong, Y.-B. Synthesis of Chiral Covalent Organic Frameworks via Asymmetric Organocatalysis for Heterogeneous Asymmetric Catalysis. *Angew. Chem. Int. Ed.* **2022**, *61*, e202115044. [CrossRef] [PubMed]
93. Chen, H.; Gu, Z.-G.; Zhang, J. Chiral-Induced Ultrathin Covalent Organic Frameworks Nanosheets with Tunable Circularly Polarized Luminescence. *J. Am. Chem. Soc.* **2022**, *144*, 7245–7252. [CrossRef] [PubMed]
94. Wang, J.-C.; Kan, X.; Shang, J.-Y.; Qiao, H.; Dong, Y.-B. Catalytic Asymmetric Synthesis of Chiral Covalent Organic Frameworks from Prochiral Monomers for Heterogeneous Asymmetric Catalysis. *J. Am. Chem. Soc.* **2020**, *142*, 16915–16920. [CrossRef] [PubMed]
95. Chan, A.Y.; Perry, I.B.; Bissonnette, N.B.; Buksh, B.F.; Edwards, G.A.; Frye, L.I.; Garry, O.L.; Lavagnino, M.N.; Li, B.X.; Liang, Y.; et al. Metallaphotoredox: The Merger of Photoredox and Transition Metal Catalysis. *Chem. Rev.* **2022**, *122*, 1485–1542. [CrossRef]
96. Mateo, D.; Cerrillo, J.L.; Durini, S.; Gascon, J. Fundamentals and applications of photo-thermal catalysis. *Chem. Soc. Rev.* **2021**, *50*, 2173–2210. [CrossRef]
97. Hong, J.; Xu, C.; Deng, B.; Gao, Y.; Zhu, X.; Zhang, X.; Zhang, Y. Photothermal chemistry based on solar energy: From synergistic effects to practical applications. *Adv. Sci.* **2022**, *9*, 2103926. [CrossRef] [PubMed]
98. Ma, H.-C.; Zhao, C.-C.; Chen, G.-J.; Dong, Y.-B. Photothermal conversion triggered thermal asymmetric catalysis within metal nanoparticles loaded homochiral covalent organic framework. *Nat. Commun.* **2019**, *10*, 3368. [CrossRef]
99. Ma, H.-C.; Chen, G.-J.; Huang, F.; Dong, Y.-B. Homochiral covalent organic framework for catalytic asymmetric synthesis of a drug intermediate. *J. Am. Chem. Soc.* **2020**, *142*, 12574–12578. [CrossRef]
100. Ma, H.-C.; Sun, Y.-N.; Chen, G.-J.; Dong, Y.-B. A BINOL-phosphoric acid and metalloporphyrin derived chiral covalent organic framework for enantioselective a-benzylation of aldehydes. *Chem. Sci.* **2022**, *13*, 1906–1911. [CrossRef]
101. Li, X.-T.; Zou, J.; Wang, T.-H.; Ma, H.-C.; Chen, G.-J.; Dong, Y.-B. Construction of Covalent Organic Frameworks via Three-Component One-Pot Strecker and Povarov Reactions. *J. Am. Chem. Soc.* **2020**, *142*, 6521–6526. [CrossRef]
102. Kan, X.; Wang, J.-C.; Chen, Z.; Du, J.-Q.; Kan, J.-L.; Li, W.-Y.; Dong, Y.-B. Synthesis of Metal-Free Chiral Covalent Organic Framework for Visible-Light-Mediated Enantioselective Photooxidation in Water. *J. Am. Chem. Soc.* **2022**, *144*, 6681–6686. [CrossRef] [PubMed]
103. He, T.; Liu, R.; Wang, S.; Keng, I.; On, W.; Wu, Y.; Xing, Y.; Yuan, W.; Guo, J.; Zhao, Y. Bottom-Up Design of Photoactive Chiral Covalent Organic Frameworks for Visible-Light-Driven Asymmetric Catalysis. *J. Am. Chem. Soc.* **2023**, *145*, 18015–18021. [CrossRef] [PubMed]
104. Jin, C.; Li, N.; Lin, E.; Chen, X.; Wang, T.; Wang, Y.; Yang, M.; Liu, W.; Yu, J.; Zhang, Z.; et al. Enzyme Immobilization in Porphyrinic Covalent Organic Frameworks for Photoenzymatic Asymmetric Catalysis. *ACS Catal.* **2022**, *12*, 8259–8268. [CrossRef]

molecules

MDPI

Review

Transition Metal Catalysis for the Asymmetric Synthesis of 2-Arylethylamines: A Review of the New Millennium

Alejandro Manchado, Ángel García-González, Carlos T. Nieto, Nieves G. Ledesma, David Díez and Narciso M. Garrido *

Department of Organic Chemistry, Faculty of Chemical Sciences, University of Salamanca, Pl. Caídos, s/n, 37008 Salamanca, Spain; alex92mc@usal.es (A.M.); u149349@usal.es (Á.G.-G.); eneas@usal.es (C.T.N.); nievesgarcial11@usal.es (N.G.L.); ddm@usal.es (D.D.)

* Correspondence: nmg@usal.es

Abstract: The 2-arylethylamine motif is very well-known in medicinal chemistry because of its interesting properties when it comes to interacting with the Central Neural System thanks to its ability to pass the blood–brain barrier. This nitrogen-containing family of compounds is of great interest in synthetic organic chemistry and, when it comes to its asymmetric synthesis, great challenges can be faced in order to obtain the chiral purity required in the drug industry. Thus, we provide a concise transition metal review presenting the recent advances in the synthesis of chiral 2-arylethylamines using transition metals as the main catalysts in the introduction of chirality. Both conventional and photocatalysis methods will be covered, considering the main transition metal used in the studies.

Keywords: 2-phenethylamine; 2-arylethylamine; asymmetric synthesis; transition metal catalysis; metal catalysis; photocatalysis; transition metal-catalyzed asymmetric catalysis and synthesis; kinetic resolution; bioactive compounds; molecular design; organic synthesis

1. Introduction

Phenethylamine derivatives constitute a well-established class of bioactive compounds with significant pharmacological relevance [1,2]. Their relatively low molecular weight and amphiphilic nature facilitate their passage through the blood–brain barrier, allowing direct interaction with dopaminergic neurons and modulating key physiological processes such as motor control, stress response, and mood regulation [3]. These scaffolds, whether of natural origin (e.g., dopamine) or synthetically designed (e.g., amphetamine and its analogs), serve as the backbone for a wide range of central nervous system stimulants used in the treatment of neurological and psychiatric disorders (Figure 1).

Our research group has previously reviewed the medicinal chemistry of 2-phenethylamines and related scaffolds, examining key pharmacophores and their therapeutic relevance [4], as well as extending this analysis to bioactive 2-heteroarylethylamines [5]. More recently, we examined metal-free strategies for the asymmetric synthesis of 2-arylethylamines (AEA), covering chiral induction catalysis, organocatalysis, organophotocatalysis, and enzymatic approaches [6].

Previous reviews on this topic have discussed multiple aspects of the AEA syntheses, emphasizing different strategies and catalytic approaches. Some have focused on the advances made in asymmetric direct hydrogenation of C=N bonds with metal catalysts, such as those by Wencel-Delord et al. in 2021 [7] and Riera et al. in 2022 [8]. Other approaches have also been explored, such as enzymes and chemoenzymatic cascades making use of amine transaminases by Rueping et al. in 2023 [9], as well as the broader review by Moran et al. [10] encompassing both metal-catalyzed and metal-free synthetic methods towards β-(hetero)arylethylamines. From a substrate-based perspective, alkene functionalization has emerged as a powerful strategy for AEA synthesis, as highlighted in reviews by Wang et al. in 2022 [11] and Zhang et al. in 2025 [12]. Additionally, aziridines have been recently investigated by Lee et al. [13] as key intermediates for the synthesis of medicinally relevant chiral amines, while Xiaowei et al. reviewed the asymmetric photocatalytic synthesis of azaarene derivatives in 2022 [14]. Also, collectively, these studies examine the plethora of options available in the synthesis of AEAs, each providing some unique advantages and scope within the field.

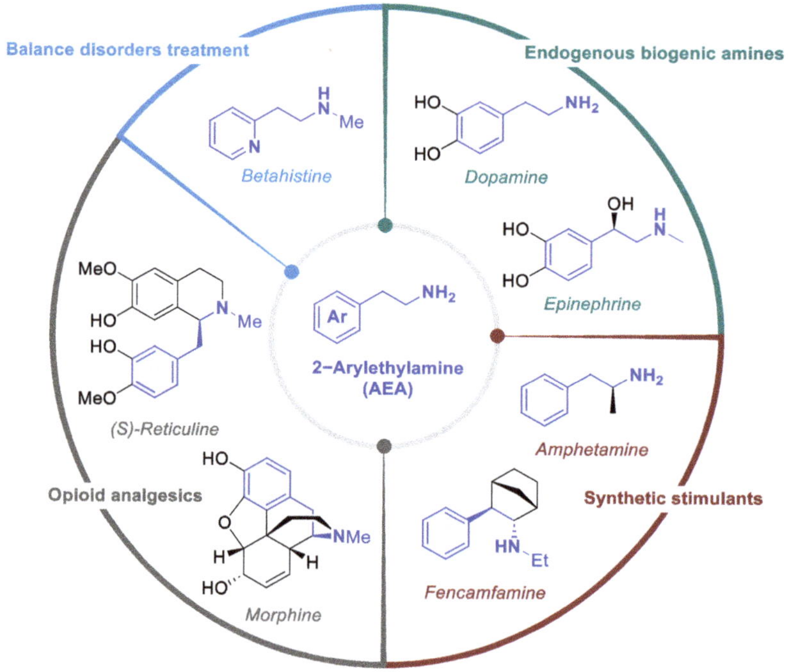

Figure 1. Representative examples of 2-arylethylamine derivatives.

The use of transition metal catalysis in the synthesis of enantioenriched compounds has been proved to be great tool in the synthesis of bioactive compounds which are suitable for the pharmaceutical industry [7–9]. In comparison with non-transition metal methodologies, this field has a large number of potential applications, both when it comes to suitable substrates and the moieties synthesized [6]. As for the synthesis of chiral AEAs, using transition metal catalysis has proved to be a path towards the development of interesting strategies, such as asymmetric hydrogenation of C=N bonds, asymmetric ring-openings, or C–N cross-coupling reactions, among others. These key methodologies, with which it is difficult to achieve success in their organocatalytic counterpart [6], are the main core of the synthetic advances in the synthesis of asymmetric 2-arylethylamine.

Due to the lack of reviews considering both methodology and substrate scopes, the current work aims to contribute to this body of knowledge by providing a comprehensive review of asymmetric metal catalysis for chiral AEA synthesis that, together with our previous metal-free review, completes the current domain of asymmetric synthesis of the aforementioned family of compounds in the new millennium.

Thus, we report the main advances in the last millennium in asymmetric synthesis of 2-AEA when using transition metal catalysis, covering the chiral synthesis of AEA skeletons considering different substituents, functional groups, and aryl/heteroaryl rings and their combinations, such as those shown in Figure 2A. Chirality is considered either in C-1, C-2, or in both carbon atoms. Also, different moieties in which nitrogen is two carbons apart from an aryl group are not covered, as shown in Figure 2B.

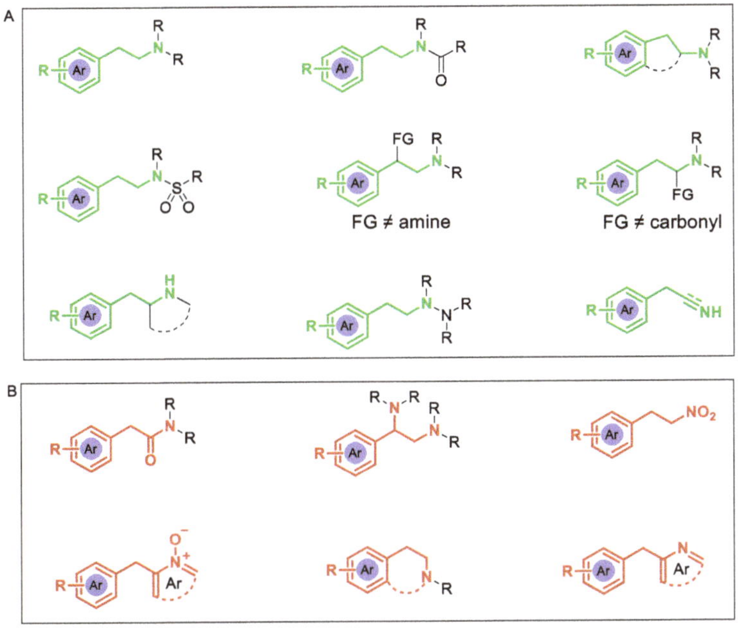

Figure 2. (**A**) 2-arylethylamines covered in this review. (**B**) Compounds not covered in this review.

This review is organized based on the transition metal used to enable the main step in the production of asymmetric AEA, with the transition metals organized based on the number of studies on the topic (Figure 3). Also, photocatalysis-based methodologies will be presented separately, and main and novel catalytic cycles will be presented and discussed.

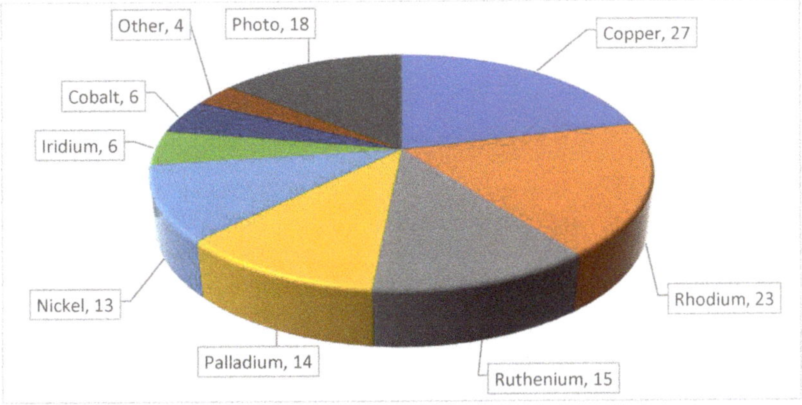

Figure 3. Number of studies on chiral AEA based on the transition metal used to perform the corresponding methodology, as well as photocatalysis.

2. Conventional Metal Catalysis

2.1. Copper

In 2007, Chemler et al. [15] reported a Cu(II)-catalyzed enantioselective intramolecular carboamination of alkenes, involving intramolecular addition of arylsulfonamides across terminal alkenes to provide chiral sultams. Chirality was enabled by a chiral copper–oxazolinyl complex and the corresponding products could be derivatized in order to obtain chiral AEA, in which the nitrogen atom is embedded into a pyrrolidine ring (Scheme 1).

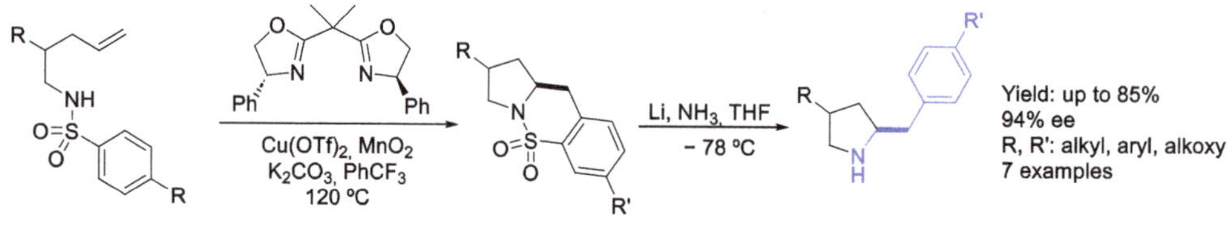

Scheme 1. Chiral synthesis of AEA from sultam derivatives by Chemler et al. in 2017.

In 2011, Hajra et al. [16] reported a catalytic enantioselective one-pot aziridoarylation reaction of aryl cinnamyl ethers followed by an intramolecular arylation (Friedel–Crafts) reaction towards the synthesis of an interesting chiral AEA based on arylchromans. In situ-generated tethered aziridine provides easy access to *N*-sulfonyl-protected *trans*-3-amino-4-arylchromans in moderate yields and good ee by using the chiral organo-copper complex as a catalyst (Scheme 2).

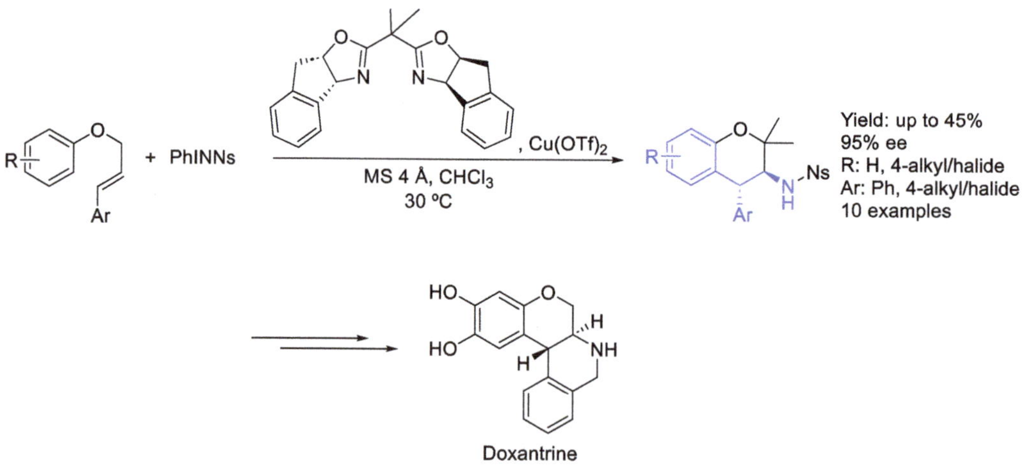

Scheme 2. Chiral arylchorman-based AEA synthesized by Hajra et al. in 2011 with copper catalysis.

Three years later, in 2014, Miura et al. [17] developed the asymmetric synthesis of (borylmethyl)cyclopropylamines by aminoboration of methylenectclopropanes using chiral-copper catalysis. This highly regio- and stereoselective process could afford interesting AEA in excellent yields and dr, which could be derivatized in order to obtain versatile *trans*-2-phenylcyclopropylamine derivatives, interesting potential moieties in medicinal chemistry (Scheme 3A). Two years later [18], they reported the synthesis of β-boryl-α-aminosilanes. This copper-catalyzed regioselective and stereospecific aminoboration of vinylsilanes with bis(pinacolato)diboron(pinB-Bpin) and hydroxylamines afforded asymmetric silyl boron AEA in good yields and dr (Scheme 3B).

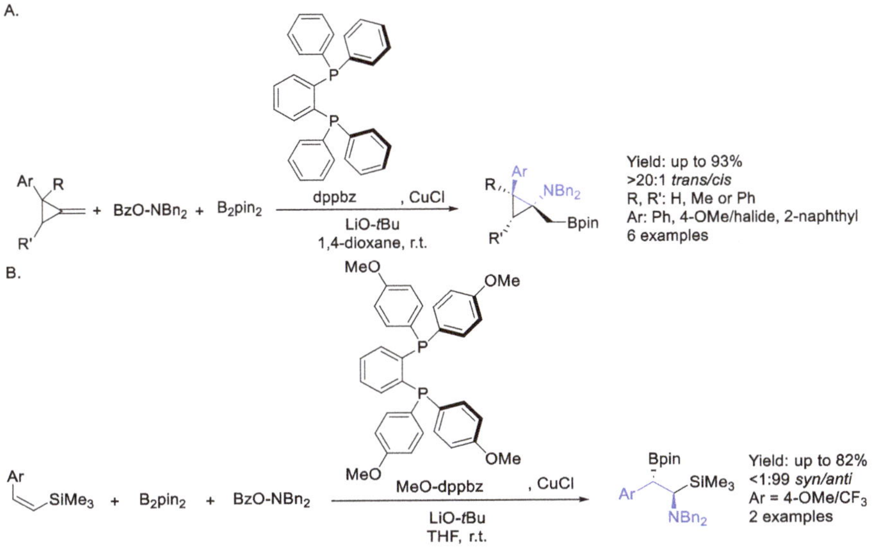

Scheme 3. Cyclopropane-based chiral AEA (**A**) and β-boryl-α-aminosilane-based AEA (**B**) synthesized in 2014 and in 2016 by Miura et al., respectively.

In 2014, Tortosa et al. [19] reported the first asymmetric synthesis of cyclopropylboronates with a quaternary stereocenter by copper-catalyzed diastereo- and enantioselective desymmetrization of cyclopropenes. Trapping the cyclopropylcopper intermediate with electrophilic amines allowed for the synthesis of chiral AEA with good yields and ee (Scheme 4).

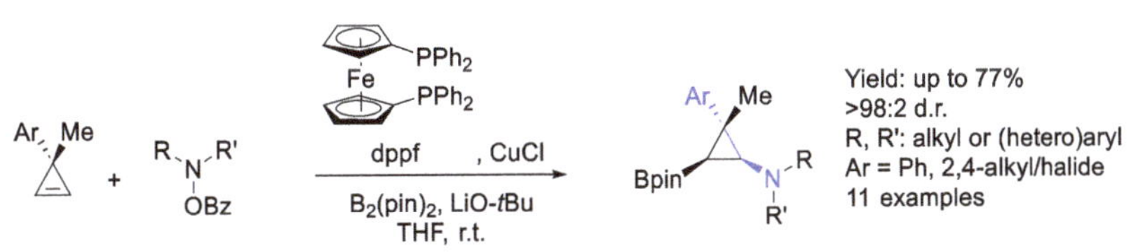

Scheme 4. Asymmetric synthesis of AEA based on cyclopropylboronates by Tortosa et al. in 2014.

Two years later, Buchwald et al. [20]. developed the enantioselective synthesis of an interesting AEA by copper catalysis. Starting from styrene-derived nucleophiles and imine derivatives, this intermolecular enantioselective addition of styrenes to imines afforded highly enantiomerically enriched amines bearing contiguous stereocenters in excellent yields and ee. This method relied on the use of styrenes as latent carbanion equivalents via the intermediacy of catalytically generated benzyl-copper derivatives. Interestingly, mechanistic studies remarked on the preferential styrene hydrocupration in the presence of an imine with the Ph-BPE-derived copper catalyst (Scheme 5).

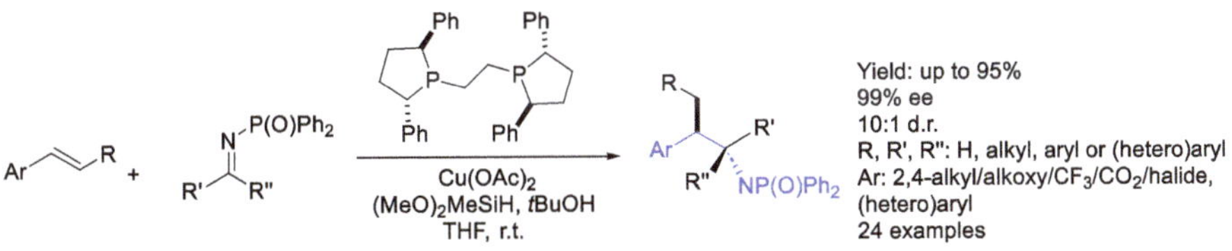

Scheme 5. Chiral AEA by copper catalysis synthesized in 2016 by Buchwald et al.

Also in 2016, Jia et al. [21] reported a Cu(I)-catalyzed enantioselective Friedel–Crafts alkylation of indoles with 2-aryl-*N*-sulfonylaziridines (Scheme 6). This process afforded indole-substituted chiral AEA in good yields and ee. Starting from racemic substituted aziridine, asymmetric addition by ring-opening reaction with indoles could be possible due to chiral Cu(I)/(*S*)-Segphos catalyst.

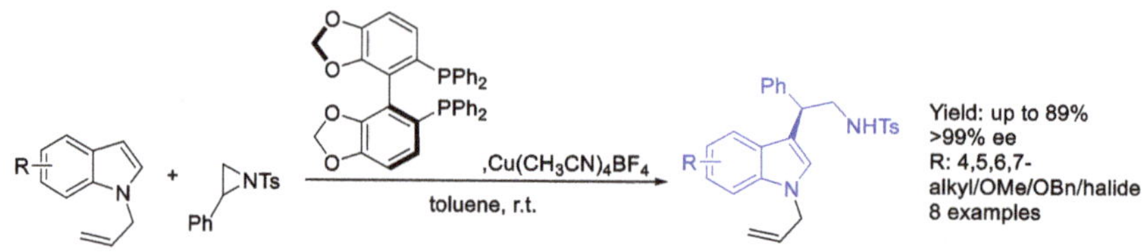

Scheme 6. Asymmetric synthesis of indolethylamines by Jia et al. in 2016.

Ma et al., also in 2016 [22], developed the asymmetric synthesis of a-silyl *N*-tosylamides by using [2.2]paracyclophane-based *N*-heterocyclic carbene in combination with a copper complex. The dual role of the new carbene precursor as an organocatalyst and as a ligand of the copper catalyst was crucial in the synthesis of the desired products. Starting from the corresponding imines, this methodology could afford a particular chiral α-silyl AEA in good yield and ee (Scheme 7), explained by the donor capacity of the carbene center through the transannular electronic effects, which significantly alters the activity of the resulting copper catalyst.

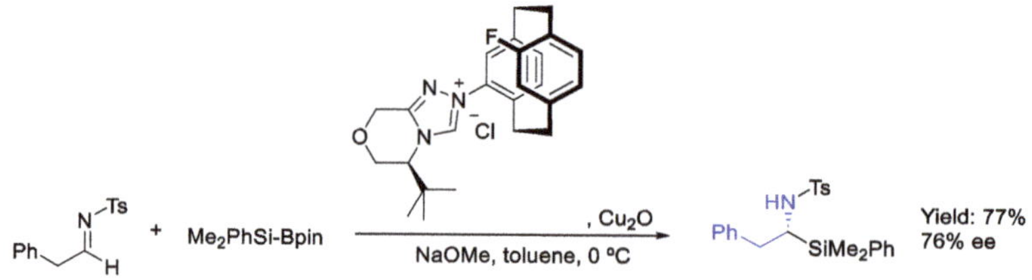

Scheme 7. Chiral AEA synthesis via transannular organo-copper catalyst by Ma et al. in 2016.

In the same year, Carretero et al. [23] reported the direct synthesis of polisubstituted chiral pyrrolidines by catalytic asymmetric 1,3-dipolar cycloaddition reactions of azomethine ylides, enabled by chiral Cu(I)/(*R*)-DTBM-Segphos or Ag(I)/(*R*)-Fesuphos complexes. This methodology could allow total control of the stereocenters by using different copper or silver complexes, resulting in either *exo*- or *endo*-4-aryl substituted pyrrolidine with high diastereoselectivity and excellent enantioselectivity (Scheme 8).

In 2017, Chung et al. [24] reported the formal synthesis of Omarigliptin, a long-acting DPP-4 inhibitor. The key step in the methodology was the asymmetric synthesis of a β-phenethylnitro compound from the corresponding benzaldehyde derivative and nitromethane, enabled by the chiral copper–dibenzylamine complex (Scheme 9). The subsequent treatment of the substrate led to chiral AEA Omarigliptin in excellent yield and ee.

Liu et al. have shown great success when it comes to the synthesis of chiral AEA using organo-copper complexes as transition metal catalysts. In 2017 [25], they reported the asymmetric copper-catalyzed intermolecular aminoarylation of styrenes. Starting from the corresponding styrene derivative and *N*-fluoro-*N*-alkylsulfonamide, several chiral AEA were synthesized in good yields and ee by its addition to styrene, so the generated benzylic radical could couple with a chiral L*Cu(II)Ar complex (Scheme 10A). Also, they reported the enantioselective copper-catalyzed intermolecular amino- and azidocyanation of styrenes in order to synthesize an interesting AEA in which the nitrogen functional group consisted of either a sulfonamide, a nitrile, or an azide [26]. By using organo-copper complexes, a variety of enantiomerically enriched β-amino/azido alkylnitriles were efficiently synthesized in excellent yields and ee. Also, the corresponding cyano-azido compounds could be derivatized in order to obtain bioactive compounds as antibacterials (Scheme 10B). Finally, in 2021 [27], following their studies in synthetic applications of styrene derivatives when it comes to organo-copper catalysis, they reported an enantioselective intermolecular aminoalkynylation using *N*-fluoro-*N*-alkylsulfonamides as nitrogen-centered radical precursors and alkynyltrimethoxysilanes as alkynylating reagents. This methodology tolerates wide substrate scope, high functional group tolerance, and mild conditions, and its products can be derivatized in order to obtain interesting synthons when it comes to both synthetic and medical applications (Scheme 10C).

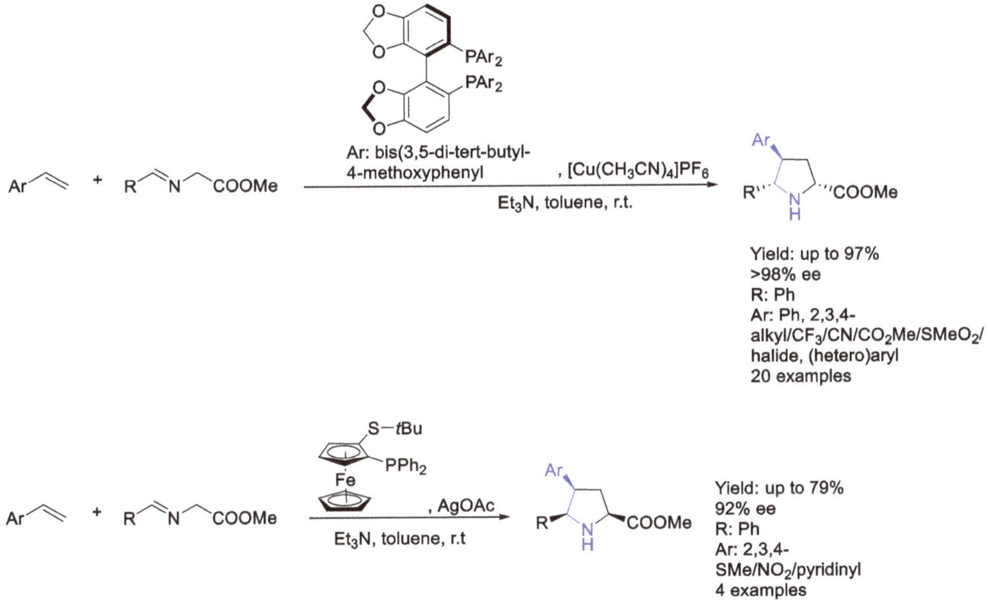

Scheme 8. Asymmetric synthesis of pyrrolidine-based AEA by Carretero et al. in 2016.

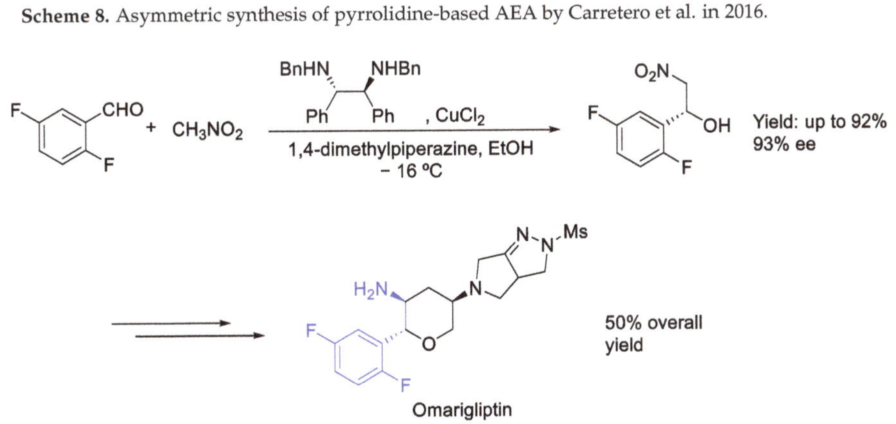

Scheme 9. Omarigliptin synthesis in 2017 by Chung et al.

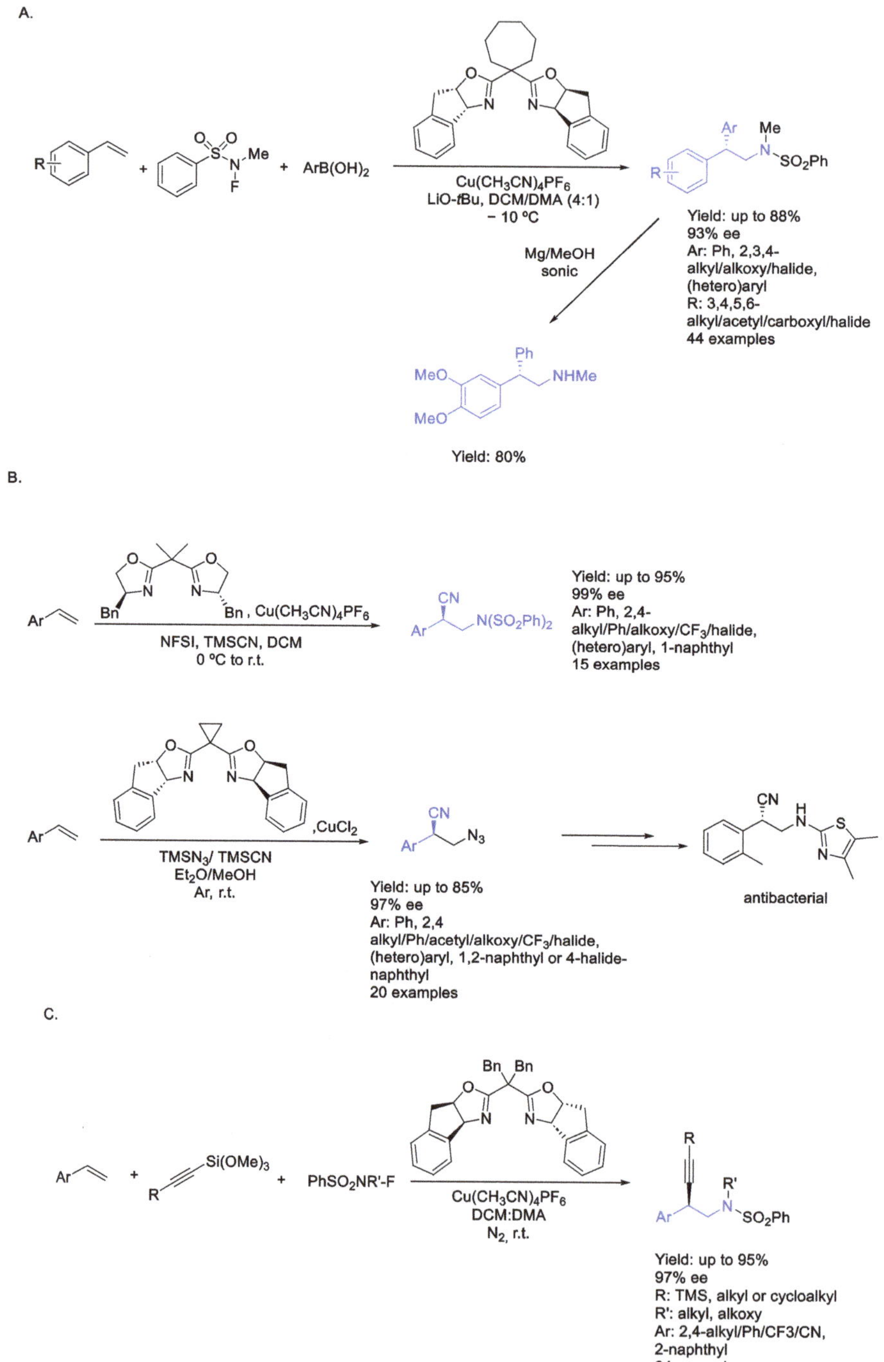

Scheme 10. (**A**) *N*-sulfonyl-based AEA via copper catalysis by Liu et al. in 2017. (**B**) Amino- and azidocyanation towards the synthesis of an interesting AEA with high synthetic potential synthesized in 2017 by Liu et al. (**C**) Intermolecular aminoalkynylation towards the synthesis of an asymmetric AEA by Liu et al. in 2012.

In 2018, Chai et al. [28] reported the synthesis of a chiral AEA by copper-catalyzed dynamic kinetic asymmetric transformation of racemic *N*-sulfonylaziridines. Interestingly, different products were afforded when starting from either aryl aldehyde derivatives or indol derivatives, via C–N bond cleavage with nucleophiles, including [3 + 2] annulations with (hetero)aromaticaldehydes and 1,3-disubstituted indoles, an asymmetric Friedel–Crafts-type reaction with electron-rich (hetero)arenes and asymmetric aminolysis with amines, resulting in a chiral AEA in which the nitrogen atom is embedded into an oxazolidine or into a fused dihydroindolopyrrolidine, respectively (Scheme 11). Thus, excellent yields and ee were achieved.

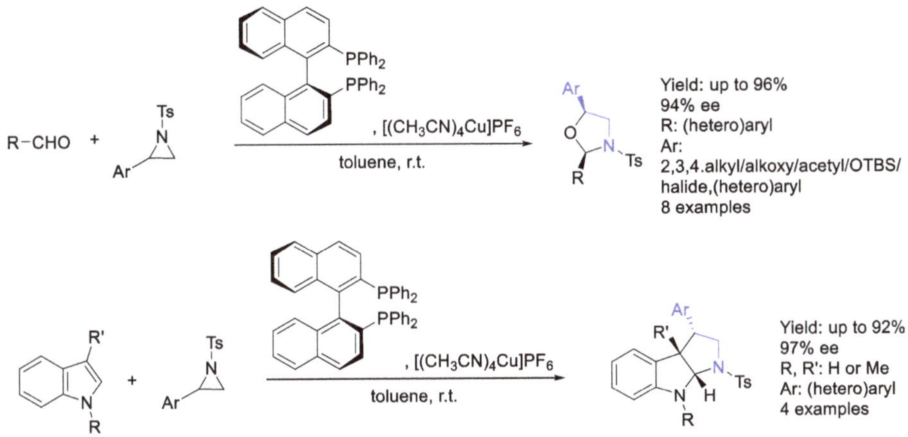

Scheme 11. Oxazolidines and dihydroindolopyrrolidines synthesized by Chai et al. in 2018.

In 2019, Zhao et al. [29]. developed the synthesis of chiral 2-arylcyclopropylamines enabled by copper-catalyzed cyclopropene carbometallation with an organo-boron reagent. This three-component synthesis afforded a polisubstituted AEA in excellent yields and ee. By using different chiral organic complexes, a large variety of compounds could be synthesized, representing the first example of highly enantioselective multicomponent cyclopropane synthesis (Scheme 12A). Mechanistic insights proposed that oxidative trapping of copper complex D by *O*-benzoyl hydroxylamine 3 delivered the desired products, while N insertion in early produced copper complex B did not occur, as shown in the proposed catalytic cycle (Scheme 12B).

In 2021, Wang et al. [30] developed the synthesis of an interesting polycyclic chiral AEA by Cu(I)-catalyzed asymmetric 1,3-dipolar cycloaddition of azomethine ylides in excellent yields and ee. The resulting asymmetric AEA consisted of pyrrolidine substrates and could be obtained in gram-scale, as well as derivatized in order to obtain key intermediates in the synthesis of bioactive compounds, such as HCV inhibitors. Various β-substituted alkenyl heteroarenes were successfully employed as dipolarophiles for the first time (Scheme 13). Also, DFT calculations proposed uncommon dual activation/coordination of both the dipole and dipolarophile substrates by the metal, in which a sterically bulky, rigid, and monodentate phosphoramidite ligand with triplehomoaxial chirality played a pivotal role in providing an effective chiral pocket around the metal center. The additional coordination of the heteroatom in the dipolarophile substrate to copper is also critical for the exclusive diastereoselectivity and enhanced reactivity.

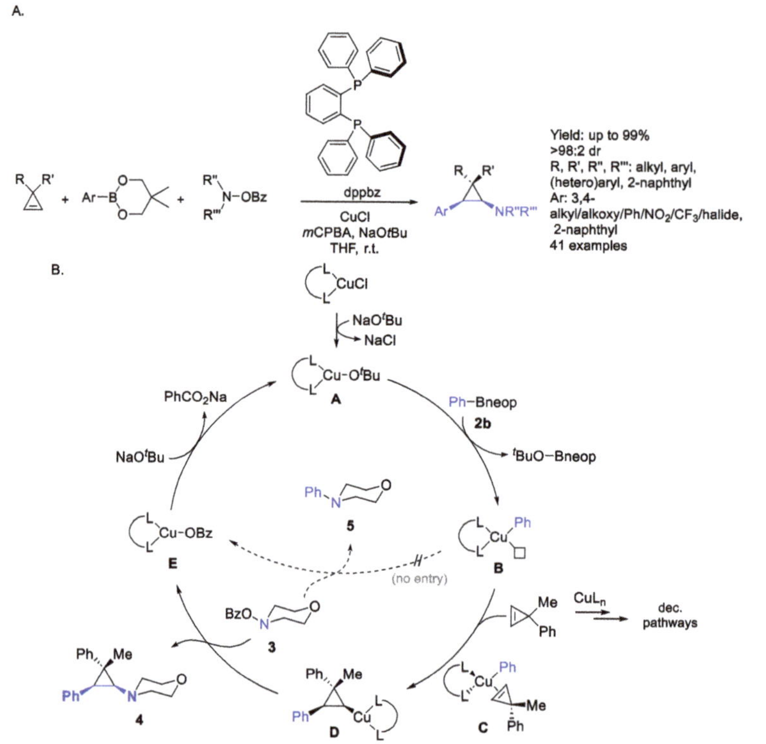

Scheme 12. (**A**) Chiral AEA by Zhao et al. in 2019. (**B**) Proposed mechanism for the copper-catalyzed reaction.

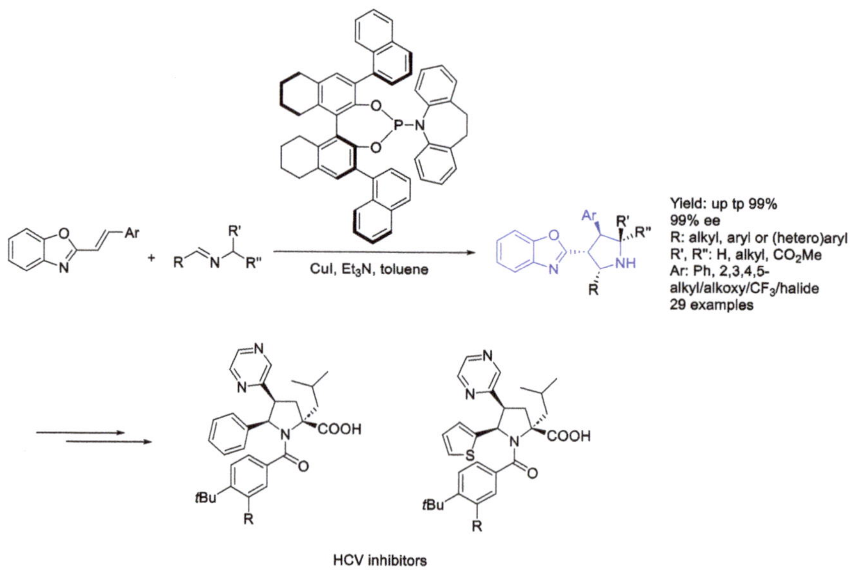

Scheme 13. HCV derivatized from chiral AEA by Wang et al. in 2021.

Following this copper-catalyzed asymmetric synthesis of AEA, Hirano et al. have shown great success in recent years. Starting in 2021 [31], they developed the regioselective synthesis of α-aminoacid by asymmetric hydroamination of α,β-unsaturated carbonyls enabled by an organo-copper complex. This resulted in an interesting AEA that incorporated ester groups into its structure (Scheme 14A). The key to regioselectivity control was the use of hydroxylamine as an umpolung, electrophilic amination reagent, and remote steric hindrance by both the chiral catalyst and DTBM-dppbz ligand. One year later [32], they published back-to-back studies based on the synthesis of α-amino acids by either borylamination or silylamination reations enabled by carefully chosen bulky chiral organo-copper complexes, synthesizing a large variety of chiral AEAs and achieving great success in both yields and ee, with α-substitution of either boron or silane derivatives (Scheme 14B). Finally, in 2023 [33], they reported the asymmetric synthesis of α-aminophosphonates by copper-catalyzed regioselective hydroamination of α,β-unsaturated phosphonates. Following their previous strategies, an umpolung, electrophilic amination with the hydroxylamine resulted in an interesting chiral AEA in great ee and moderate dr (Scheme 14C).

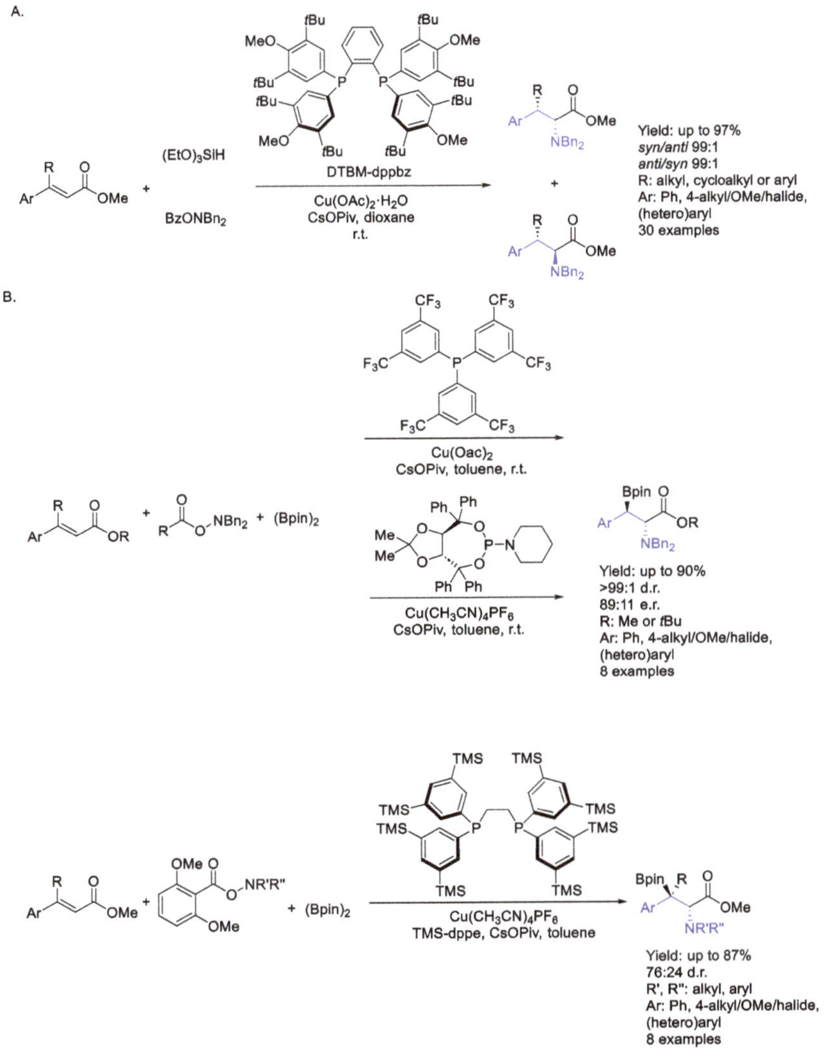

Scheme 14. *Cont.*

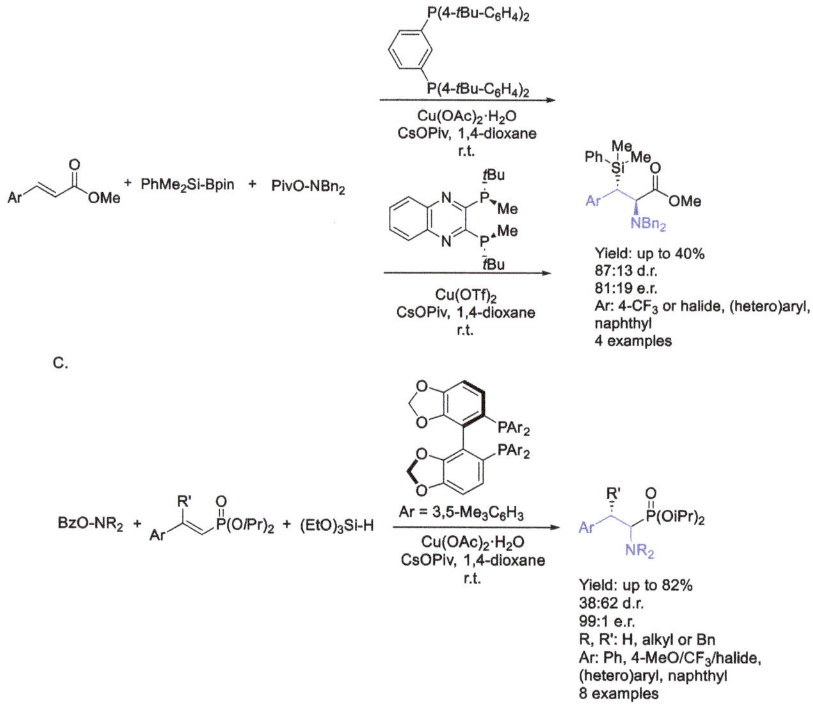

Scheme 14. Chiral synthesis of AEA by copper catalysis by Hirano et al. in 2021 (**A**), 2022 (**B**), and 2023 (**C**).

Following that year's studies, Yu et al. [34]. developed the synthesis of interesting β-chiral amines with quaternary stereocenters by copper-catalyzed reductive aminomethylation of 1,3-dienes with *N*,*O*-acetals, resulting in a chiral AEA with high chemo-, regio-, *E*/*Z*- and enantioselectivities, as shown in Scheme 15.

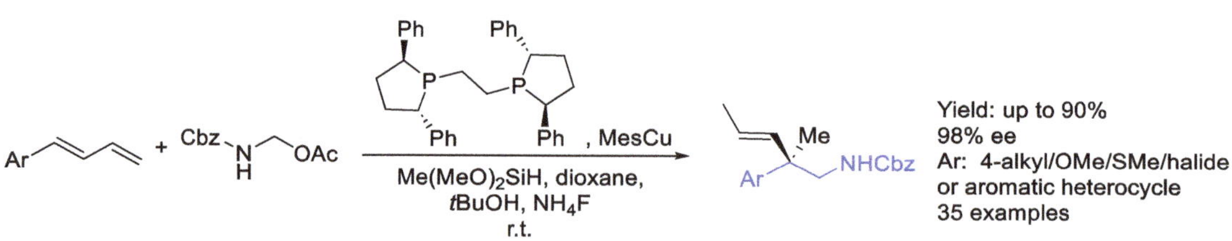

Scheme 15. Chiral AEA from 1,3-dienes and *N*,*O*-acetals synthesized in 2023 by Yu et al.

In 2023, Zhou et al. [35] developed a methodology in which an intermolecular enantioselective benzylic C(sp^3)-H amination by cationic copper catalysis occurred, synthesizing useful α-phenethylamines (Scheme 16). Mechanistic studies suggested that the amination likely proceeded via a radical mechanism in which C–H bond cleavage was the rate-limiting step. In this process, an interesting chiral β-phenyl phenethylamine was produced in moderate yield and great ee.

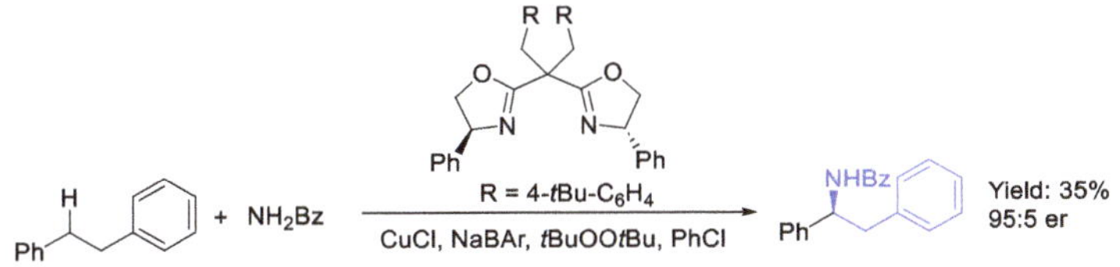

Scheme 16. Single AEA synthesized by Zhou et al.'s methodology in 2023.

Also in 2023, Liu et al. [36]. reported a copper-catalyzed enantioconvergent radical C(sp^3)–N cross-coupling of activated racemic alkyl halides with (hetero)aromatic amines, resulting in asymmetric quaternary carbon AEA (Scheme 17). Copper catalysis in combination with multidentate anionic ligands led to excellent yields and ee, as this kind of ligand could not only enhance the reducing capability of the copper catalyst to provide an enantioconvergent radical pathway but also avoid coordination with other coordinating heteroatoms, thereby overcoming catalyst poisoning and/or chiral ligand displacement.

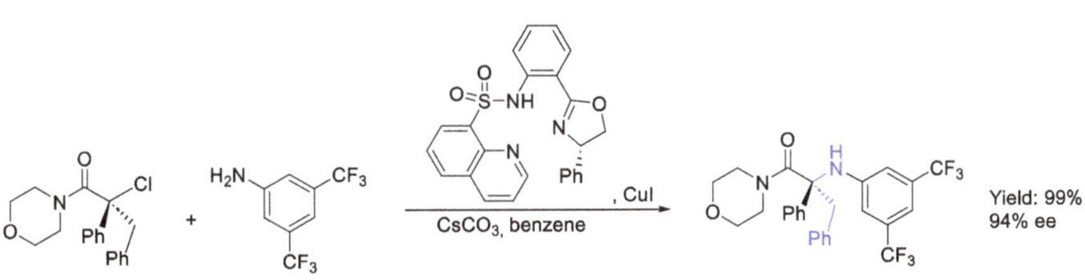

Scheme 17. Chiral quaternary carbon containing AEA by Liu et al. in 2023.

Ending that year's research, Sieber et al. [37] reported an interesting synthesis of chiral 1,2-aminoalcohols through enantioselective copper-catalyzed reductive coupling of aldehydes and allenamides, overcoming the problematic competitive reduction of the aldehyde electrophile by a CuH catalyst. Enantiopure vinyl AEAs were synthesized in great ee and dr. Also, subsequent derivatization of the products could lead to key intermediate in the synthesis of eliglustat, a treatment for Gaucher's disease (Scheme 18).

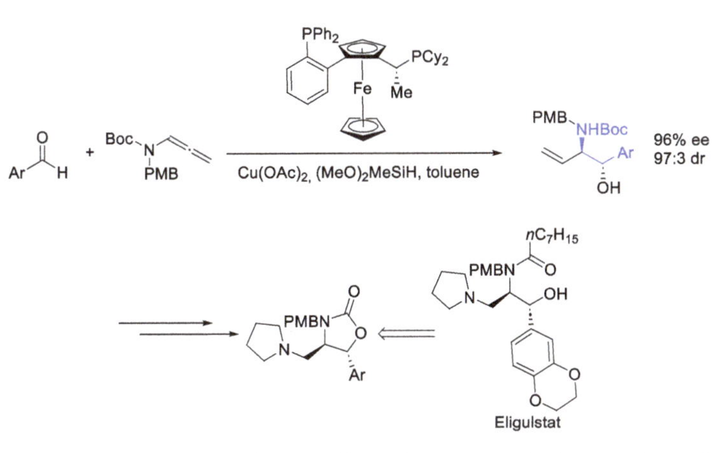

Scheme 18. Asymmetric synthesis of AEAs and their derivatization towards an eliglustat precursor by Sieber et al.

Later, in 2024, Guo et al. [38] reported the synthesis of an interesting chiral AEA by dynamic kinetic stereodivergent transformation of racemic propargylic ammonium salts enabled by dual copper and nickel catalysis. The resulting α-quaternary amino ester containing an optically active AEA was synthesized via C–N bond cleavage in good yields and ee (Scheme 19A). A plausible mechanism was proposed, in which nickel catalysis intertwined with copper catalysis for the DyKAT of racemic propar-gylic ammonium salts with prochiral aldimine esters. Thus, chiral nickel complex cleavaged C–N bonded to generate the electrophilic allenylnickel(II) intermediates. Subsequent nucleophilic addition of the copper-coordinated azomethine ylides onto the allenylnickel(II) intermediates via a catalytic enantio–convergent pathway would provide a novel stereoselective route to access chiral α-tertiary amines through the propargylation process (Scheme 19B).

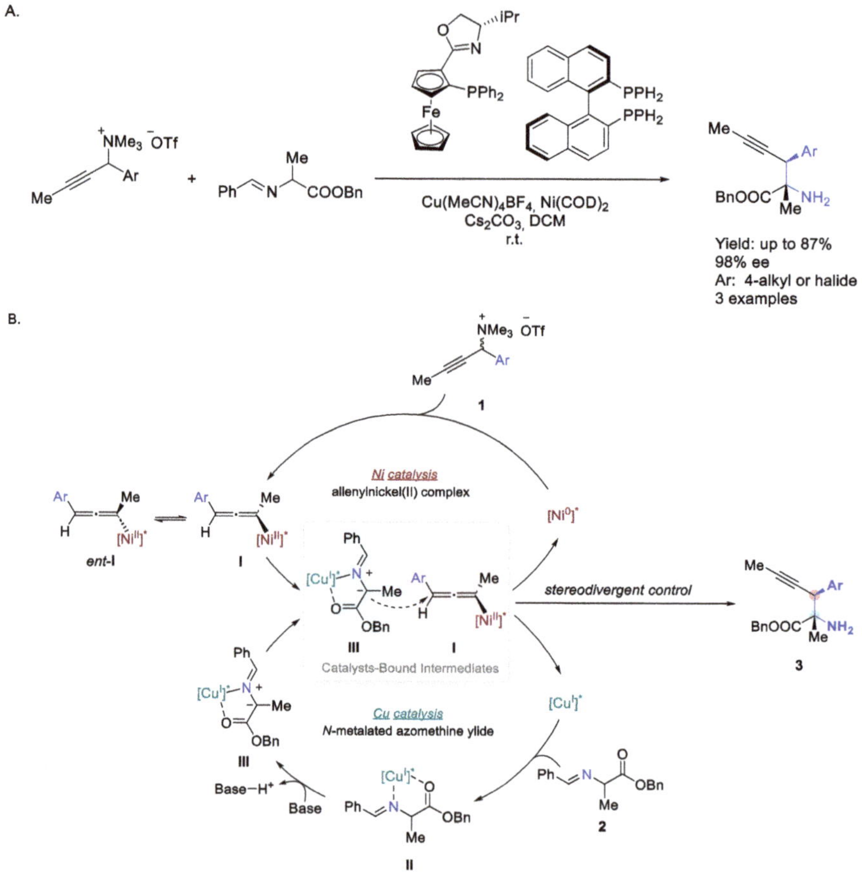

Scheme 19. Chiral quaternary carbon containing AEA (**A**) and proposed dual catalysis mechanism (**B**) by Guo et al.

Also in 2024, Malcolmson et al. [39] developed a methodology for the (*Z*)-selective aminoallylation of a range of ketones to prepare allylic 1,2-amino tertiary alcohols. Thus, copper-catalyzed reductive couplings of 2-azatrienes with aryl/alkyl and dialkyl ketones proceed with Ph-BPE as the supporting ligand, generating *anti*-amino alcohols with >98% (*Z*)-selectivity under mild conditions. Thus, subsequent derivatization of the corresponding products afforded a range of chiral AEAs in excellent yields and ee (Scheme 20).

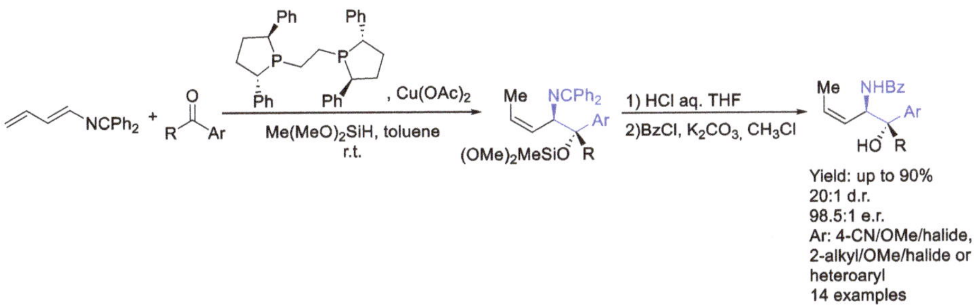

Scheme 20. Chiral 1,2-aminoalcoholes by Macolmson et al. in 2024.

The same year, Lee at al. [40] reported an interesting regioselective and stereospecific aryl Grignard addition for the copper-catalyzed trifluoromethyl aziridine opening. Desired products could be derivatized in order to obtain interesting azaindoles. An achiral copper complex was used for the ring opening of optically active activated aziridines with aryl iodides in order to synthesize an asymmetric AEA in good yields and maintaining the ee (Scheme 21).

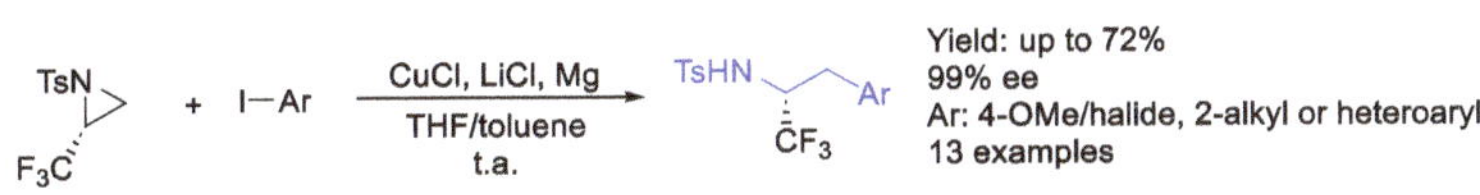

Scheme 21. Single chiral AEA through copper catalysis by Lee et al. in 2024.

Finally, Zhao et al. [41] reported the enantioselective aminative difunctionalization of alkenes via copper-catalyzed electrophilic addition with nitrogen sources, synthesizing cyclic hydrazine derivatives via either [3 + 2] cycloaddition or intramolecular cyclization in high chemo-, regio-, enantio-, and diastereoselectivities. By using a chiral organo-copper complex, interesting polisubstituted chiral AEAs were produced in excellent yields and ee. Also, subsequent derivatization could lead to potent orexin receptor antagonist 2-Epi-CP-99,994 (Scheme 22).

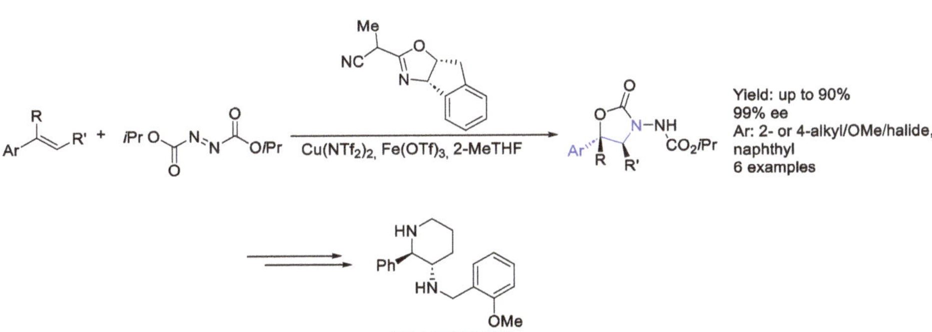

Scheme 22. Polisubstituted chiral AEAs and their derivatization by Zhao et al. in 2024.

2.2. Rhodium

Starting in 2009, Wallace et al. [42] reported the asymmetric synthesis of the bioactive compound taranabant, a CB1R inverse agonist in the treatment of obesity. The key step in the introduction of chirality was the asymmetric hydrogenation of the corresponding substrate enabled by an organo-rhodium catalyst based on a ferrocene moiety, leading to a couple of chiral AEAs in great yields and ee (Scheme 23).

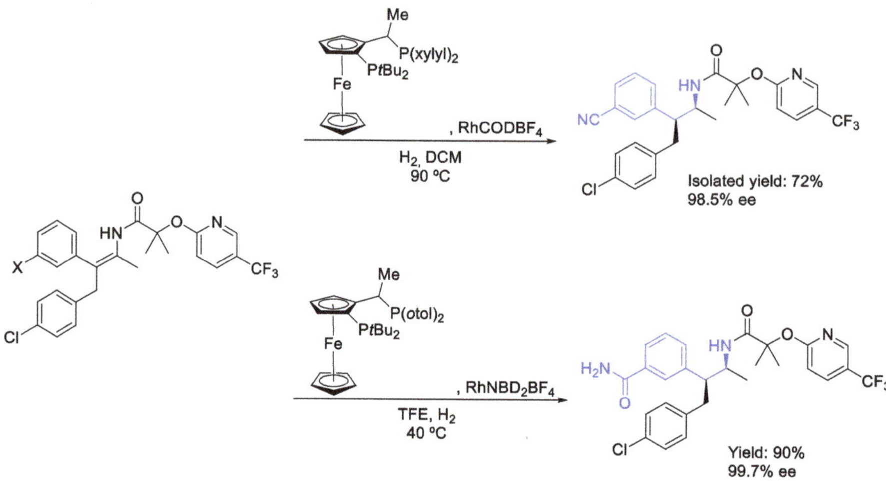

Scheme 23. Chiral AEA by rhodium catalysis by Wallace et al. in the synthesis of taranabant.

In the same year, Leighton et al. [43] developed the asymmetric synthesis of an interesting chiral AEA in which the nitrogen atom is included in a hydrazine group. This rhodium–zinc dual catalyzed tandem asymmetric aza-arzens/ring-opening reaction led to excellent yields and ee, affording interesting α-amino esters as products that could be derivatized to obtain the corresponding free amine substrates (Scheme 24). Interestingly, the chiral silane Lewis acid performs two distinct functions, activating the initially formed aziridine toward ring-opening reactions with either chloride or arene nucleophiles to deliver complex amino acid derivatives in a simple one-pot process.

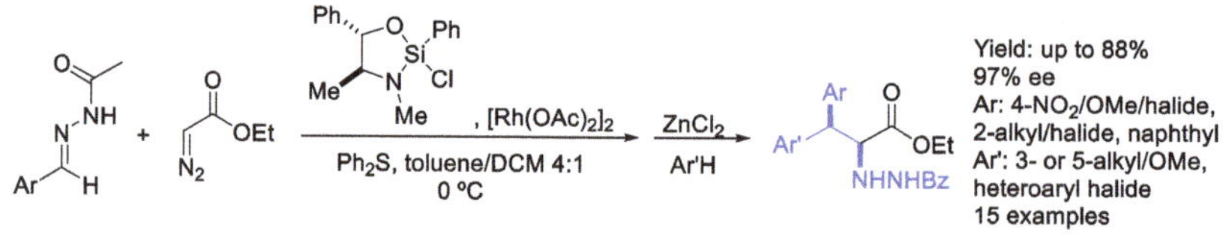

Scheme 24. Asymmetric biaryl AEA by Leighton et al. in 2009.

Two years later, in 2011, Praquin et al. [44] developed an asymmetric hydrogenation route to (*S*)-*N*-Boc-2,6-dimethyltyrosine. The discovery of a new chiral organo-rhodium complex based on a ferrocene moiety was the key step in the improvement of the previously known synthetic routes. Thus, the corresponding substrate was hydrogenated in excellent ee and yield, leading to an interesting tyrosine derivative (Scheme 25).

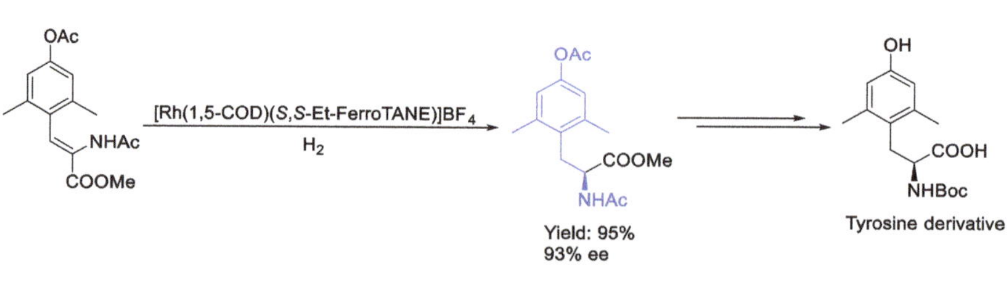

Scheme 25. Asymmetric tyrosine derivative through rhodium catalysis by Praquin et al. in 2011.

In the same year, Fox et al. [45] reported large-scale synthesis of a phenylalanine derivative by asymmetric hydrogenation using a chiral organo-rhodium catalyst from the corresponding enamine substrate. This multi-gram reaction could afford the desired product in excellent yield and ee (Scheme 26)

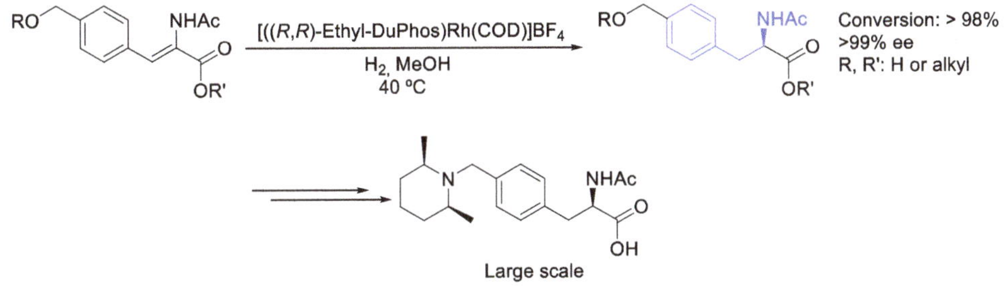

Scheme 26. Large-scale synthesis of phenylalanine derivative by Fox et al. in 2011.

One year later, Anderson et al. [46] reported the direct addition of arylboronic acids to unsaturated esters containing basic γ-amino groups. A Miyaura–Hayashi-based synthetic route was developed using asymmetric rhodium catalysis, obtaining chiral AEAs as γ-aminobutirc acid derivatives in good yields and ee (Scheme 27).

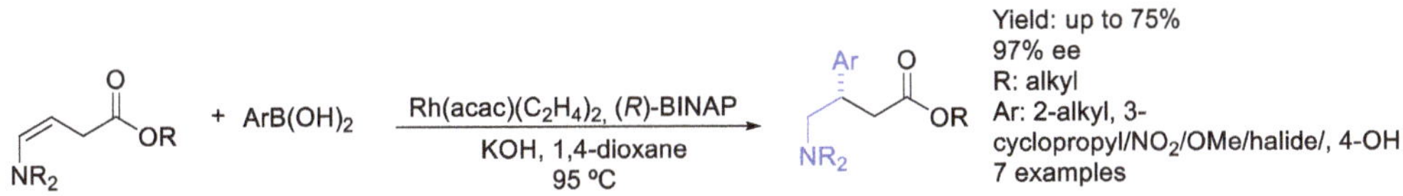

Scheme 27. Asymmetric synthesis of γ-aminobutirc acid derivatives by Anderson et al. in 2012.

Also in 2012, Ramsden et al. [47] developed a multi-kilogram reaction towards the synthesis of (*S*)-*N*-Boc-bis(4-fluorophenyl)alanine, by asymmetric hydrogenation of the corresponding substrate using rhodium catalysis (Scheme 28). They achieved the manufacture of 900 kg of the desired product in excellent yield and ee by subsequent isolations.

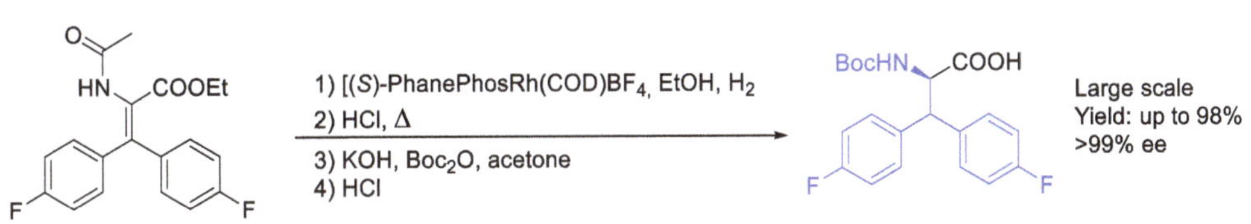

Scheme 28. Multi-kilogram scale symmetric synthesis of AEA by Ramsden et al. in 2012.

In 2015, Xiao et al. [48] reported an asymmetric synthesis of an α-substituted chiral AEA. This asymmetric hydroaminomethylation of styrenes was enabled by a rhodium–phosphine species and a chiral phosphoric acid and afforded good yields and ee (Scheme 29A). The transformation was made possible by combining metal- and organocatalysis, with the former catalyzing the hydroformylation while the latter catalyzed reductive amination via DKR, as shown in Scheme 29B.

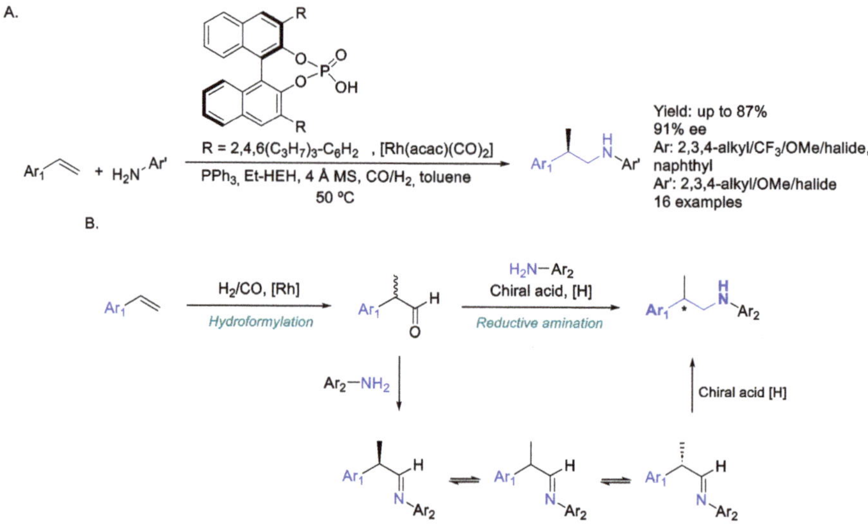

Scheme 29. Asymmetric synthesis of AEA (**A**) by DKR process (**B**) by Xiao et al. in 2015.

In 2016, Zhang et al. [49] reported the first interrupted asymmetric hydroaminomethylation reaction, obtaining an interesting chiral AEA in which the nitrogen atom is embedded into a pyrrolidine ring. Starting from *trans*-1,2-disubstituted olefins, the resulting enantiopure products were obtained in excellent yields and ee using a chiral organo-ruthenium complex (Scheme 30). The key step in the synthesis was trapping the resulting hemiaminal intermediate by either oxidation with PCC or reduction with boron species, ending in the desired compounds.

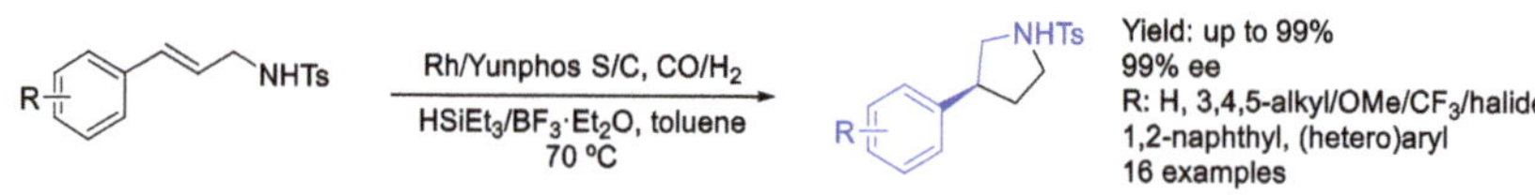

Scheme 30. Pyrrolidine-containing chiral AEA by Zhang et al. in 2016.

In 2017, Feng et al. [50] reported a rhodium-catalyzed chemoselective and regioselective allylic alkylation of hydroxyarenes with vinyl aziridines. Starting from the corresponding optically active aziridine substrates, this methodology afforded an interesting wide range of chiral 2-vinyl-2-arylethylamine derivatives in good yields and ee by taking advantage of chirality-transfer strategy. Thus, this methodology demonstrated that hydroxyarenes can serve as a *C*-nucleophiles rather than *O*-nucleophiles in rhodium-catalyzed regioselective ring-opening reactions of vinylaziridines (Scheme 31).

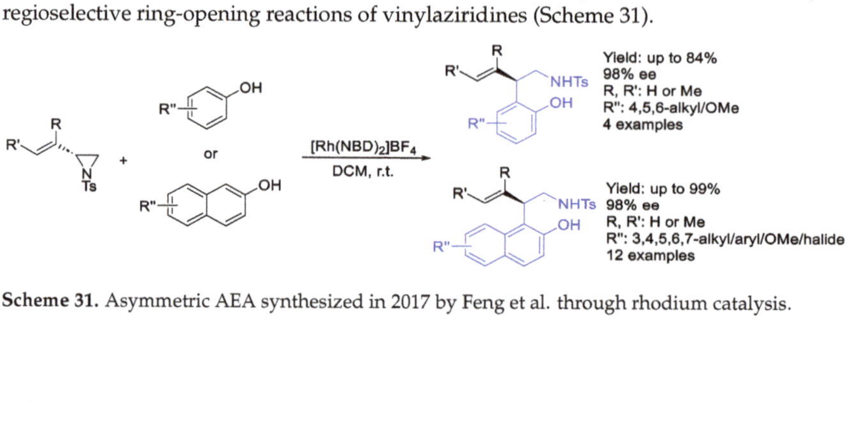

Scheme 31. Asymmetric AEA synthesized in 2017 by Feng et al. through rhodium catalysis.

Han et al., in 2017 [51], reported the synthesis of chiral α-methyl AEA in excellent yields and ee by using a rhodium catalyst with a commercial ligand and a phosphoric acid catalyst by asymmetric hydroaminomethylation enabled by rhodium-complex formation, which has been proved to be a very good strategy in order to synthesize chiral AEAs (Scheme 32). The relay catalytic reaction consisted of a rhodium-catalyzed hydroformylation step and Brønsted acid-catalyzed subsequent dynamic kinetic reductive amination process.

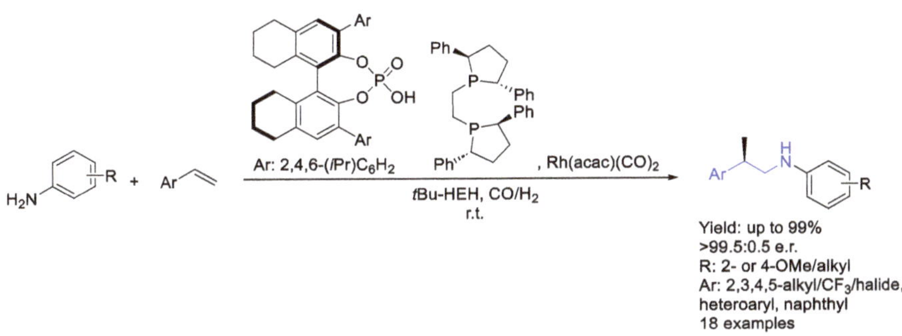

Scheme 32. Chiral AEA enabled by rhodium catalysis by Hen et al. in 2017.

One year later, Zhang et al. [52] reported the asymmetric hydrogenation of γ-lactams, introducing great values of enantiopurity by using an organo-rhodium complex based on a ferrocene moiety. This methodology could afford a chiral AEA in which the nitrogen atom is embedded into a chiral γ-lactam ring (Scheme 33). Also, they proved the methodology's applicability by the synthesis of drug molecules such as rolipram, baclofen, and α-2-adrenoceptor antagonist.

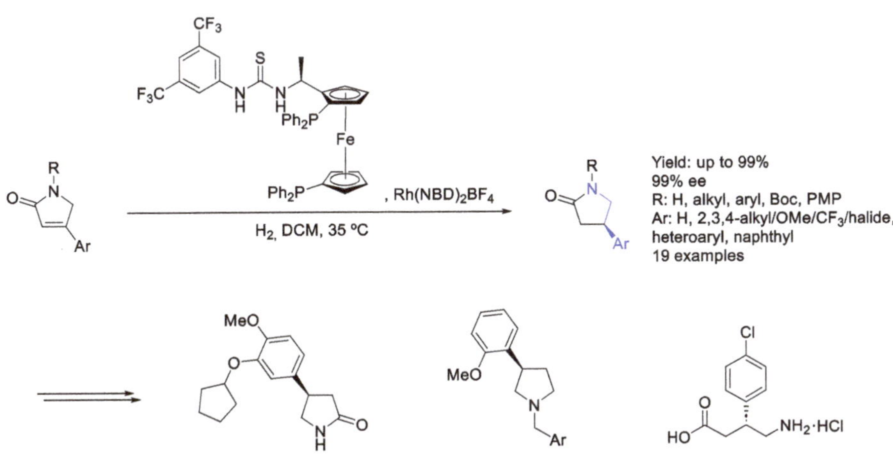

Scheme 33. Chiral γ-lactam and their derivatization towards bioactive compounds by Zhang et al. in 2018.

Also in 2018, Li et al. [53] developed a highly enantioselective synthesis of propargyl amides enabled by rhodium-catalyzed asymmetric hydroalkynylation of enamides. This process afforded a great family of chiral AEAs in excellent yields and ee (Scheme 34). Starting from the corresponding silyl alkyne substrate, this methodology relied on stereospecific hydroalkynylation of the electron-rich alkene, a completely different mechanism with conventional alkynylations that occurred under proton-transfer conditions.

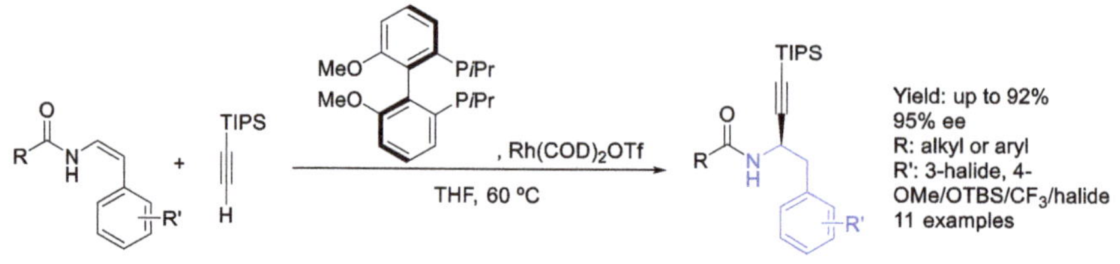

Scheme 34. Chiral α-propargyl amide synthesis by Li et al. in 2017.

In 2020, Blakey et al. [54] developed the synthesis of a new organo-rhodium-based catalyst for regio- and enantioselective allylic C–H amidation. Thus, a planar chiral rhodium indenyl complex was successfully used in order to synthesize a chiral AEA by the asymmetric allylic C–H amidation of unactivated olefins, delivering a wide range of high-value enantioenriched AEAs in excellent yields and excellent ee (Scheme 35). Computational studies suggested that C–H cleavage is rate- and enantio-determining, while reductive C–N coupling from the Rh(V)-nitrenoid intermediate is regio-determining.

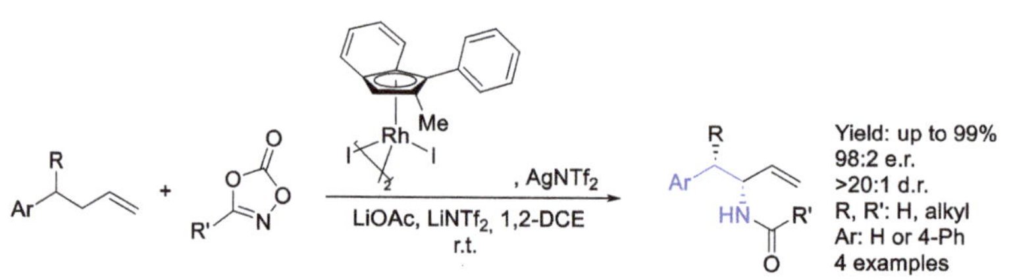

Scheme 35. Asymmetric synthesis of β-vinyl AEA by Blakey et al. in 2020.

In 2021, Wang et al. reported [55] a rhodium-catalyzed enantioselective C–H activation three-component coupling of arene, diene, and dioxazolone towards the synthesis of a chiral allylic AEA, in 1,2-*ortho*-selectivity, *E*-selectivity, and enantioselectivity. This reaction proceeded via a C–H activation pathway en route to thermodynamically stable as well as kinetically reactive π-ally intermediates. Thus, excellent yields and ee were achieved (Scheme 36).

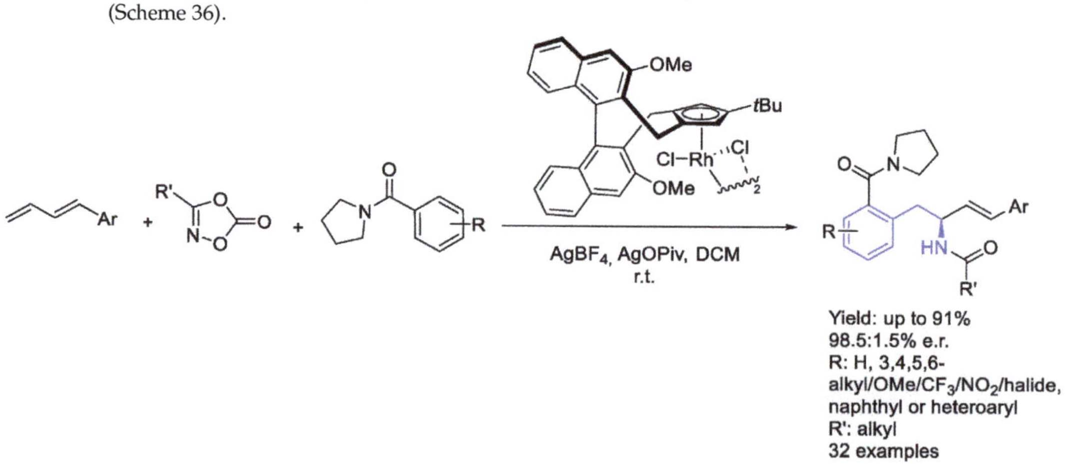

Scheme 36. Rhodium-catalysed chiral AEA by Wang et al. in 2021.

The same chiral allylic amine structural motif was also synthesized at the same time and enabled by the same type of organo-rhodium catalyst by Yi et al. [56]. Thus, starting from the corresponding *N*-phenoxy amides, intermolecular aryl C–H activation and intramolecular amide transfer were achieved by CpXRh(III)-catalyzed enantioselective intermolecular carboamination (Scheme 37). Sequential formation of a completely regioselective C–C bond and a highly enantioselective C–N bond occurred in a catalytic pathway in which an alkene insertion/π-allylation/intramolecular nucleophilic substitution cascade was involved.

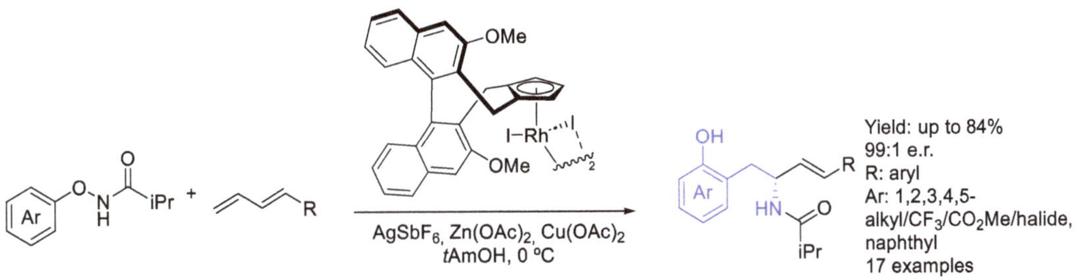

Scheme 37. CpXRh(III)-catalyzed asymmetric AEA by Yi et al. in 2021.

One year later, Xu et al. [57] developed an asymmetric three-component reaction towards the synthesis of chiral α-alkoxy-β-amino-carboxylates by rhodium-catalysis, resulting in a polisubstituted asymmetric AEA (Scheme 38A). Thus, asymmetric formal aminohydroxylation of diazo compounds with *O*-benzyl hydroxylamines involving N–O bond activation, fragment modification, and a reassembly cascade process was achieved. This cascade reaction forms multiple bonds in one pot, including C=N, C–O, and C–C bonds, providing a potent complement for the aminohydroxylation, as shown in the proposed double catalytic cycle (Scheme 38B), thus resulting in good yields and ee.

Also in 2022, Franciò et al. [58] kept pushing the synthetic methodology of styrene derivatives by hydroaminomethylation reactions enabled by rhodium catalysis when starting from the corresponding styrenes and hydrazides. This methodology could afford a chiral AEA in which the nitrogen atom forms either hydrazones, hydrazides, hydrazines, or amines, in excellent yields and ee. Subsequent derivatization of the products could lead to the synthesis of interesting bioactive compounds, such as a potassium channel inhibitor (Scheme 39).

Novel developments in the synthesis of unnatural peptides came from the studies of Rovis et al. In 2022 [59], they achieved the modular synthesis of peptides by a diastereoselective three-component carboamidation reaction enabled by an organo-rhodium complex. Starting from the corresponding dioxazolones, arylboronic acids, and acrylamide derivatives, new amide bonds were formed towards the synthesis of an interesting chiral AEA embedded into a peptide chain in good yields and ee (Scheme 40A). Two years later, in 2024 [60], they achieved the same level of success by using this methodology to obtain unnatural peptide macrocycles, starting from four to fifteen amino acid chains, incorporating chiral AEA motives towards three-component macrocyclation, achieving 15-mer macrocyclic substrate reaction (Scheme 40B).

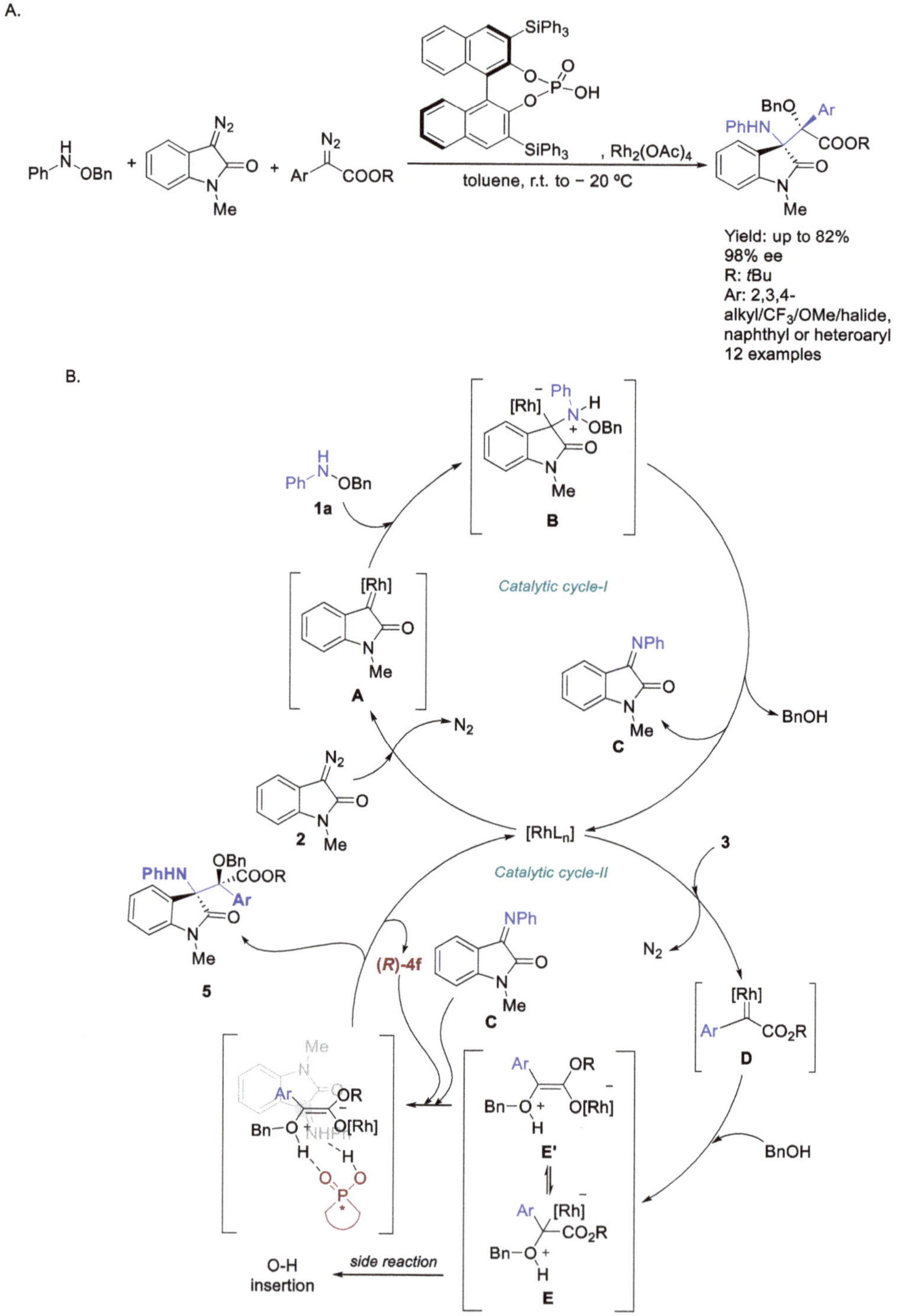

Scheme 38. (**A**) Asymmetric synthesis of AEA by rhodium catalysis. (**B**) Proposed dual mechanism. Xu et al., 2020.

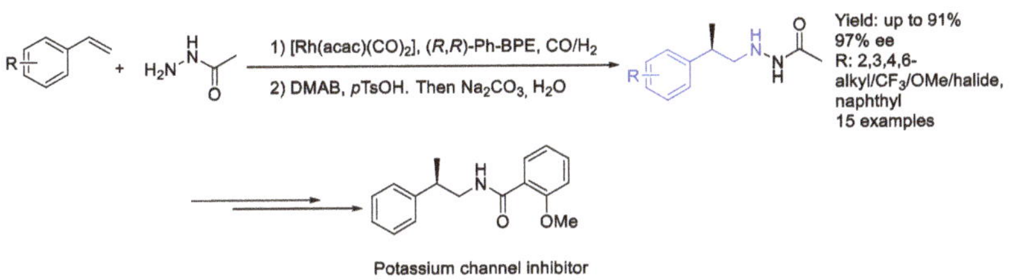

Scheme 39. Chiral AEA precursor of potassium channel inhibitor by Franciò et al. in 2022.

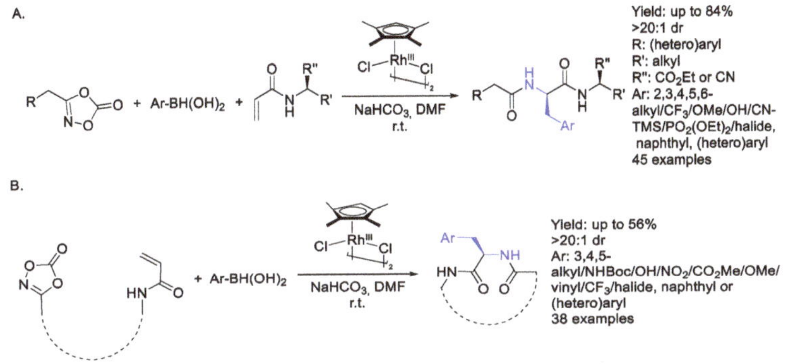

Scheme 40. Chiral AEA containing polypeptide synthesis through rhodium catalysis by Rovis et al. in 2022 (**A**) and 2024 (**B**).

In 2023, Li et al. [61] reported the asymmetric synthesis of an AEA which included axially and centrally chiral amino alcohols starting from interesting *O*-allylhydroxyamine derivatives. The key step was the C=C bond in *O*-allylhydroxyamine activation by the intramolecular electrophilic amidating moiety, as well as a migrating directing group. This methodology could afford a polisubstituted chiral AEA controlling the stereochemistry of the process by switching between chiral organo-rhodium complex; and, by the employment of axially prochiral or axially racemic heteroarenes, it afforded amino alcohols with both axial and central chirality in excellent enantio- and diastereoselectivity (Scheme 41).

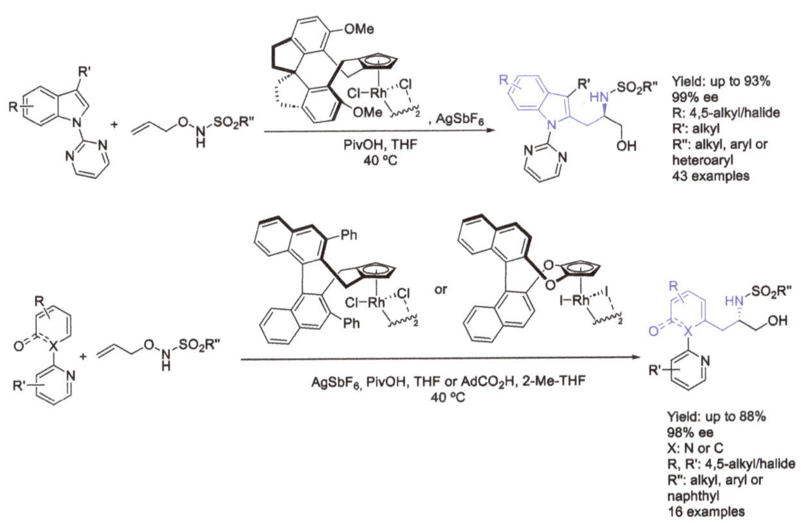

Scheme 41. Rhodium catalysts used by the asymmetric synthesis of AEA by Li et al. in 2023.

In 2024, Seyedsayamdost et al. [62] developed the synthesis of non-canonical tryptophan derivatives by C-H functionalization of anilines enabled by organo-rhodium catalysis. This regioselective synthesis afforded C4-C7 substituted tryptophan derivatives enabled by catalyzed annulation between structurally diverse *tert*butyloxycarbonyl(Boc)-protected anilines and alkynyl chlorides, achieving great success in yields and ee (Scheme 42).

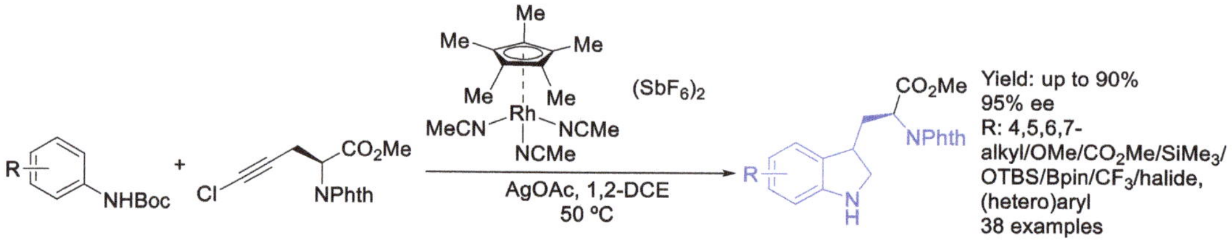

Scheme 42. Non-Canonical tryptophan synthesis by Seyedsayamdost et al. in 2024.

Also in 2024, Lv et al. [63] developed the synthesis of chiral β-aminoamides by asymmetric hydrogenation of the corresponding tetrasubstituted α,β-unsaturated amide. Using an organo-rhodium complex, the resulting chiral AEAs with two contiguous chiral centers were synthesized in excellent yields and ee. Also, gram-scale reaction and efficient transformation of β-amino amide to β-amino acid and β-amino cyanide were achieved (Scheme 43).

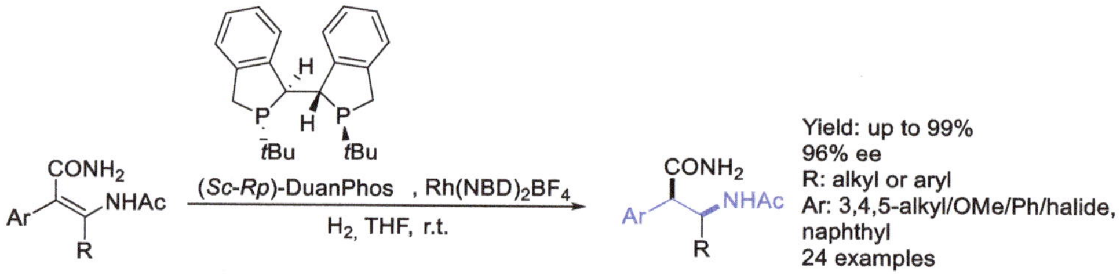

Scheme 43. Asymmetric synthesis of AEA through rhodium catalysis by Lv et al. in 2024.

Finally, Wang et al. [64] reported a diastereo- and enantioselective regiodivergent (hetero)arylamidation of (homo)allylic sulfides enabled by Rh(III) catalysis. This three-component reaction methodology could afford chiral AEAs in excellent yields, ee and dr starting from the corresponding homoallyl sulfide, arylboronic acid, and dioxazolone derivatives (Scheme 44A). Although an asymmetric AEA could be synthesized by following one of the two possible catalytic pathways, a sulfide moiety plays an important role in the reaction mechanism, promoting migratory insertion and controlling the regioselectivity of the reaction, as shown in Scheme 44B.

2.3. Palladium

Starting in 2010, Jackson et al. [65] developed the synthesis of interesting chiral AEA derivatives by a Negishi cross-coupling reaction starting from enantiopure alkyl iodide and aryl iodide substrates using an organo-palladium complex. The resulting chiral AEAs were synthesized, maintaining the stereochemistry in good yields (Scheme 45).

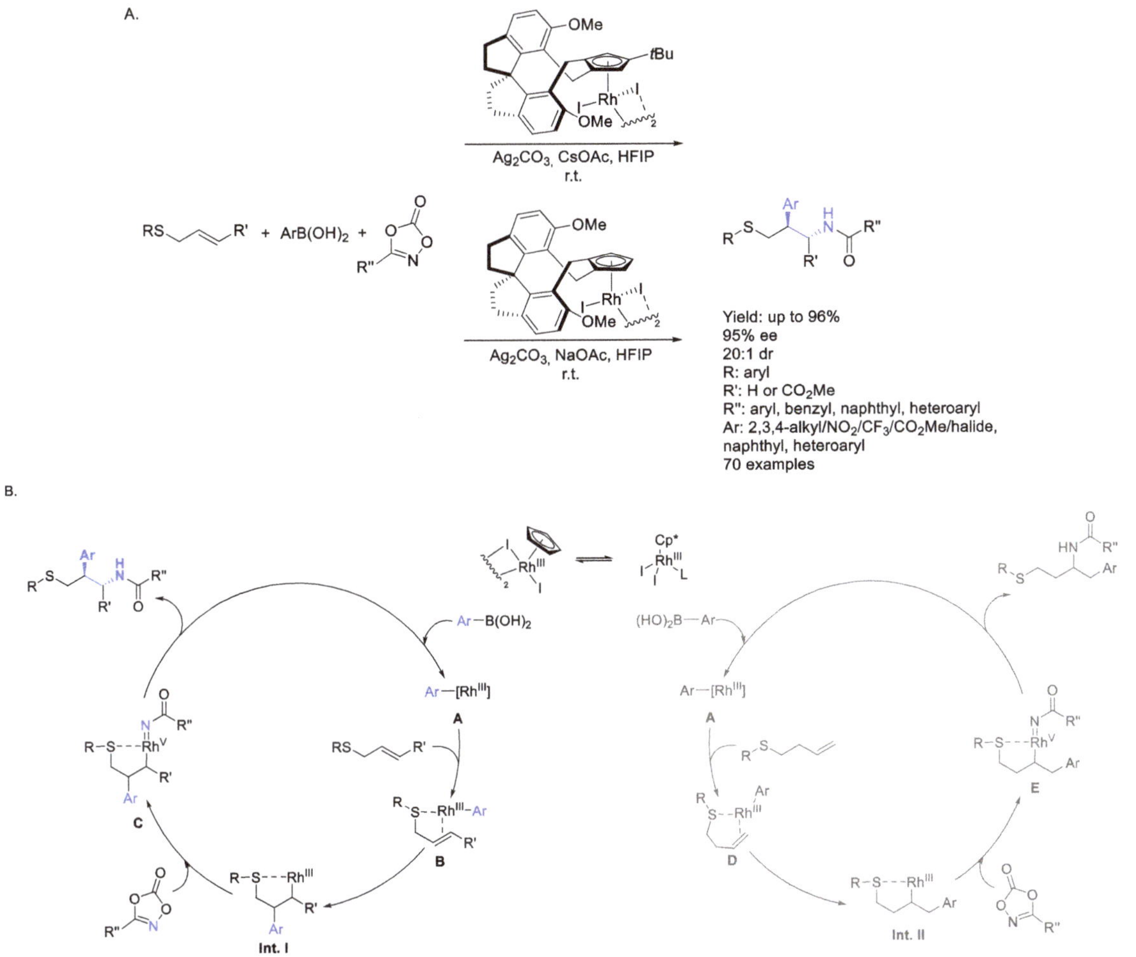

Scheme 44. Amide-containing chiral AEA synthetized by Wang et al. in 2024 (**A**) and its proposed mechanism (**B**).

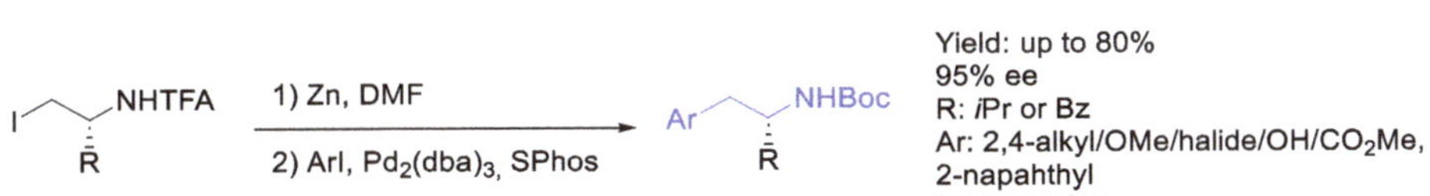

Scheme 45. Chiral AEA by Jackson et al. in 2010 through palladium catalysis.

In the same year, Trost et al. [66] reported a palladium-catalyzed dynamic kinetic asymmetric alkylation of vinyl aziridines. Starting from both substituted 1*H*-pyrroles and 1*H*-indoles, this high regio-, chemo-, and enantioselective process could afford an interesting chiral AEA in excellent yields and ee. Interestingly, the obtained products could be derivatized to obtain bioactive compounds (Scheme 46).

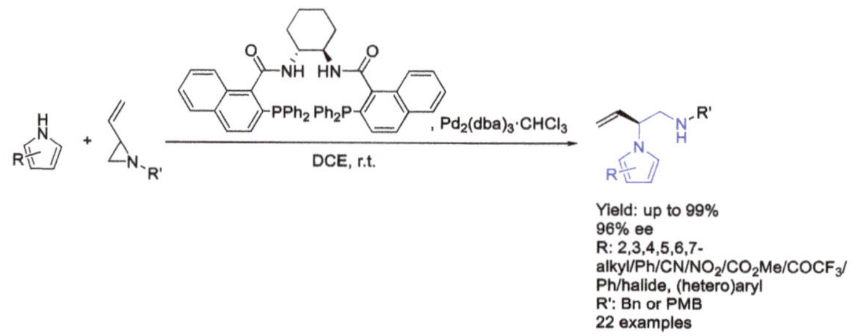

Scheme 46. Interesting asymmetric AEA by Trost et al. in 2010.

Also in 2010, Wolfe et al. [67] achieved the synthesis of an enantioenrich AEA in which the nitrogen atom is embedded into a tetrahydropyrrole ring. Starting from readily available alkenyl or aryl bromides and *N*-boc-pent-4-enylamines and using a chiral organopalladium complex, they reported high yields and ee (Scheme 47A). Five years later [68], they developed the asymmetric synthesis of 2-aminoindane derivatives by asymmetric palladium-catalyzed alkene carboamination reactions. In this way, interesting chiral AEAs were synthesized in excellent yields and ee (Scheme 47B).

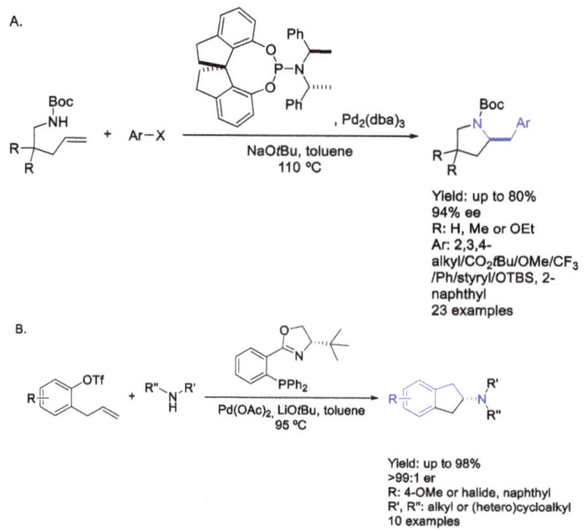

Scheme 47. Palladium-catalysed chiral AEA by Wolfe et al. in 2010 (**A**) and 2015 (**B**).

In 2013, Michael et al. [69] developed an interesting methodology to synthesize substituted chiral phenethylamines by a cross-coupling reaction of unsubstituted and 2-alkyl-substituted aziridines with arylboronic acid nucleophiles using a palladium complex. Starting from enantiopure activated aziridines, this reaction afforded an AEA that maintained the stereochemistry in good yields (Scheme 48).

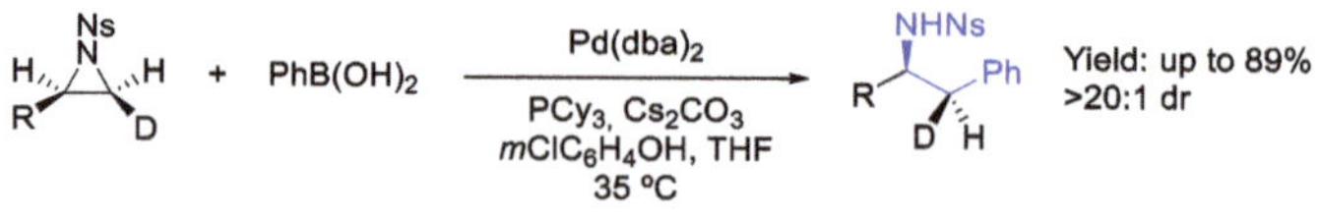

Scheme 48. Cross-coupling reaction catalyzed by a palladium complex towards the synthesis of a chiral AEA by Michael et al. in 2013.

When it comes to the asymmetric synthesis of AEAs enabled by organo-palladium catalysis, Minakata et al. have shown great success in this field. Starting in 2014 [70], they developed a methodology towards the synthesis of chiral 2-arylphenethylamine derivatives by Pd/NHC-catalyzed enantiospecific and regioselective Suzuki–Miyaura arylation of 2-arylaziridines (Scheme 49A). Starting from enantiopure substrates, they achieved great numbers when it came to yields, maintaining the enantiopurity. Then, in 2016 [71], they increased the substrate scope and developed a methodology starting from enantioenrich aziridine and bis(pinacolato)diboron towards palladium-catalyzed regioselective and stereo-inversive ring-opening borylation of 2-arylaziridines, resulting in interesting products that could be derivatized in order to obtain a chiral AEA in good yield and ee (Scheme 49B). Later, in 2019 [72], by synergistic palladium/copper dual catalysis, starting from chiral activated aziridines and silylborane, they achieved a regiodivergent and stereospecific ring-opening C(sp^3)–Si cross-coupling reaction leading to an interesting silicon-containing chiral AEA in good yields and ee (Scheme 49C). Also in 2019 [73], they developed a palladium-catalyzed enantiospecific and regioselective ring-opening Suzuki–Miyaura arylation of aziridine-2-carboxylates by a cross-coupling reaction. Starting from enantioenriched substrates, a ring-opening reaction with arylboronic acids led to interesting chiral AEAs in excellent yields and ee (Scheme 49D).

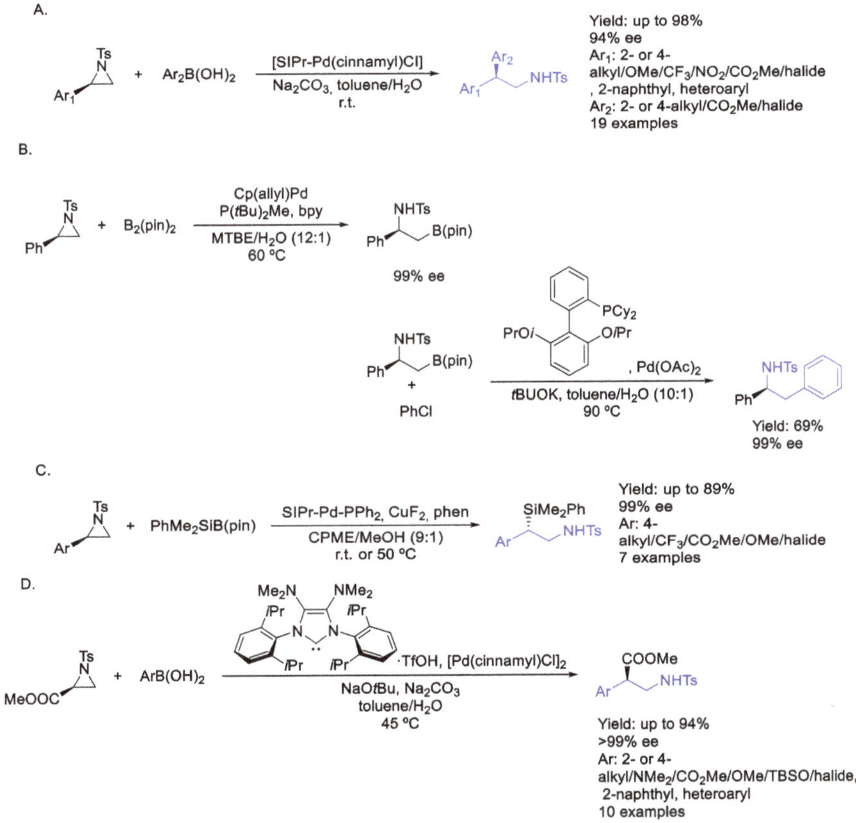

Scheme 49. Chiral palladium catalysis methodology by Minakata et al. towards the synthesis of biarylethylamines in 2014 (**A**), phenethylamines in 2016 (**B**) and silyl-containing AEA in 2019 (**C**) and AEA based on beta-amino esters (**D**) in 2019.

In 2015, Yu et al. [74] introduced chirality in an already-formed racemic AEA by palladium(II)-catalyzed highly enantioselective C–H arylation of cyclopropylmethylamines. Using aryl iodides as substrates and mono-*N*-protected amino acid ligands, they synthesized several chiral AEAs in excellent yields and ee (Scheme 50).

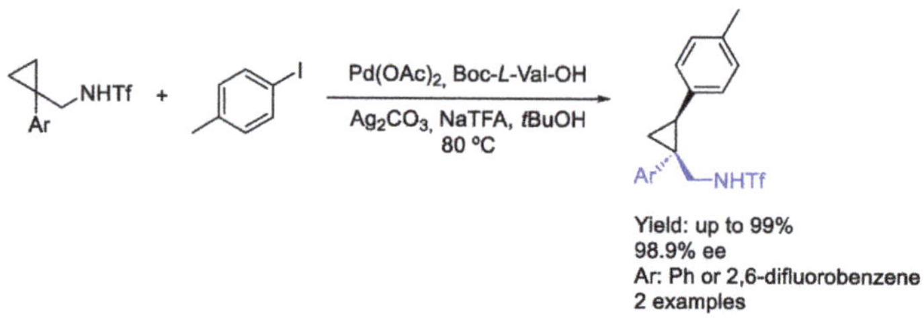

Scheme 50. Cyclopropane-based chiral AEA by Yu et al. in 2015.

Then, in 2019, Zhao et al. [75] used aliphatic aziridines to enable late-stage functionalization of aromatic acids, leading to an interesting AEA by using an organo-palladium-complex via C–H activation (Scheme 51). When it comes to asymmetric synthesis, starting from enantiopure methyl substituted aziridine, the synthesis of a chiral AEA was achieved in good yield and ee.

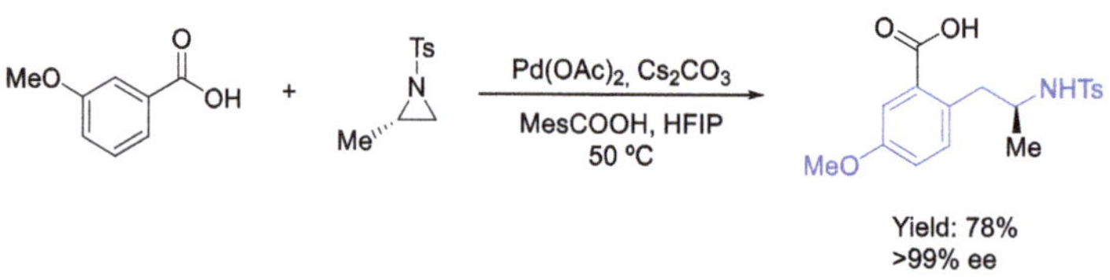

Scheme 51. Asymmetric AEA by Zhao et al. in 2019 through palladium catalysis.

Later, in 2022, Koley et al. [76] developed a methodology towards the synthesis of β-indolylethylamines by using a palladium complex as a catalyst by a cross-coupling reaction of aliphatic aziridines. Starting from the corresponding indole and enantioenriched aziridine, they could synthesize a single AEA in which the aryl group was an indole, in good ee and yield (Scheme 52).

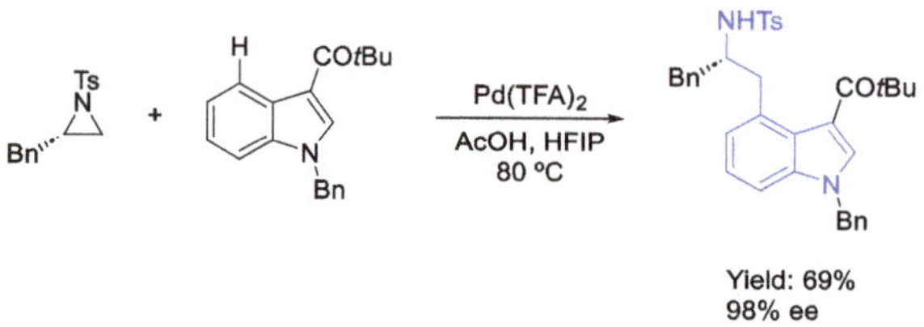

Scheme 52. Single chiral AEA synthesized by Koley et al. in 2022.

In 2023, Stecko et al. reported a two-step approach to α-amino ketones involving cross-coupling and an oxidative cleavage sequence, synthesizing a wide variety of chiral AEAs in the process, as key intermediates in the synthesis of α-aminoketones (Scheme 53). Using a palladium complex as a catalyst and starting from chiral amines, they achieved great numbers when it came to yields and ee.

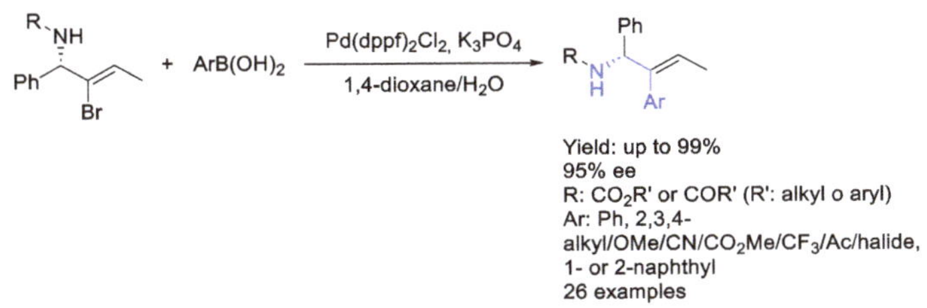

Scheme 53. Asymmetric AEA achieved through palladium catalysis by Stecko et al. in 2023.

In the same year, Guo et al. [77] developed a methodology using modular chiral-aldehyde/palladium catalysis towards the synthesis of an interesting chiral AEA. Starting from racemic α-amino esters and phenylallene, they achieved excellent yields and ee (Scheme 54).

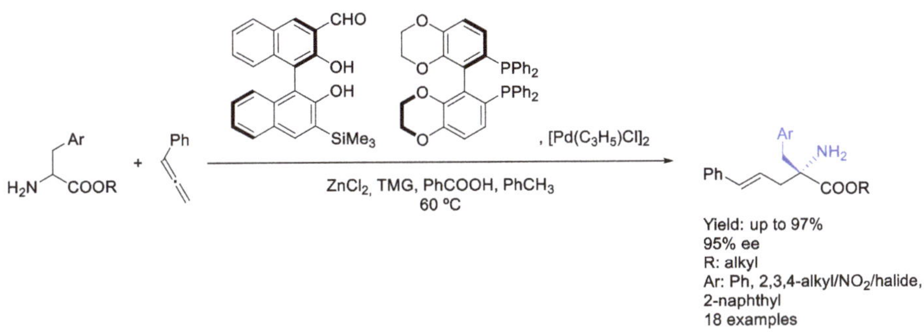

Scheme 54. Quaternary-center AEA by Guo et al. in 2023.

Finally, also in 2024, Chen et al. [78] reported a Pd(0)-catalyzed diastereoselective and enantioselective intermolecular Heck–Miyaura borylation of internal enamides. By using a chiral organo-palladium complex, three-component reactions were achieved, resulting in a polisubstituted boron-containing chiral AEA in excellent yields and ee (Scheme 55).

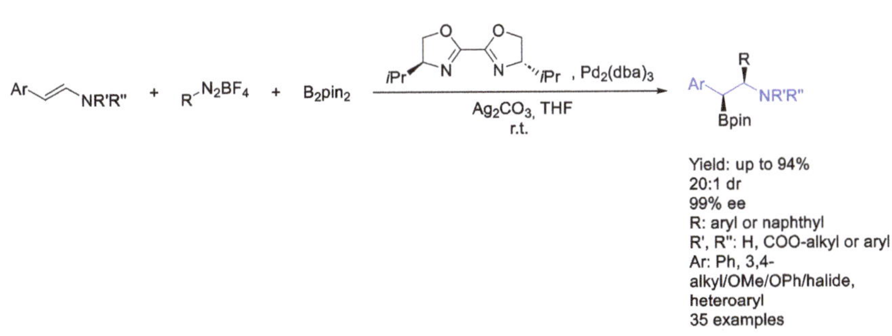

Scheme 55. Chiral AEA by Chen et al. in 2024 obtained through palladium catalysis.

2.4. Ruthenium

In 2009, Xu et al. [79] developed a promising methodology to synthesize an interesting chiral AEA, sitagliptin, a bioactive molecule used in the treatment of type 2 diabetes, reducing the total waste generated per kilogram and eliminating aqueous waste streams. The process involved asymmetric enamine hydrogenation enabled by an organo-ruthenium complex based on a ferrocene moiety, yielding the desired product in excellent yield and ee (Scheme 56).

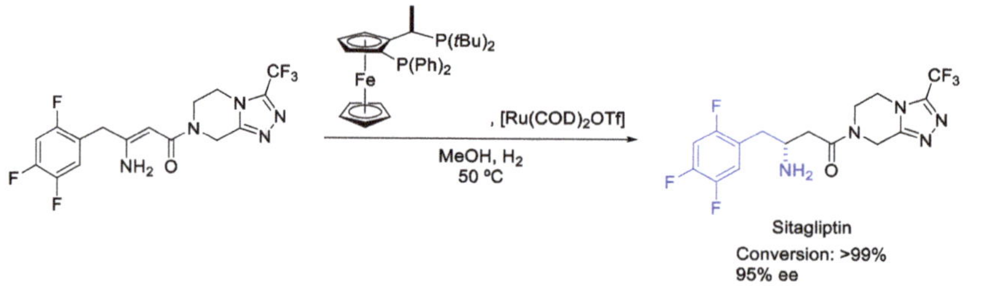

Scheme 56. Sitagliptin synthesis by Xu et al. in 2009.

In this sense, Steinhuebel et al. also obtained sitagliptin by reductive amination of a β-keto amide derivative in a new, high-yield, highly enantioselective synthetic route using a chiral organo-ruthenium complex, also capable of being used in different β-keto amide compounds (Scheme 57A) [80]. Following this study, one year later [81], they also reported the asymmetric hydrogenation of protected allylic amines by ruthenium catalysis, leading to interesting chiral AEAs in excellent yields and ee. Also, a remarkable application of the methodology was shown, as the medicinally relevant bioactive compound telcagepant could be synthesized through this route (Scheme 57B).

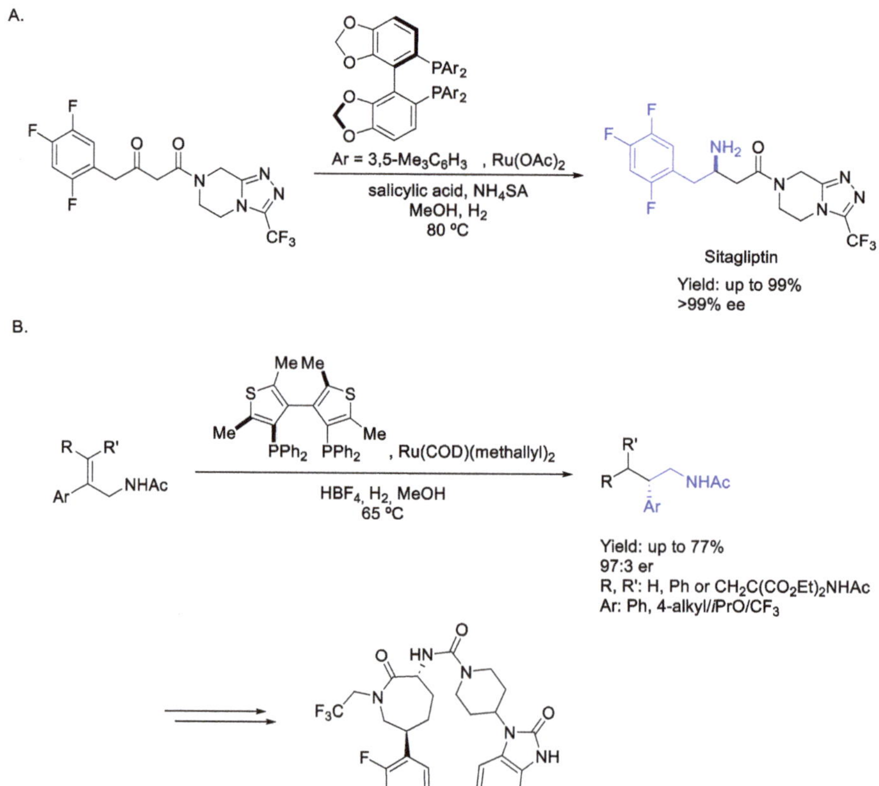

Scheme 57. Sitgliptin (**A**) and telgacepant precursor (**B**) by Steinhuebel et al. in 2009 and 2010, respectively.

Later, in 2012, Shibata et al. [82] achieved the asymmetric addition of amines to substituted styrenes by cationic ruthenium complexes combined with diphosphine ligands, with this addition being both β- and enantioselective. Arene-exchange from benzene to α-methylstyrene and subsequent nucleophilic attack towards a previously formed complex were the key steps in the regio- and enantioselection of the reaction. The resulting chiral phenethylamine was afforded in moderate yields and good ee (Scheme 58).

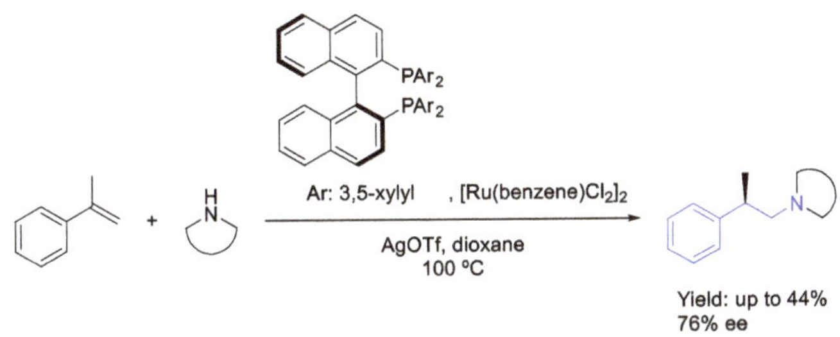

Scheme 58. Single chiral AEA by Shibata et al. in 2012.

Two years later, in 2014, Komiyama et al. [83] reported the asymmetric reduction of a key intermediate in the synthesis of β2-adrenergic receptor agonist, involving the reduction of a ketone group by a scalable ruthenium-catalyzed asymmetric transfer hydrogenation, enabled by both the arene/aryl interaction and potential repulsive interaction of the complex between catalyst and substrate (Scheme 59). This process could afford the desired chiral AEA in good yield and ee on a multi-kilogram scale (up to 4.2 kg).

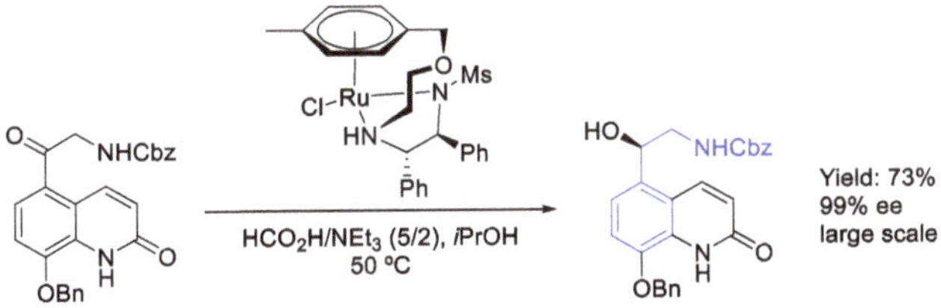

Scheme 59. Large-scale synthesis of chiral AEA by Komiyama et al. in 2014.

Later, in 2015, Wysocki et al. [84] developed an asymmetric homogeneous hydrogenation of 2-pyridones using a ruthenium complex with a NHC ligand. The resulting chiral AEAs were synthesized in good yields and moderate ee, leading to compounds in which the nitrogen group was embedded into a piperidone ring (Scheme 60).

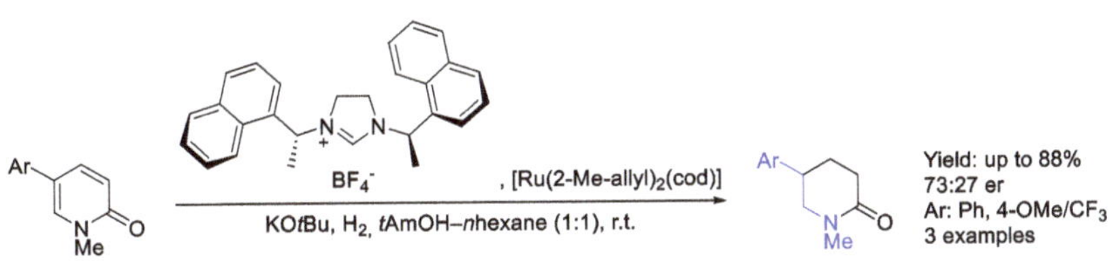

Scheme 60. Piperidone-based chiral AEA by Wysocki et al. in 2015.

One year later, Beliaev et al. [85] achieved multi-kilogram-scale asymmetric synthesis of a key intermediate in the production of etamicastat, a chiral AEA peripherally selective dopamine β-hydroxylase inhibitor. The process involved asymmetric hydrogenation of an enamide moiety enabled by several ruthenium-based complexes, affording the desired product in excellent yields and ee (Scheme 61). Their studies allowed identification of four ruthenium-based catalytic systems for asymmetric hydrogenation: two isolated catalysts, Ru(*R*)-Xyl-PPhos(acac)$_2$ and Ru(*R*)-C3-TunePhos(acac)$_2$, and two preformed non-isolated catalytic systems, CatASiumT$_3$/[Ru(pcymene)Cl$_2$]$_2$ and (*R*)-TolBINAP/[Ru(p-cymene)Cl$_2$]$_2$. They concluded that the most evident advantage of the isolated ones was lower catalyst loading, as preformation of the catalyst in solution in most cases produced less reactive species, although isolated catalysts were less stable and more oxygen- and moisture-sensitive than the corresponding ligand and metal precursor. On the other hand, considered in terms of commercial efficacy, using low loading of an isolated catalyst may be as convenient as larger amount of a non-isolated one, they concluded.

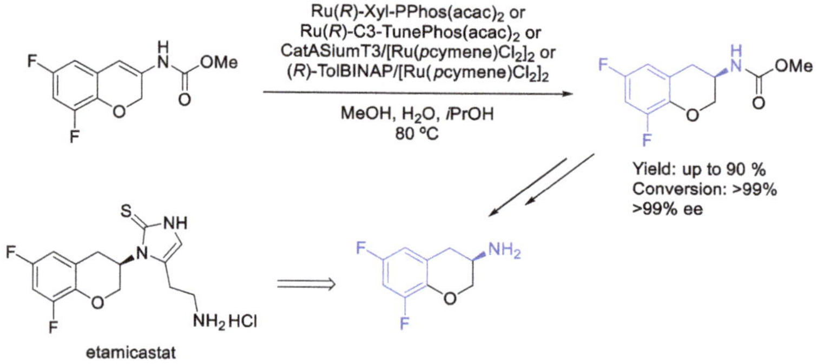

Scheme 61. Chiral etimicastat precursor obtained by ruthenium catalysis by Beliaev et al. in 2016.

In 2018, Schaub et al. [86] reported the direct asymmetric reductive amination of alkyl-aryl ketones enabled by ruthenium catalysis with ammonia and hydrogen. The corresponding amines were synthesized in excellent yields and ee. In this process, an interesting phenethylamine was produced (Scheme 62).

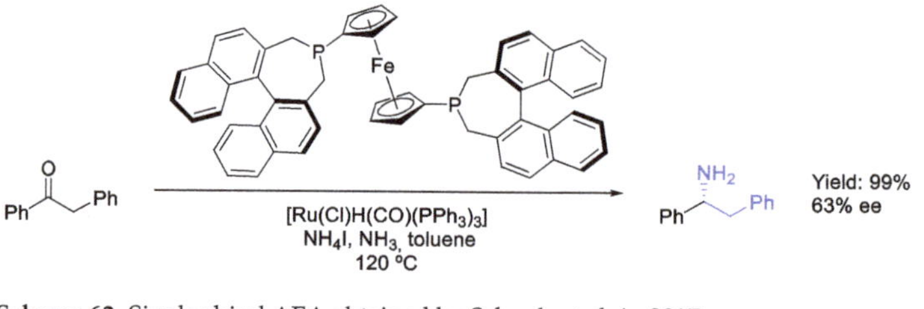

Scheme 62. Single chiral AEA obtained by Schaub et al. in 2017.

Two years later, in 2020 [87], Yin et al. published a study in which they performed an asymmetric reductive amination of diaryl and sterically hindered ketones using ammonium salts and hydrogen, enabled by a chiral ruthenium complex. Thus, synthetically useful chiral primary diarylmethylamines and sterically hindered benzylamines were synthesized in excellent yields and ee. In this process, an interesting chiral AEA was produced (Scheme 63).

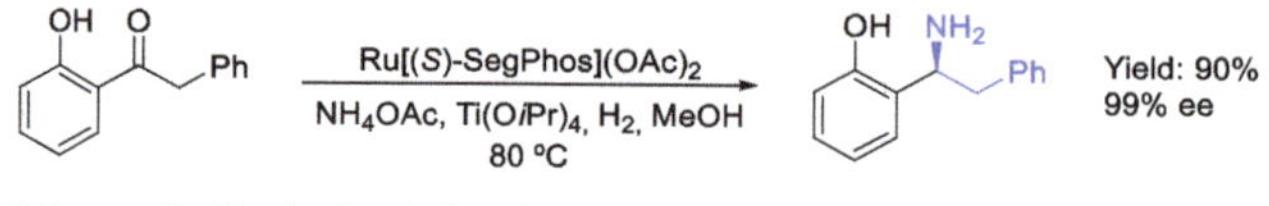

Scheme 63. Single chiral diaryl AEA obtained by Yin et al. in 2020.

In 2021, Ellman et al. [88] developed a three-component 1,2-carboamidation of bicyclic alkenes, synthesizing an interesting complex chiral AEA. By using a chiral Ru(III) complex, they achieved excellent yields and ee starting from aryl, dioxazolone, and bicyclic substrates. The AEA synthesized in this study consisted of polycyclic compounds in which the nitrogen atom formed an amide group (Scheme 64).

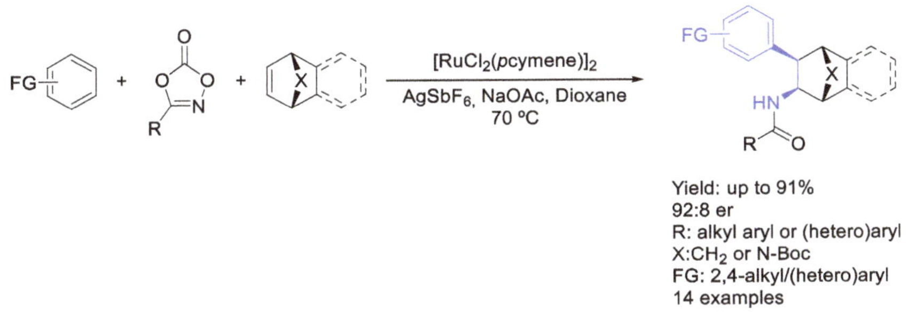

Scheme 64. Chiral bicyclic-containing AEA synthesized by Ellman et al. in 2021.

Later, in 2022, Zhang et al. [89] reported the synthesis of a highly substituted chiral AEA through an unprecedented highly enantioselective ruthenium-catalyzed direct asymmetric reductive amination of α-keto amides with ammonium salts, achieving excellent yields and ee. Using ammonium salts as the nitrogen source, this methodology allowed the synthesis of medicinally interesting *N*-unprotected β-branched α-amino acids containing two contiguous stereogenic centers. Also, they proved the utility of this method through its application in the efficient synthesis of chiral drug intermediates, peptides, and organocatalysts/ligands (Scheme 65).

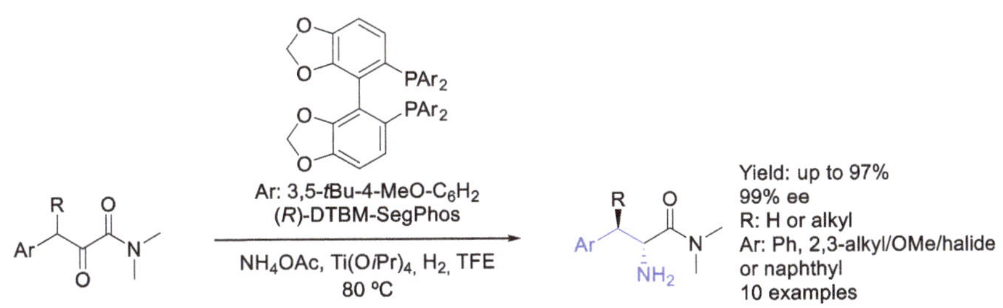

Scheme 65. Interesting chiral AEA synthesized in 2022 by Zhang et al.

Later, in 2023, Meggers et al. [90] achieved the synthesis of chiral γ-aminomethyl-γ-lactones, producing an interesting chiral AEA in the process. This methodology involved a nitrene-mediated enantioselective intramolecular olefin oxyamination enabled by a chiral ruthenium complex. Achieving great success in yields and ee, they were also able to derivatize the products, producing interesting building blocks to synthesize bioactive compounds (Scheme 66A). DFT calculations supported the possibility of both a singlet (concerted oxyamination of the alkene) and a triplet pathway (stepwise oxyamination) for the formation of the predominant stereoisomer, as shown in the proposed catalytic cycle. Amine–imine equilibrium led to the two possibilities, resulting in the same enantiomer by either a concerted or a stepwise pathway (Scheme 66B).

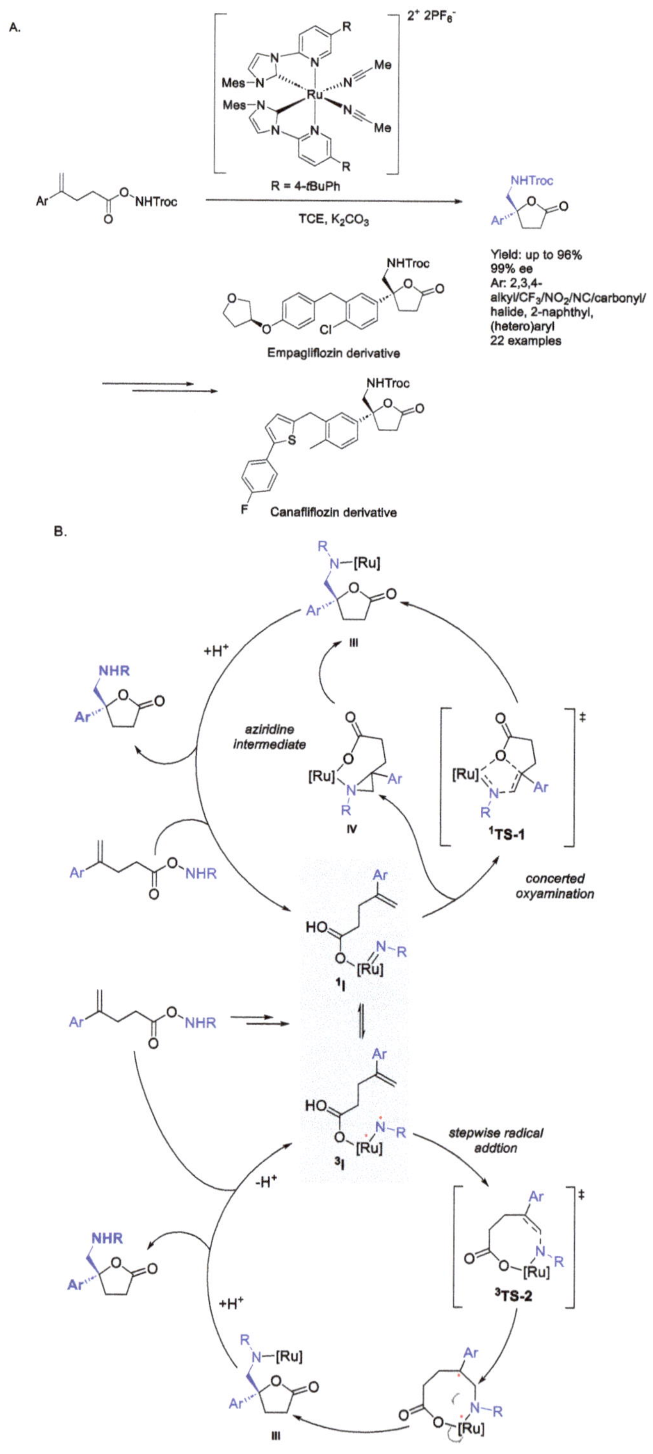

Scheme 66. Ruthenium-catalyzed chiral AEA as building block by Meggers et al. (A) and its proposed mechanism (B).

In the same year, Chen et al. [91] developed an asymmetric hydrogenation of dibenzo-fused azepines using a cationic ruthenium–diamine complex. In this process, an interesting chiral AEA was synthesized, in which the amine group was embedded into an azipine ring, achieving excellent yields and ee (Scheme 67). Interestingly, the enantioselectivities could be regulated by the counteranion of the catalyst in the asymmetric hydrogenation of dibenzo [b,f][1,4]thiazepines and dibenzo[b,e]azepines.

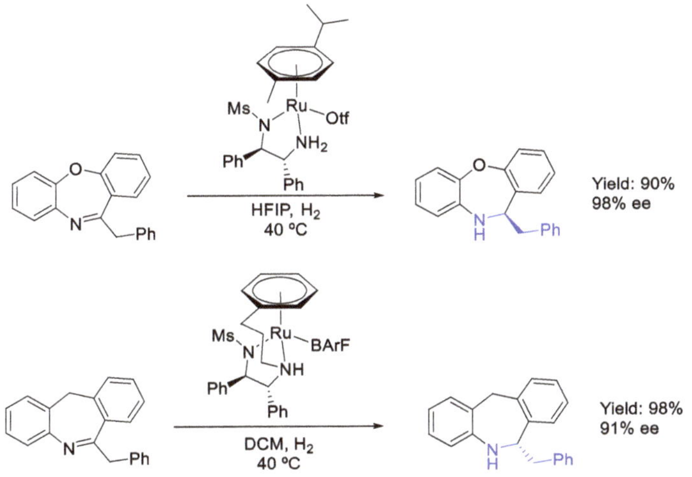

Scheme 67. Benzoazepine-based AEA obtained by Chen et al. in 2023.

Also in 2023, Liu et al. [92] achieved the asymmetric addition of hydrazones to aryl imines in water, by using a ruthenium–diamine–diphosphine system in the presence of crown ether. The methodology used a Ru(II)-catalyzed "umpolung" asymmetric addition of non-substituted hydrazones as an alkyl carbanion equivalent. Thus, poliaryl-substituted chiral AEAs were synthesized in good yields and ee (Scheme 68).

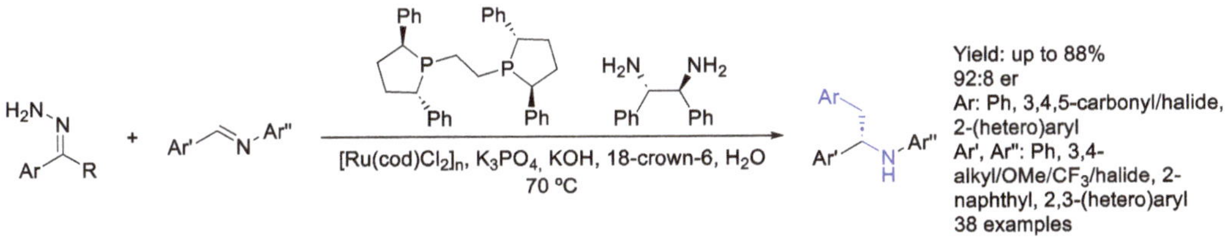

Scheme 68. Poliaryl-substituted chiral AEA by Liu et al. in 2023.

Finally, in 2024, Zhang et al. [93] used a chiral ruthenium complex for the asymmetric hydrogenation of α,β-unsaturated γ-lactams. Thus, when aryl-substituted γ-lactam was used as the starting material, interesting chiral AEAs were synthesized in excellent yields and ee. Also, subsequent derivatization of the products could be undertaken in order to synthesize interesting bioactive molecules, such as rolipram and an α-2-adrenoceptor antagonist (Scheme 69).

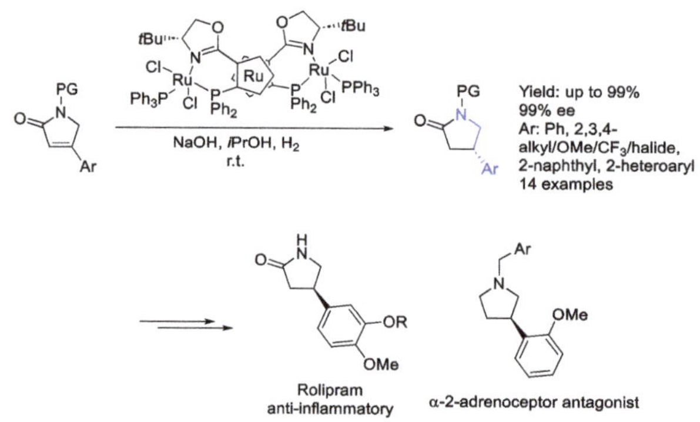

Scheme 69. Chiral pyrrolidinone-based AEAs as bioactive compound precursors by Zhang et al. in 2024.

2.5. Nickel

Doyle et al. have been developing interesting ring-opening reactions from aziridines in order to synthesize, among others, enantiopure AEA. In 2012 [94], they achieved a nickel-catalyzed Negishi alkylation of styrenyl aziridines starting from optically active substrates, maintaining the enantiomeric excess towards the products, resulting in chiral AEAs in good yields and ee (Scheme 70A). Later, in 2015 [95], they used an organo-nickel complex as the catalyst for the ring opening for enantiopure quaternary carbon-containing aziridines, synthesizing an interesting substituted chiral AEA in moderate ee (Scheme 70B). Finally, in 2017 [96], they used a chiral organo-nickel complex to introduce chirality in the ring-opening reaction of racemic aziridines by direct asymmetric arylation, resulting in chiral diaryl ethyl amines in good yields and ee (Scheme 70C).

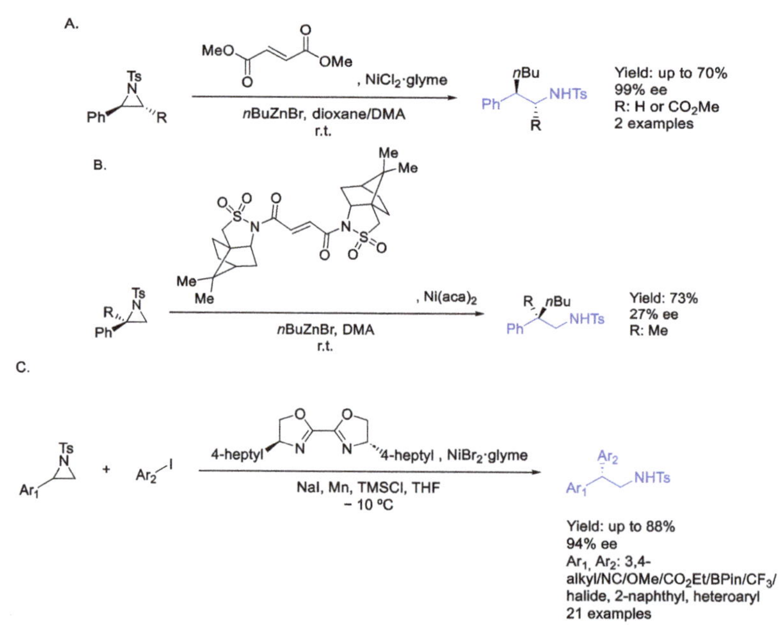

Scheme 70. Aziridine ring-opening methodology by Doyle et al. throughout the years, synthesizing interesting chiral AEAs: (**A**) Negishi alkylation. (**B**) Chiral sultam auxiliaries ligand for Ni. (**C**) Chiral bioxazoline ligand for Ni.

Then, in 2020, Zhu et al. [97] developed an interesting synthesis of a chiral AEA in which the aryl group is fused either with a dihydrofurane or a cyclopentane ring (Scheme 71). Starting from diverse alkene-tethered aryl iodides and O-benzoyl-hydroxylamines, an asymmetric reductive 1,2-carboamination of unactivated alkenes was performed, enabled by nickel catalysis with a β-chiral diamine organo-complex, resulting in good yields and ee.

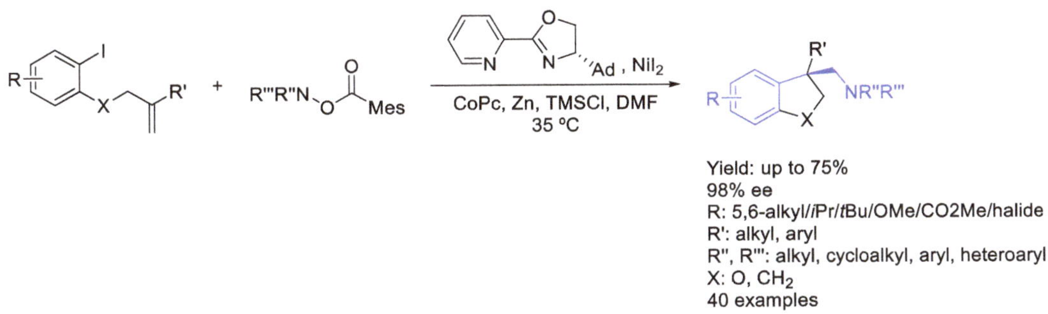

Scheme 71. Chiral AEA based on benzofused rings by Zhu et al. in 2020.

Later, in 2021, an asymmetric 1,2-carboamination of alkenes was also achieved by Engle et al. [98]. Enabled by nickel catalysis, interesting chiral AEAs were synthesized through a three-component reaction from unactivated alkenes with arylboronic esters and electrophilic aminating agents (Scheme 72). This methodology relied on tailored O-(2,6-dimethoxybenzoyl)hydroxylamine electrophiles that suppressed competitive processes, including undesired β-hydride elimination and transesterification between the alcohol substrate and electrophile. Thus, interesting chiral AEAs were reported in good yields and dr.

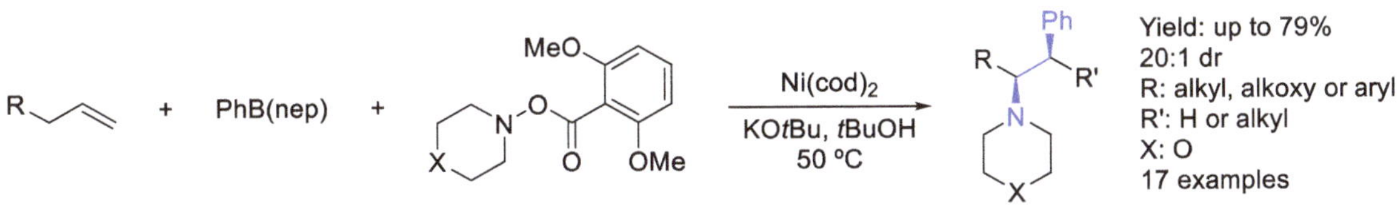

Scheme 72. Three-component reaction towards the synthesis of chiral AEAs by Engle et al. in 2021.

Back to asymmetric aziridine ring-opening reactions, in 2023 [99], Nevado et al. reported the use of a chiral organo-nickel complex to synthesize interesting chiral AEAs in excellent yields and ee. Starting from activated aziridines and the corresponding bromide substrates, an electrochemically driven nickel-catalyzed enantioselective reductive cross-coupling reaction was achieved, leading to a great variety of enantiopure AEAs (Scheme 73A). Interestingly, this electroreductive strategy proceeds in the absence of heterogeneous metal reductants and sacrificial anodes by employing constant current electrolysis in an undivided cell with triethylamine as a terminal reductant, as shown in the proposed catalytic cycle (Scheme 73B), making this method more atom-economical and scalable for synthetic applications.

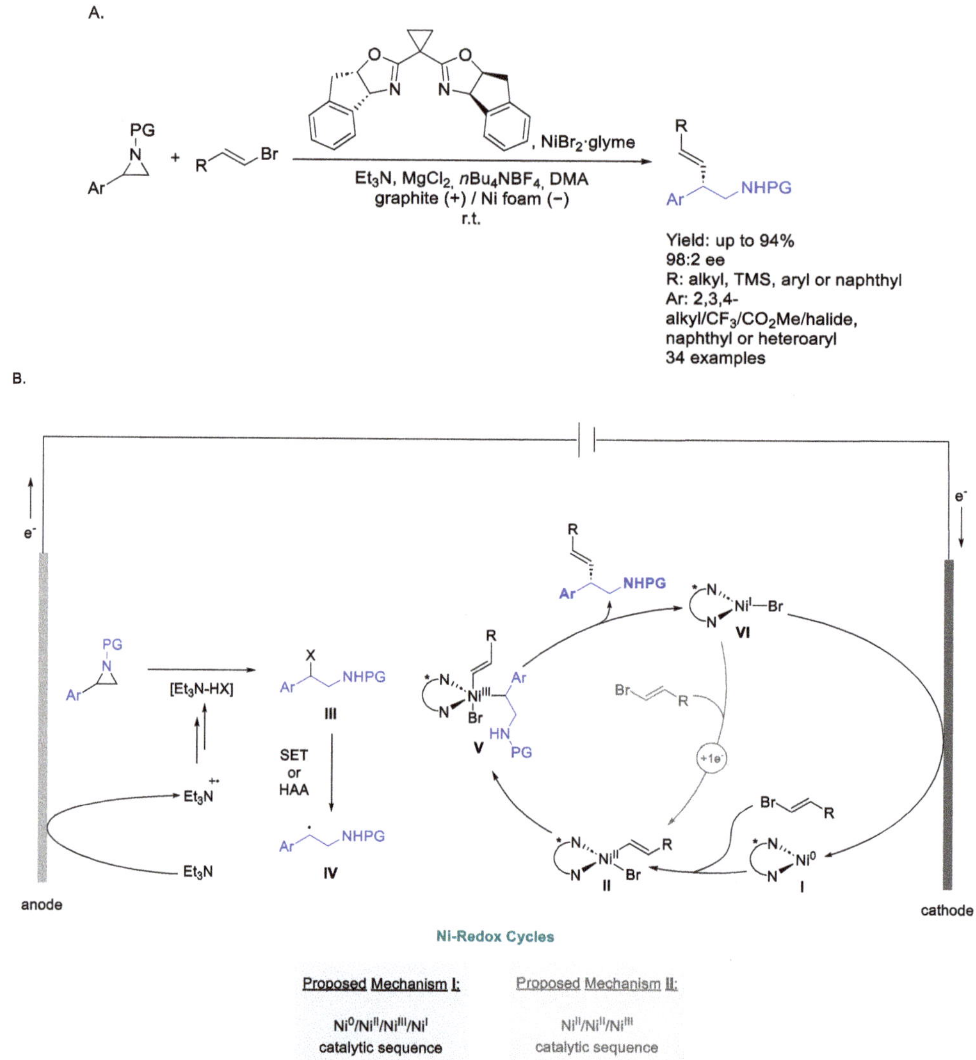

Scheme 73. (**A**) Asymmetric synthesis of AEA by electronickel catalysis by Nevado et al. in 2023. (**B**) Proposed mechanism.

Also in this field, in the same year, Mei et al. [100] also reported an asymmetric electrochemical reductive aziridine ring-opening reaction by the use of a chiral nickel–biimidazole catalyst complex. A reductive cross-coupling reaction from the corresponding aziridines and aryl iodide substrates led to excellent yields and ee when it came to the synthesis of a chiral AEA (Scheme 74). Similarly to Nevado et al.'s work, this reaction occurred in an undivided cell, allowing the electroreduction-mediated turnover of the nickel catalyst instead of a metal reductant-mediated turnover. Interestingly, later derivatization of the products could afford SCH 12679, a dopamine D2 receptor antagonist.

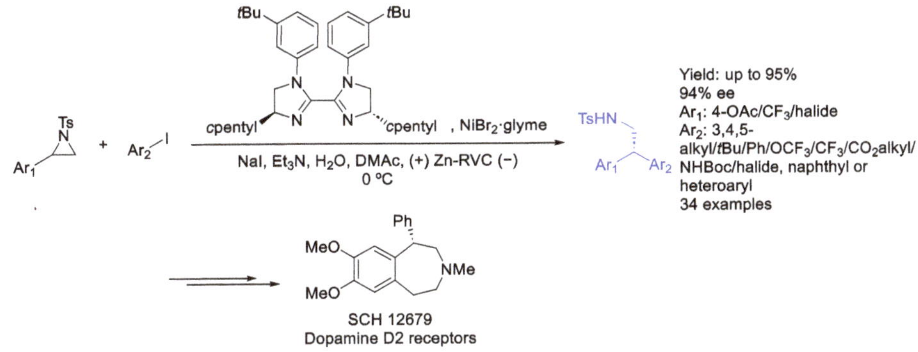

Scheme 74. Chiral synthesis of AEAs as precursors of dopamine D2 receptor antagonist by Mei et al. in 2023.

In the same year, Chang et al. [101] developed an asymmetric synthesis of β-lactams leading to interesting chiral AEAs in excellent yields and ee. By intramolecular hydroamination enabled by a chiral organo-nickel catalyst, this methodology afforded an enantiopure AEA via proximal C–N bond formation in which the nitrogen atom was embedded into a β-lactam ring. Later derivatization could afford interesting bioactive compounds, such as clofibric or estrone derivatives (Scheme 75).

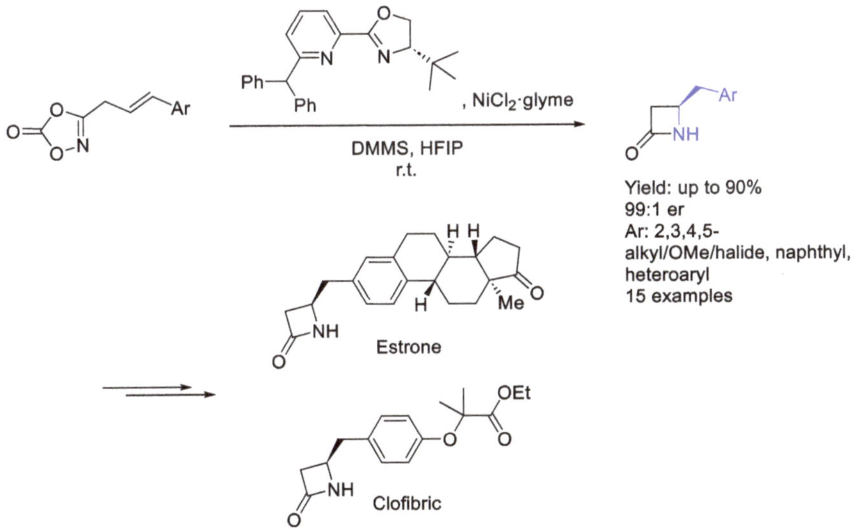

Scheme 75. β-lactam-based chiral AEA by Chang et al. in 2023.

Following studies published in the same year, Liu et al. [102] used nickel catalysis to produce chiral benzylamines, synthesizing an interesting chiral AEA in the process in moderate yield and ee. An asymmetric decarboxylation of NHP esters via a reductive cross-coupling reaction was enabled by an organo-nickel complex starting from racemic benzylamine and aryl iodide substrates (Scheme 76).

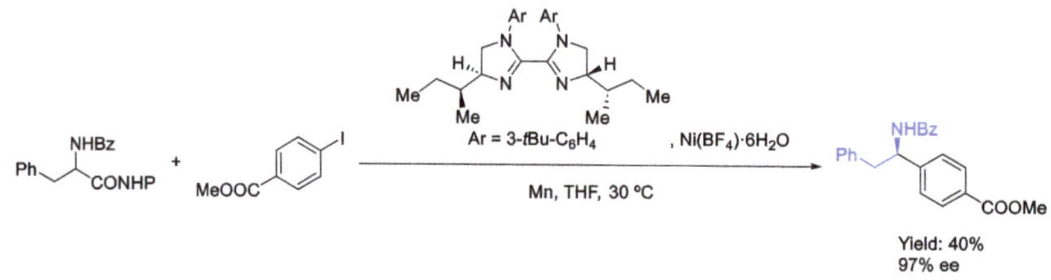

Scheme 76. Single chiral AEA enabled by nickel catalysis by Liu et al. in 2023.

Later in 2024, Powers et al. [103] reported the asymmetry ring-opening reaction of racemic *N*-pyridinium aziridines towards the synthesis of an interesting chiral AEA. Different organo-nickel complexes were used to afford *N*-pyridinium phenethylamine salts in excellent yields and ee with alkylzinc substrates, and further derivatization could afford the corresponding substituted β-phenethylamines (Scheme 77).

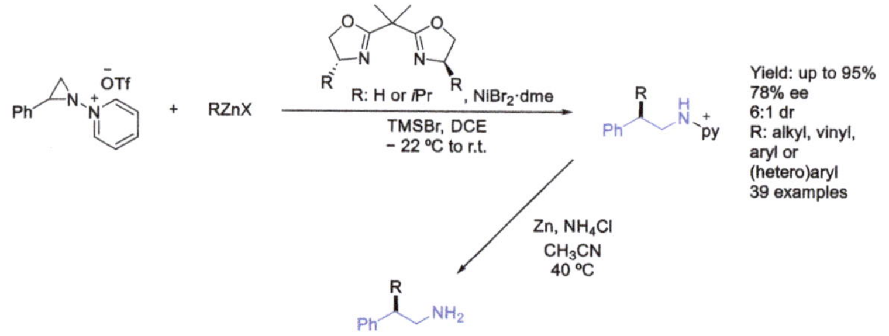

Scheme 77. *N*-pyridinium-based AEA by Powers et al. in 2024.

Also in 2024, Fürstner et al. [104] reported a study involving the obtention of a chiral AEA by Ni catalysis. This enantioselective synthesis of *vic*-aminoalcohol derivatives by nickel-catalyzed reductive coupling of aldehydes with protected amino-pentadienoates could afford a polisubtituted asymmetric AEA in excellent yields and ee (Scheme 78). The methodology relied on reductive coupling conveniently performed with a bench-stable Ni(0) precatalyst and Et_3B as the promoter to obtain interesting aminoalcohols.

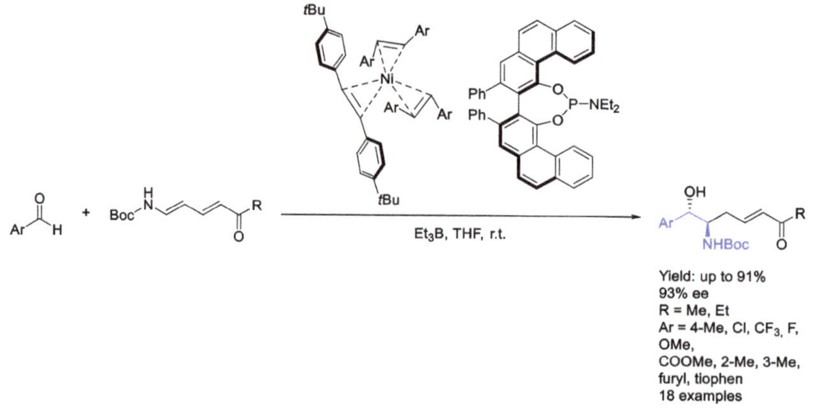

Scheme 78. Chiral aminoalcohols as AEAs by Fürstner at al. in 2024.

Liu et al. [105] reported, also in 2024, another asymmetric aziridine ring-opening reaction starting from racemic substrates. A C(sp^3)-C(sp^3) cross-coupling reaction from racemic *N*-sulfonyl styrenyl aziridines primary alkyl bromides was enabled by a nickel/pyridine-imidazoline complex, leading to phenethylamine derivatives with excellent yields and ee (Scheme 79).

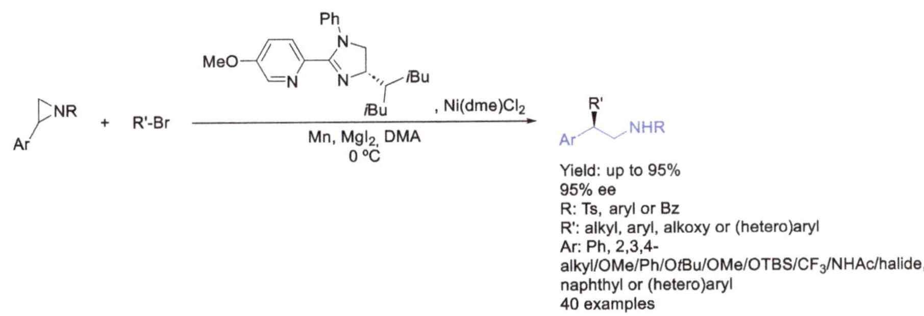

Scheme 79. Aziridine ring-opening towards the synthesis of a chiral AEA by Liu et al. in 2024.

Finally, in 2025, Chang et al. [106] achieved the asymmetric synthesis of b-arylamides by nickel-catalyzed homobenzylic hydroamidation of aryl alkenes. By employing a transposed NiH catalysis approach, this method facilitated the formation of key chiral nickel-amido intermediates, enabling asymmetric insertion into alkenes to produce the desired compounds with excellent enantioselectivity and yields. Interestingly, subsequent studies could afford corresponding bioactive compounds, such as clobenzorex, Didrex, selegiline, or tamsulosin analog (Scheme 80).

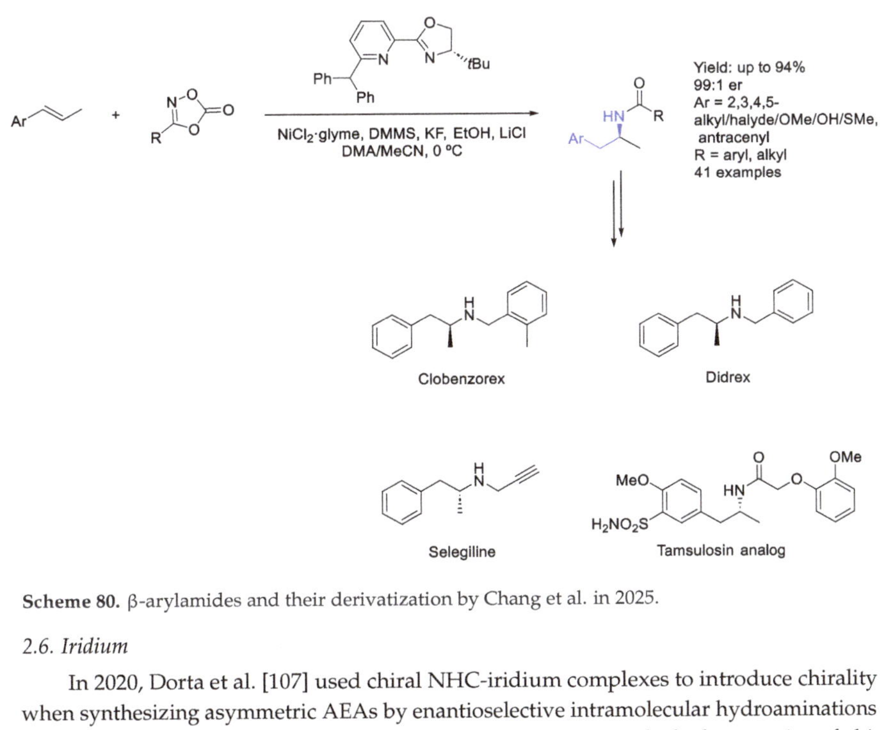

Scheme 80. β-arylamides and their derivatization by Chang et al. in 2025.

2.6. *Iridium*

In 2020, Dorta et al. [107] used chiral NHC-iridium complexes to introduce chirality when synthesizing asymmetric AEAs by enantioselective intramolecular hydroaminations and ring-opening aminations. The resulting products consisted of a large series of chiral AEAs where the corresponding aryl group is always fused with a six-member ring, achieving excellent yields and ee in the process (Scheme 81).

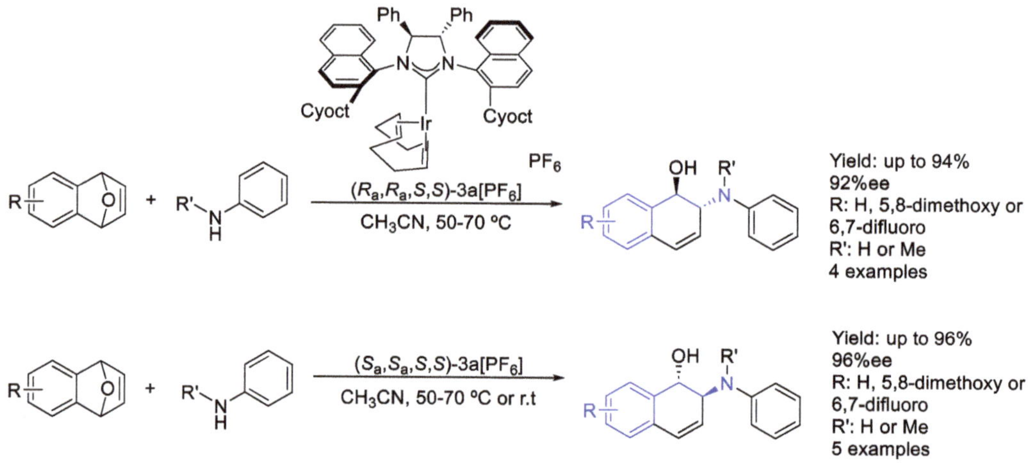

Scheme 81. NHC-iridium catalysis towards the synthesis of asymmetric AEAs by Dorta et al. in 2020.

Later, in 2020, Jouffroy et al. [108] developed the synthesis of interesting polycyclic chiral products through direct reductive amination of ketones and secondary amines enabled by an in situ-generated organo-iridium complex based on a ferrocene moiety. In the process, a particular chiral polisubstituted AEA was synthesized in low yield and excellent dr (Scheme 82).

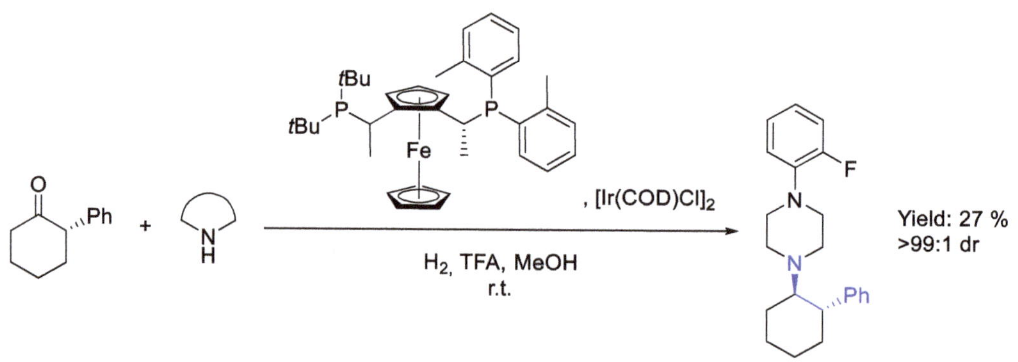

Scheme 82. Single chiral AEA by Jouffroy et al. methodology, in 2020.

Asymmetric hydroamination was also achieved by Hartwig et al. in 2022 [109], synthesizing an interesting chiral AEA from unactivated terminal alkenes by a chiral organo-iridium complex. This methodology used equimolar amounts of alkene and amine and, in order to achieve N-H addition, reversibility of the addition, reversible oxidation of the product amine, competing isomerization of the alkene reactant, and unfavorable replacement of sacrificial ligands in standard catalyst precursors by the chiral bisphosphine proved to be crucial in the process. Thus, asymmetric AEA synthesis was achieved in excellent yields and ee (Scheme 83).

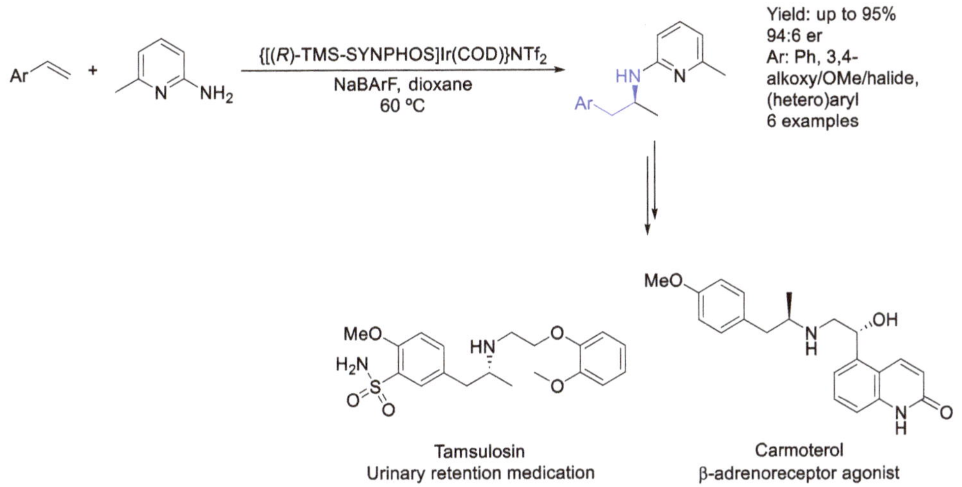

Scheme 83. Hydroamination of unactivated alkenes towards the synthesis of a chiral AEA by Hartwig et al. in 2022.

One year later, in 2023, Li et al. [110] developed a methodology in which an enantiosolective group repositioning reaction was enabled by a chiral iridium complex, synthesizing an interesting chiral AEA in which the corresponding aryl group was a hetero-five-member ring, including furan, benzofuran, and thiophene (Scheme 84). Starting from corresponding heteroarene and enamide substrates, the C–H bond at the C-2 position of the heteroarene is site-selectively cleaved and added regioselectively to the β-position of an enamide, resulting in a polisubstituted AEA in excellent yields and ee.

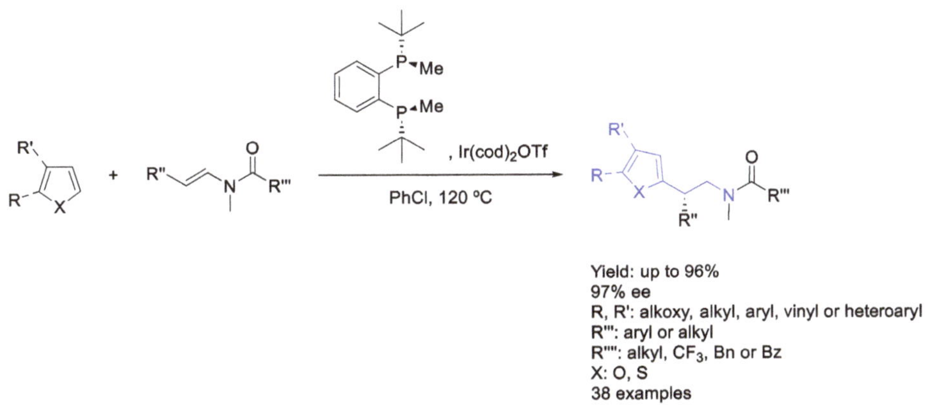

Scheme 84. Polisubstituted chiral AEA obtained by iridium catalysis by Li et al. in 2023.

Last year, Kuwata et al. [111] used a chiral organo-iridium complex for the asymmetric reductive amination of α-keto acids. Starting from optically active 2-phenyglycinol as the aminating agent, the resulting products could be derivatized by subsequent elimination of the hydroxyethyl moiety in order to obtain the corresponding chiral AEA as unprotected unnatural α-amino acids in good yields and ee (Scheme 85).

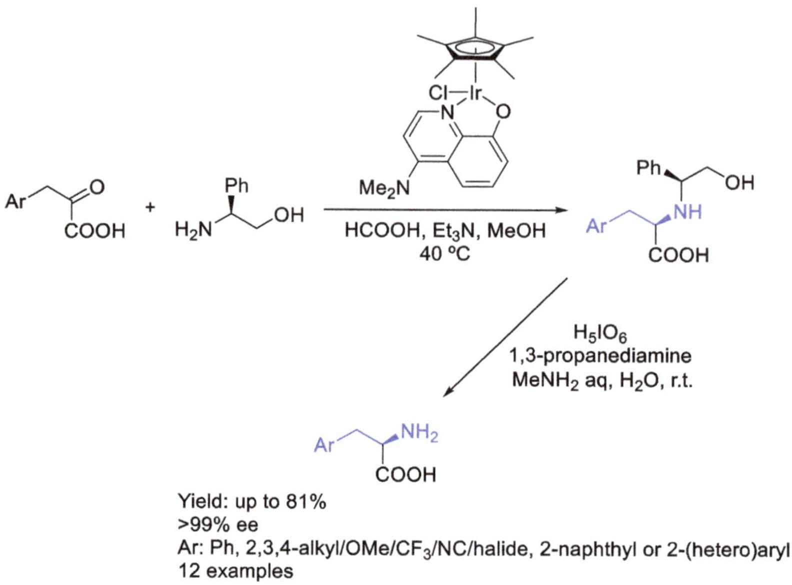

Scheme 85. Chiral AEA as unnatural α-amino acids by Kuwata et al. in 2024.

Also in 2024, Liu et al. [112] synthesized interesting δ-aminoboronic esters starting from 1,2-azaborines by enantioselective hydrogenation enabled by a chiral iridium complex as the chiral introducing agent. The resulting chiral AEAs were synthesized in good yields, ee, and dr. Also, derivatization of the products led to interesting bioactive molecules, including phenibut as a chiral AEA, among others (Scheme 86).

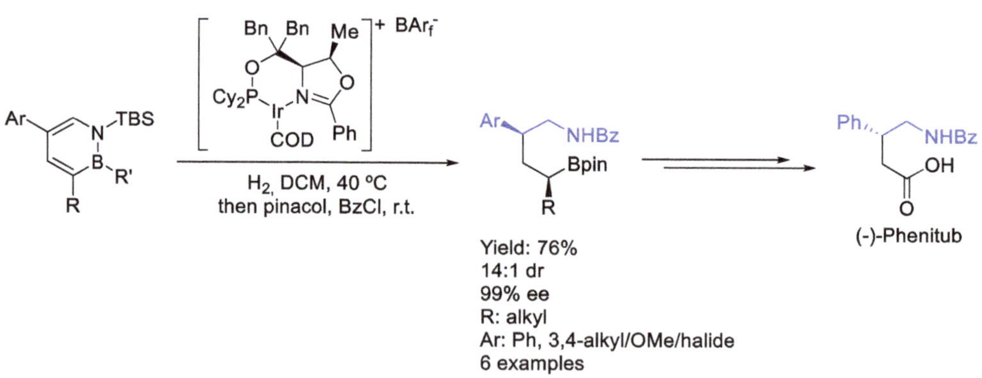

Scheme 86. Chiral AEAs and their derivatization by Liu et al. in 2024.

2.7. Cobalt

Starting in 2018, Cheng et al. [113] synthesized 1-aminoindenes via enantioselective [3 + 2] annulation enabled by a chiral cobalt complex, resulting in highly polisubstituted chiral AEAs in excellent yields and ee (Scheme 87). The desired products could be selectively prepared in either (*R*)- or (*S*)-form by the ligand-controlled synthesis, which was initiated by the cleavage of C–B, C–Br, or C–O bonds under very mild reaction conditions.

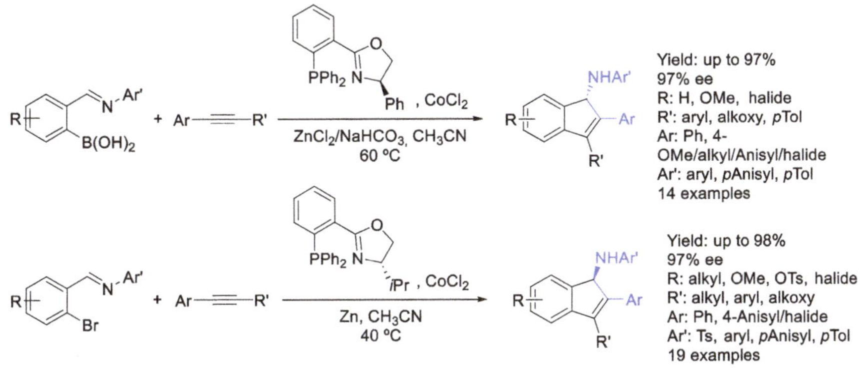

Scheme 87. Aminoindenes as chiral AEAs by Cheng et al. in 2018.

Also in 2018, Ge et al. [114] reported an asymmetric synthesis of chiral boryl-functionalized γ-lactams and tetrahydro-pyrroles containing an all-carbon quaternary stereocenter by a chiral cobalt complex. The resulting asymmetric AEAs, in which the aryl moiety was present in the substrates (Scheme 88), were synthesized by a ring-closing reaction enabled by an organo-cobalt complex.

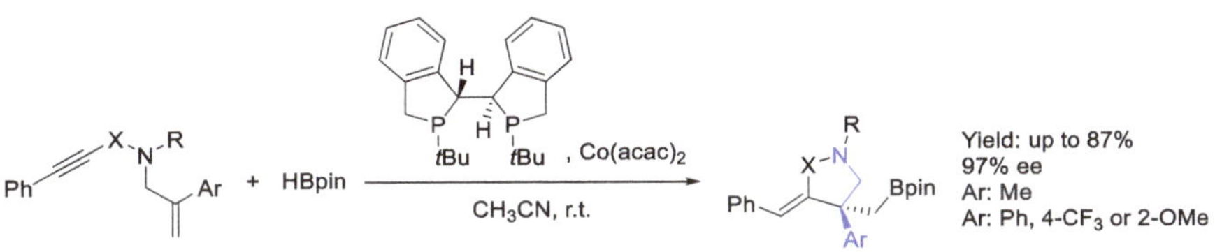

Scheme 88. Organo-cobalt-mediated reaction towards the synthesis of a chiral AEA by Ge et al. in 2018.

Then, in 2020, Cramer et al. [115] developed an enantioselective carboamination reaction by C-H activation of *N*-phenoxyamides enabled by the tailored tri-substituted chiral Cpx ligand. The resulting chiral AEAs, in which the amine group is actually an amide group, were synthesized in good yields and ee. Interestingly, products of the methodology could be derivatized in order to obtain interesting amino acid derivatives (Scheme 89).

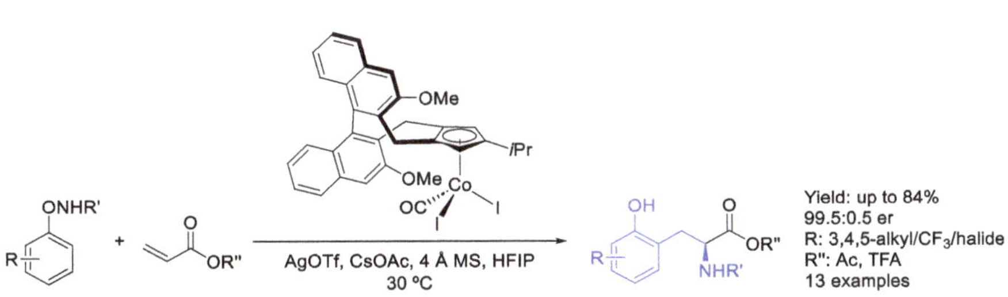

Scheme 89. Amino acid as chiral AEA by Cramer et al. in 2020.

Following the synthesis of amide-based AEAs, in 2023, Qin et al. [116] achieved an asymmetric deuteration of α-amidoacrylates, developing the synthesis of a chiral AEA by asymmetric hydrogenation enabled by a chiral organo-cobalt complex, resulting in excellent yields and ee using either HOAc or DOAc as hydrogenation/deuteration agent (Scheme 90). As a cheap deuterium source, the use of methanol enabled a concise synthesis of α,β-dideuterio *L*-DOPA as well as stereoselective syntheses of deuteralogs of drugs including nostocyclopeptide A2, a matrix metalloproteinase inhibitor, an anticancer agent bortezomib, nateglinide, and solriamfetol.

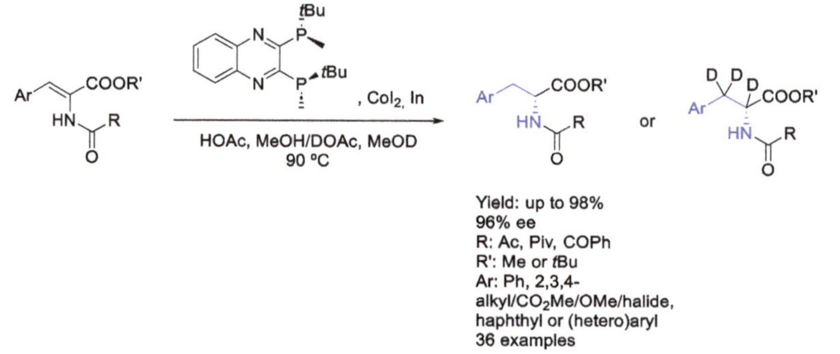

Scheme 90. Asymmetric deuterated AEAs by Qin et al. in 2023.

In the same year, Zhang et al. [117] also achieved an asymmetric hydrogenation of ketone moieties by using a cobalt-catalyzed reaction from α-primary amino ketones. The resulting chiral AEAs were obtained in excellent yields and ee and were even used as key intermediates in the synthesis of a dopamine derivative and a potent β-adrenergic agonist, both interesting (Scheme 91).

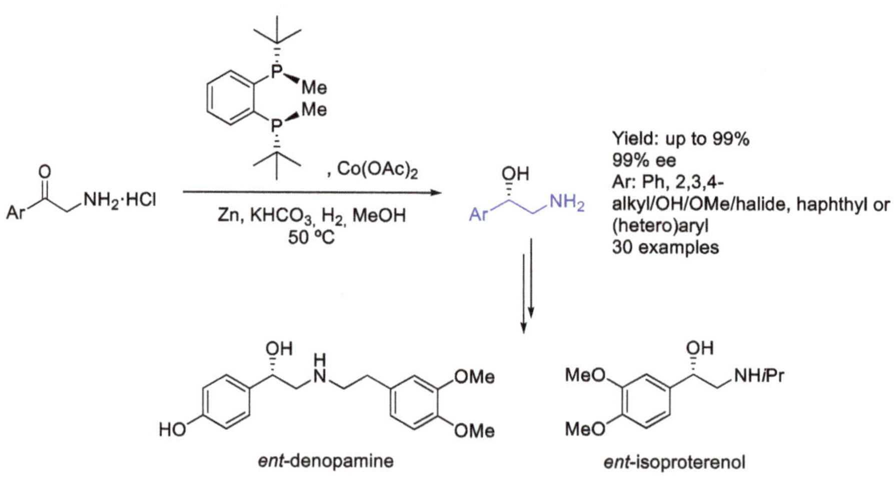

Scheme 91. Chiral AEAs as key intermediates towards the synthesis of bioactive compounds by Zhang et al. in 2023.

Finally, in 2024, Chirik et al. [118] developed an asymmetric hydrogenation of aryl-containing enamides, resulting in interesting chiral AEAs in excellent yields and ee. Cobalt-catalyzed asymmetric hydrogenation was enabled by different organo-cobalt complexes containing either Co(I) or Co(II), both of them showing great performance. Thus, indazole-containing enamides relevant to the synthesis of the calcitonin gene-related peptide (CGRP) receptor antagonist, zavegepant, approved for the treatment of migraines, could be described (Scheme 92).

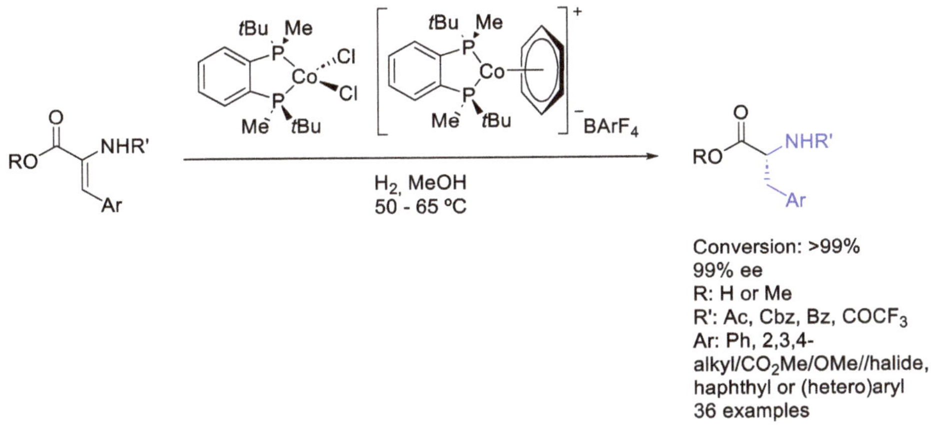

Scheme 92. Asymmetric synthesis of an AEA enabled by cobalt catalysis by Chirik et al. in 2024.

2.8. Others

2.8.1. Iron

In 2023, Meggers et al. [119] achieved the synthesis of chiral 2-imidazolidinones by using iron as the transition metal included in a chiral organic complex. By starting from branched urea-based substrates, an enantioselective ring-closing amination reaction could be undertaken, through intermediate iron nitrene species, in moderate yields and moderate ee (Scheme 93).

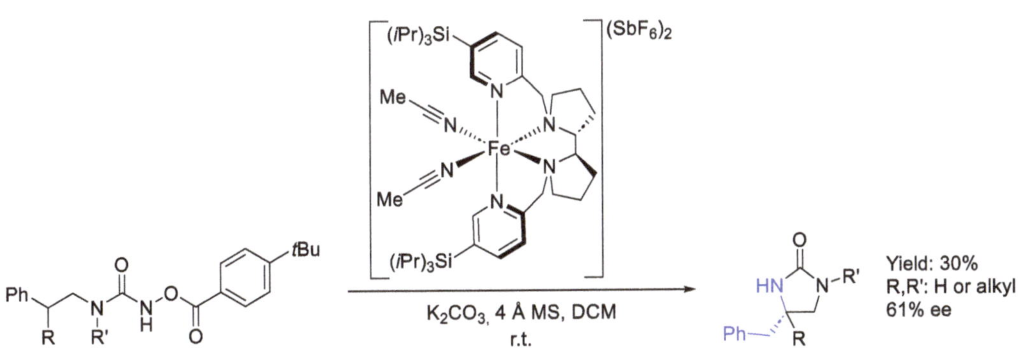

Scheme 93. Iron-catalyzed chiral AEA by Meggers et al. in 2023.

2.8.2. Titanium

In 2021, Race et al. [120] developed a methodology in which a chiral AEA could be synthesized by a phenonium ion intermediate enabled by $TiCl_4$. Starting from diastereopure aziridine substrates, high yields and dr were achieved in less than 10 min reaction time (Scheme 94A). Mechanistically, coordination of the aziridine to $TiCl_4$ showed an energetic preference for monodentate *O*-ligation to the acyl group with subsequent phenonium ion formation. Then, chloride opening took place, resulting in diasteroseletive ring opening (Scheme 94B).

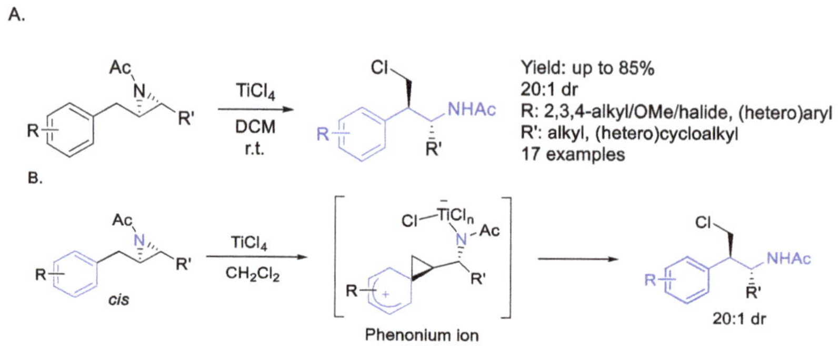

Scheme 94. Stereoselective ring-opening reaction of aziridines (**A**) and its proposed mechanism (**B**) by Race et al. in 2021.

2.8.3. Zinc

In 2013, Ghorai et al. [121] used dual catalysis of zinc and scandium Lewis acids to catalyze S_N2-type ring opening of *N*-activated aziridines with electron-rich arenes and heteroarenes. Chiral AEAs were synthesized from the corresponding activated chiral aziridines, resulting in highly functionalized 2,2-diaryl/heteroarylethylamines in excellent yields and ee (Scheme 95).

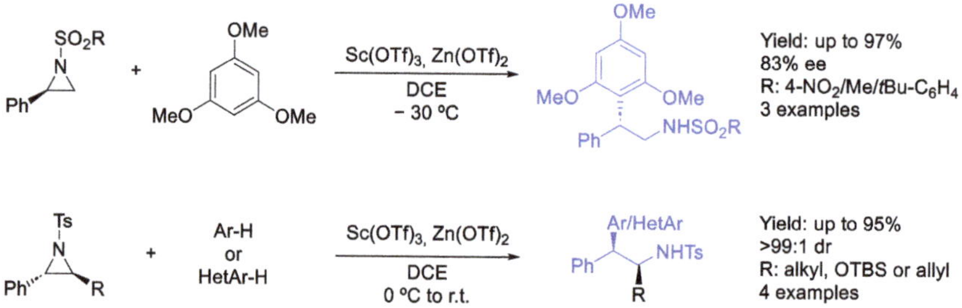

Scheme 95. Dual catalysis enabled by Zn and Sc towards the synthesis of chiral AEAs by Ghorai et al. in 2013.

Trost et al., in 2019 [122], developed an interesting synthesis of a polisubstituted chiral AEA between *N*-carbamoyl imines and α-branched ketones catalyzed by Zn-ProPhenol enabled by Mannich reactions. Key to this strategy was the introduction of unsaturation on one side of the ketone pronucleophiles, which drastically improved the reactivity and overcame the challenge of regioselectivity at the more substituted position. The resulting products with quaternary centers could be achieved in good yields and ee (Scheme 96).

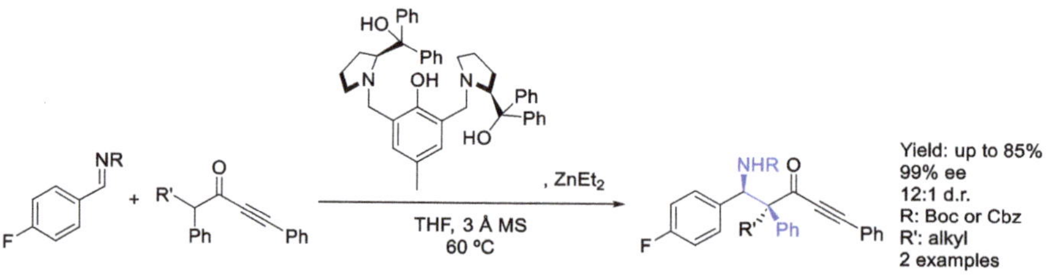

Scheme 96. Asymmetric synthesis of AEA through Mannich reaction by Trost et al. in 2019.

3. Photocatalysis

The use of visible light to promote organic transformations has shown interesting and rapid advances in the last decades. When it comes to waste management and reaction efficiency, light seems to offer great advantages to be used both in the laboratory and on an industry scale [123,124]. In this sense, progress in this field has allowed us to produce ideal cheap and energy-efficient light sources available for photocatalysis. Although ultraviolet light has been used in classical photochemistry, visible light, as well as photoredox catalysis and photosensitizers, has become much easier to be used in a typical laboratory setup [125].

Since the report by MacMillan et al. in 2008 on visible-light singly occupied molecular orbital (SOMO) photoredox catalysis allowing enantioselective α-alkylations by a combination of photoredox catalysis and organocatalysis [126], photoredox catalysis using visible light has been applied to a large range of important organic transformations [127–129].

Photocatalysts rely either on metal complexes or on organic dyes, which can be excited by light and thus activate intermediates either by electro, hydrogen, or energy transfer. The most recent uses of different photocatalysts in combination with other techniques in organic synthesis or otherwise will be discussed.

3.1. Iridium

In this regard, Phipps et al., in 2018 [130], reported a process for the addition of prochiral radicals from amino acid derivatives in order to synthesize chiral quinolines with asymmetric AEAs in their structure (Scheme 97). This method offered excellent enantio- and regioselective control enabled by a chiral Brønsted acid catalyst, which served both to activate the substrate and to induce asymmetry, while an iridium photocatalyst mediated in the required single electron transfer process.

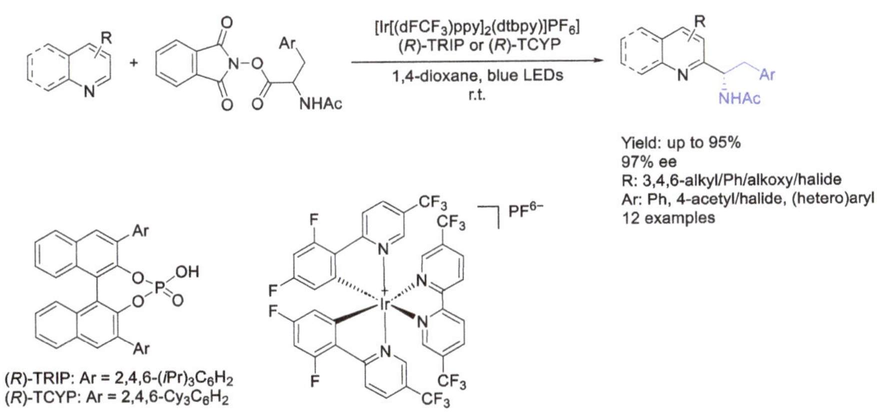

Scheme 97. Interesting polisubstituted AEA by Phipps et al. in 2018.

Also in 2018, Monos et al. [131] reported a catalytic iridium protocol for the 1,4-aryl migration of arylsulfonylacetamides across electron-rich alkenes through Smiles–Truce rearrangement. This redox-neutral, visible light-driven, single-electron alkene oxidation method achieved complete anti-Markovnikov regioselectivity and excellent diastereoselectivity, enabling the synthesis of a chiral AEA from commercially available aryl sulfonamides and unactivated alkenes under mild conditions in excellent ee and yields (Scheme 98).

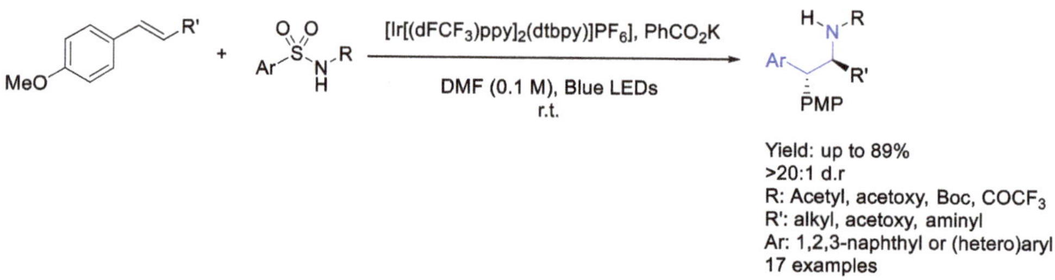

Scheme 98. Chiral AEA by obtained via Truce–Smiles rearrangement enabled by iridium photocatalysis by Monos et al. in 2018.

In 2019, Molander et al. [132] achieved a rapid and highly diastereoselective amidoarylation of unactivated olefins by a cascade process by merging, for the first time, photoredox proton-coupled electron transfer (PCET) with nickel catalysis. This novel process started with *N*-radical formation by PCET, enabled by a photoexcited iridium photocatalyst and a base, followed by an intramolecular ring-closing reaction and radical addition to nickel complex. Then, oxidative addition and reductive elimination afforded the corresponding chiral AEA in which the nitrogen atom was embedded into a pyrrolidinone ring, in good yield and dr (Scheme 99).

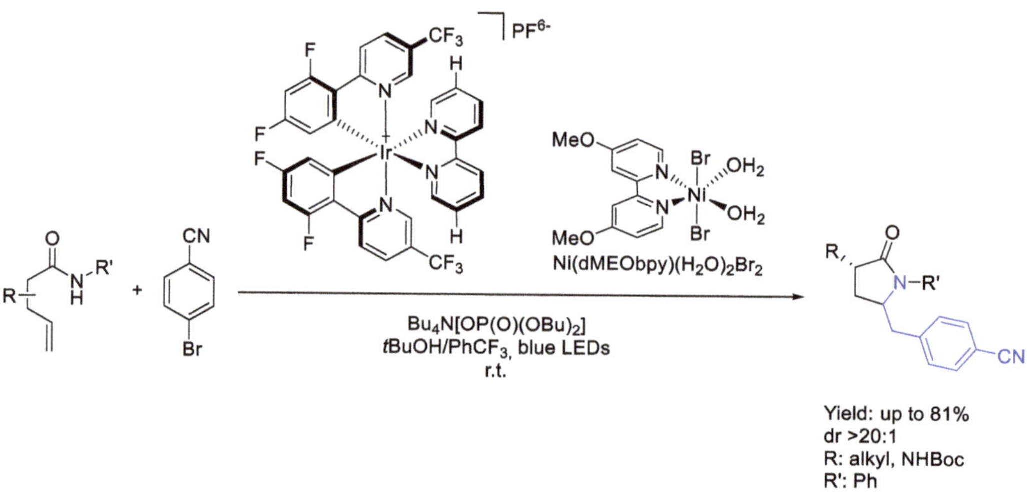

Scheme 99. Pyrrolidinone-based chiral AEA by Molander et al. in 2019.

In 2020, Greaney et al. [133] developed a new powerful decarboxylative, desulfonylative Truce–Smiles rearrangement enabled by visible light energy transfer catalysis. The reaction used starting materials derived from commercially available β-amino acids, giving extensive control over the aryl species, ethyl chain substitution, and amino protection for the synthesis of a new chiral AEA in good yields and ee (Scheme 100). This methodology relied on imine activation by energy transfer from the excited iridium catalyst, followed by homolytic N–O bond cleavage and hydrogen atom transfer by the corresponding solvent.

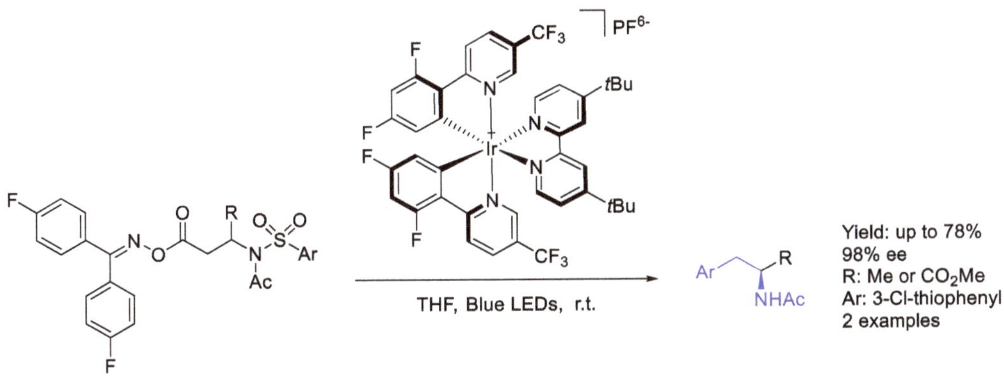

Scheme 100. Chiral AEA obtained through Truce–Smiles rearrangement by Greaney et al. in 2020.

Later, in 2022, Stephenson et al. [134] developed an intramolecular aromatic *ipso* substitution of a heteroatom leaving group by a carbon nucleophile with sulfonamides as bifunctional aryl and amine sources. This Smiles–Truce rearrangement was enabled by single electron transfer from the excited iridium catalyst to the corresponding sulfonamide substrate. Interestingly, it could only afford the desired chiral AEA when *O*-substituents were present in the aryl moiety, blocking dearomatization, giving the corresponding lactams in good yields and dr (Scheme 101A). Two years later, in 2024 [135], following their studies, they transferred the Smiles–Truce rearrangement to the aminoarylation of unactivated alkenes by lowering the oxidation state of the sulfur atom from S(VI) to S(IV), starting from the corresponding sulfinamides (Scheme 101B). The resulting change to the sulfur atom's molecular geometry contracted the C–S–N bond angle, thereby favoring the *ipso* cyclization step of the aryl migration. Using a weakly oxidizing photoredox catalyst, a sulfinamidyl radical was generated under mild conditions by single electron transfer from the iridium photocatalyst and was added to an alkene to form a new C–N bond, followed by a desulfinylative Smiles–Truce rearrangement to form a new C–C bond. Thus, interesting chiral AEAs were synthesized in good yields and dr (Scheme 101C).

In 2022, Xia et al. [136] demonstrated a methodology for a photoinitiated deaminative [3 + 2] annulation reaction of *N*-aminopyridinium salts with alkenes for the synthesis of functionalized γ-lactams enabled by a chiral iridium photocatalyst. The corresponding synthetized chiral polisubstituted AEAs had their nitrogen atoms embedded into a γ-lactam ring (Scheme 102). The mechanism relied on energy transfer from the iridium photocatalyst to the corresponding salts to enable homolytic N-N cleavage, followed by sterenyl addition, an intramolecular ring-closing process, and subsequent quenching with DMSO.

Nevado et al., in 2023 [137], described enantioenriched arylsulfinylamides as all-in-one reagents for the efficient asymmetric, intermolecular aminoarylation of alkenes. Under mild photoredox conditions, nitrogen addition of the arylsulfinylamide onto the double bond, followed by 1,4-translocation of the aromatic ring, afforded, in a single operation, production of the corresponding aminoarylation adducts in an enantiomerically enriched form (Scheme 103). Mechanistic investigations revealed the likelihood of multiple reaction pathways operating in these transformations. In the case of electron-rich styrenes, the proposed mechanism was based on radical cation formation by single-electron oxidation. In contrast, the single-electron oxidation of the deprotonated arylsulfinylamide by the excited iridium photocatalyst to form an *N*-centred radical seemed to occur in the case of poorly oxidizable olefins. This methodology could afford interesting optically pure β,β-diarylethylamines in good yields and ee.

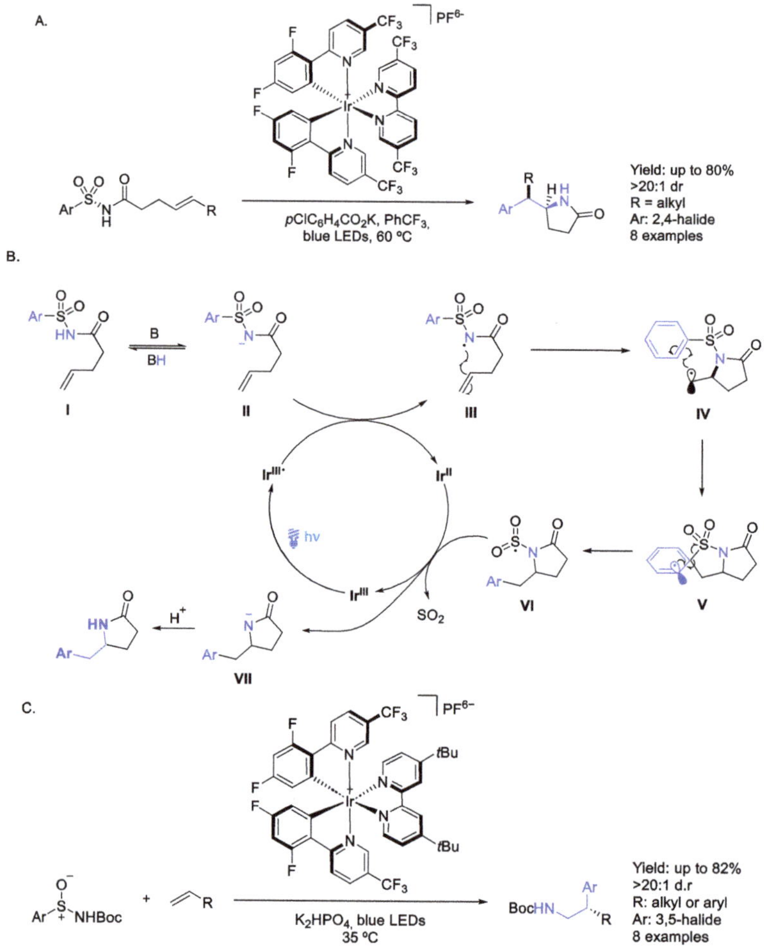

Scheme 101. Chiral AEA from high-oxidation-state sulfur present in sulfunamide moieties (**A**), its proposed catalytic cycle (**B**), and from low-oxidation-state sulfur present in sulfinimide moieties (**C**).

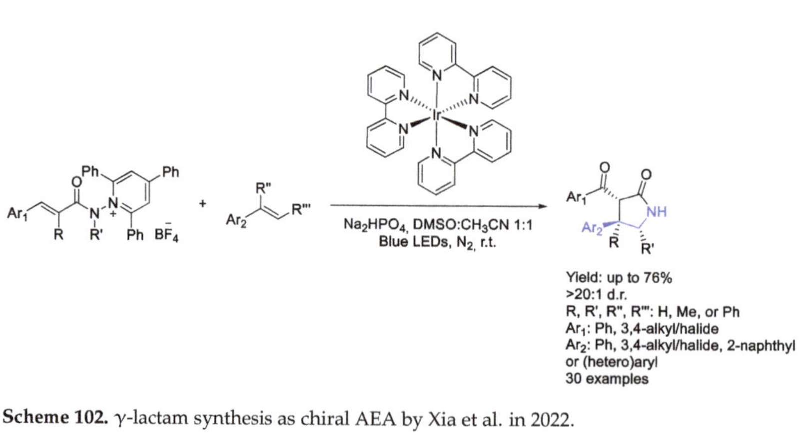

Scheme 102. γ-lactam synthesis as chiral AEA by Xia et al. in 2022.

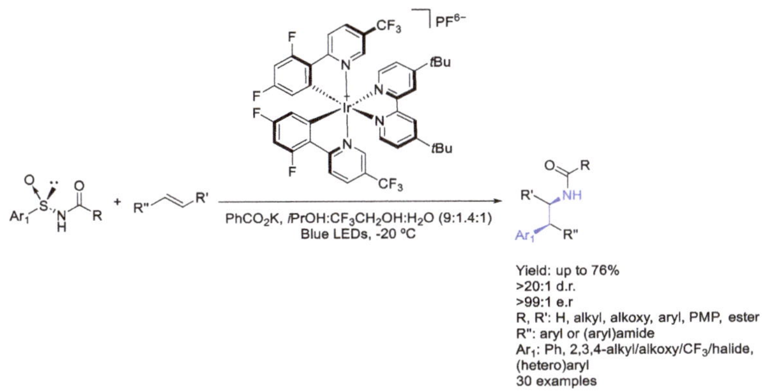

Scheme 103. β,β-diarylethylamines as asymmetric AEAs by Nevado et al. in 2023.

Wang et al., in 2024 [138], presented a novel synergistic triple-catalysis approach for the asymmetric α-C–H addition of readily available *N*-sulfonyl amines to aldehydes under mild conditions. This method allowed the efficient synthesis of a diverse variety of valuable β-amino alcohols bearing vicinal stereocenters, affording an interesting chiral AEA in good yield and ee (Scheme 104A). An iridium photocatalyst enables quinuclidine activation by the single electron transfer process, which, by hydrogen atom transfer, activated the corresponding benzylamine substrate, proceeding towards its addition to the aldehyde–chrome complex, affording the aforementioned products (Scheme 104B).

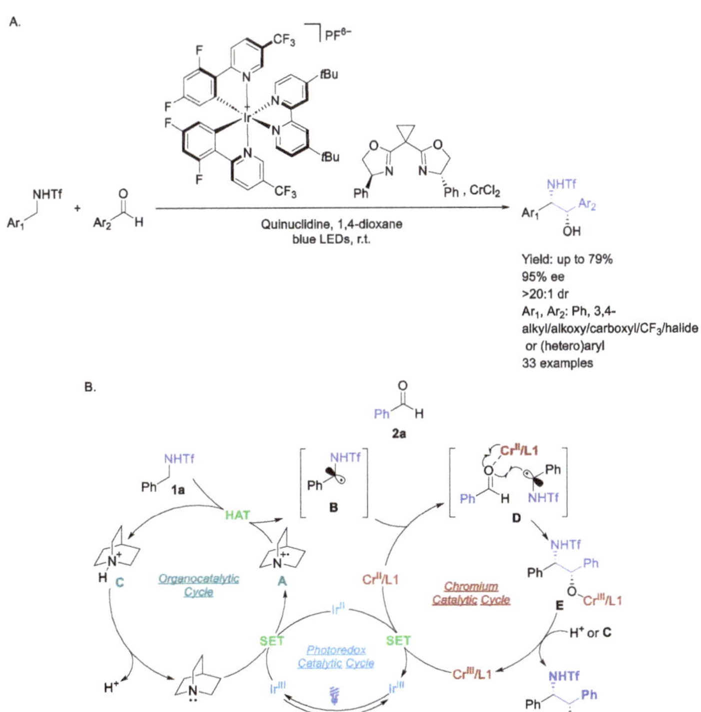

Scheme 104. Iridium-catalyzed synthesis of chiral AEA by Wang et al. in 2024 (**A**) and its proposed catalytic cycle (**B**).

3.2. Ruthenium

In 2014, Xia et al. [139] developed a mild and efficient procedure for the regioselective ring-opening nucleophilic addition reactions of aziridines via visible light photoredox catalysis, which provided practical synthetic access to 1,2-bifunctional compounds, synthesizing chiral AEAs in the process. Starting from the corresponding optically active aziridine, the mechanism relied on single electron transfer from the excited ruthenium photocatalyst towards the substrates, followed by nucleophilic ring opening and subsequent hydrogen atom transfer from the solvent, affording the desired products in excellent yields and ee (Scheme 105).

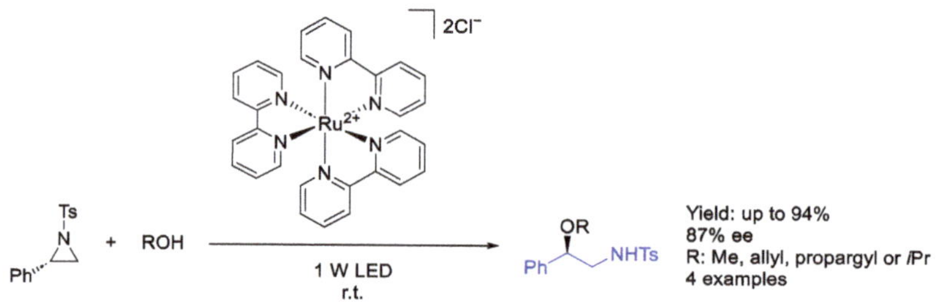

Scheme 105. 1,2-bifunctional AEA by Xia et al. in 2014.

Liu et al., in 2021 [140], developed a visible light-mediated approach for synthesizing 1,4,5,6-tetrahydropyridazines with arylethylamine motifs using *N*-radical cyclization and Smiles–Truce aryl transfer. This cascade reaction, enabled by a photoredox catalyst and a base, operated under mild, redox-neutral conditions and demonstrated excellent functional group compatibility and diastereoselectivity. Additionally, it enabled the construction of synthetically challenging all-carbon quaternary centers through alkene aminoarylation. Single electron transfer from the excited ruthenium photocatalyst towards the corresponding substrates followed by an intramolecular ring-closing reaction and the desulfonation process afforded the aforementioned chiral AEA in moderate yields and dr (Scheme 106).

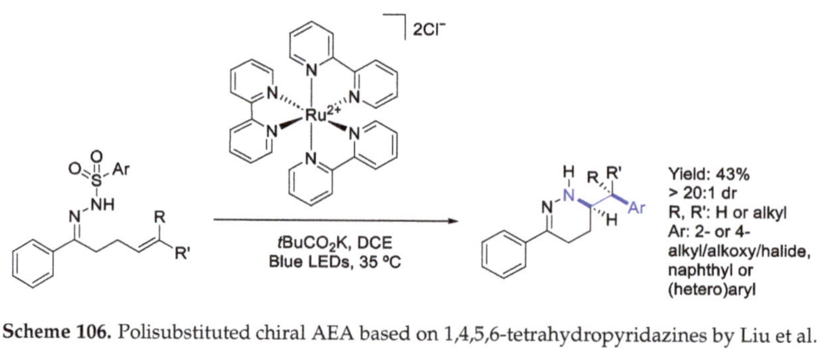

Scheme 106. Polisubstituted chiral AEA based on 1,4,5,6-tetrahydropyridazines by Liu et al. in 2021.

3.3. Nickel

Subsequently, Huo et al., in 2020 [141], achieved a direct enantioselective acylation of α-amino C(sp^3)–H bonds with carboxylic acids via the combination of a transition metal and photoredox catalysis. This straightforward protocol enabled cross-coupling of a wide range of carboxylic acids, with readily available *N*-alkyl benzamides to produce chiral AEAs in high enantioselectivities under mild conditions. A photoexcited nickel catalyst enabled Br radical formation, which prompted hydrogen atom transfer with the corresponding substrates to form radical intermediates, followed by oxidative addition of the chiral nickel organo-catalyst and reductive elimination to afford the desired products (Scheme 107).

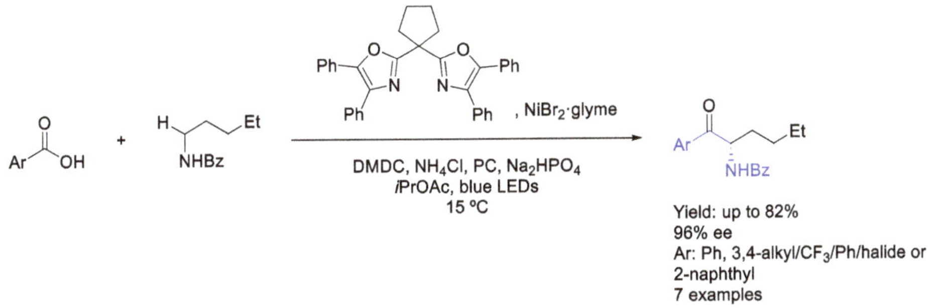

Scheme 107. Asymmetric synthesis of an AEA enabled by nickel catalysis by Huo et al. in 2020.

Doyle et al., in 2020 [142], reported a photoassisted nickel-catalyzed reductive cross-coupling between tosyl-protected alkyl aziridines and commercially available (hetero)aryl iodides. 4CzIPN as the photocatalyst enabled single electron transfer from the aminoiodine intermediate towards aziridine substrates, followed by oxidative addition and reductive elimination of the corresponding chiral organo-nickel complex, affording an interesting chiral AEA, when starting from the optically active substrate, maintaining the stereochemistry in good yield (Scheme 108).

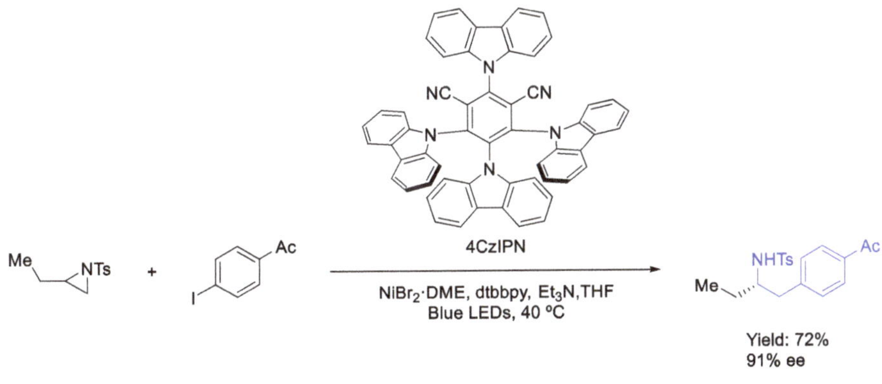

Scheme 108. *N*-tosyl-based AEA by Doyle in 2020.

Studer et al., in 2021 [143], reported a three-component stereoselective 1,2-aminoarylation of electron-rich alkenes through synergistic photoredox and nickel catalysis. Anti-Markovnikov addition of the amidyl radical to the alkene and nickel-mediated radical/transition metal crossover led to the corresponding 1,2-aminoarylation product, offering a route for the preparation of enantiopure α-arylated β-amino alcohols. In this process, interesting chiral AEAs were synthesized in moderate yields and excellent dr (Scheme 109). The mechanism involved single electron transfer from an excited organocatalyst towards the carboxylic acid substrates; sequential fragmentation of CO_2 and acetone which generates the electrophilic N-radical; followed by addition to the corresponding alkene. Then, radical trapping by the Ni-complex followed by reductive elimination afforded the desired products.

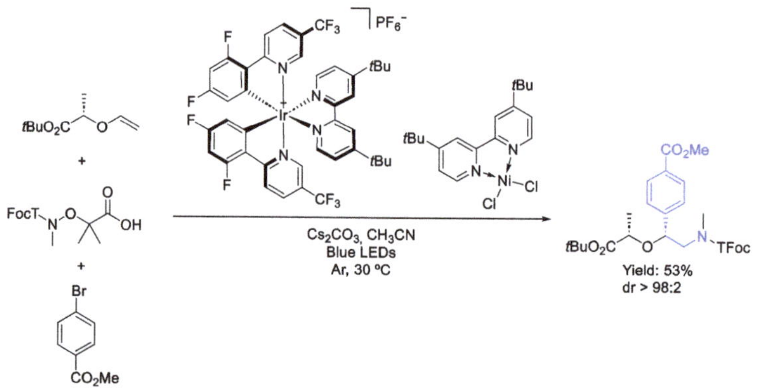

Scheme 109. Chiral AEA incorporating ether moiety by Studer et al. in 2021.

3.4. Copper

Nicewicz et al., in 2023 [144], described for the first time an enantioselective difunctionalization of olefins via a cation radical intermediate utilizing an acridinium photooxidant in conjunction with copper catalysis. The transformation could be rendered asymmetric by using a serine-derived bisoxazoline ligand. The wide array of nucleophiles in this three-component coupling allowed the synthesis of an interesting chiral α-cyano AEA in good yields and great dr and ee (Scheme 110A). The scope of amines for the aminocyanation reaction was greatly expanded by undergoing a cation radical intermediate as opposed to previous *N*-centered radical-initiated aminocyanations, as shown in the proposed mechanistic cycle (Scheme 110B).

A.

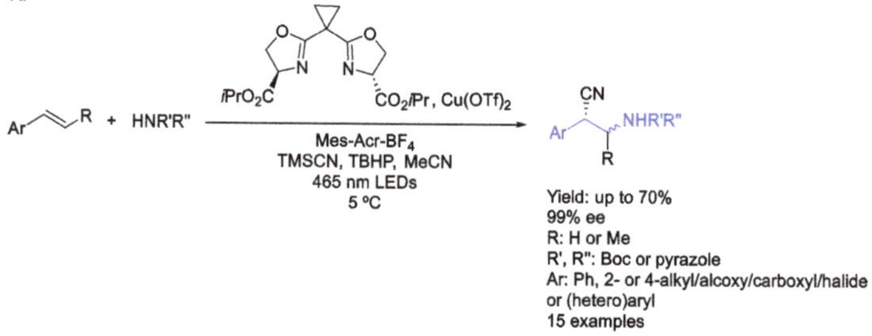

B.

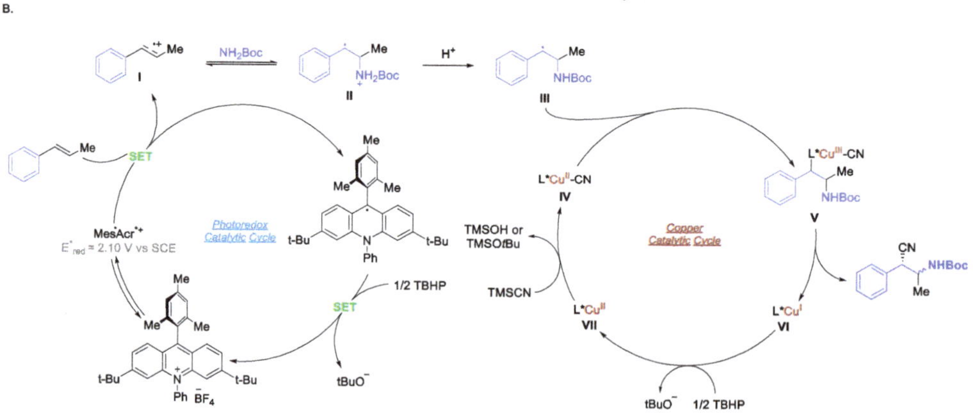

Scheme 110. α-Ciano AEA by Nicewicz et al. in 2023 (**A**) and its proposed catalytic cycle (**B**).

In 2025, Gong et al. [145] developed a methodology towards the synthesis of β,β-aminoacyl AEA (Scheme 111). Starting from glycine derivatives, this photoinduced regio-divergent and enantioselective cross-coupling relied on a catalytic system that integrated photoinduced hydrogen atom transfer (HAT) and chiral copper catalysis. This strategy facilitated regiodivergent and enantioselective cross-couplings between *N*-aryl glycine ester/amide derivatives and abundant hydrocarbon feedstocks through strong C(sp^3)-H bond activation, obtaining interesting AEAs in good yields and ee.

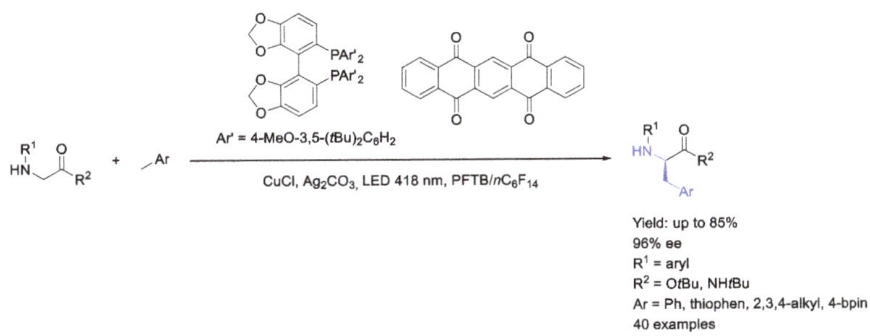

Scheme 111. Copper catalysis towards the synthesis of interesting AEAs by Gong et al. in 2025.

3.5. *Wolframium*

Masson et al., in 2024 [146], described a sustainable and efficient photocatalytic method for the stereoselective radical alkylation of chiral sulfinyl imines. By employing readily available non-prefunctionalized radical precursors and cost-effective TBADT (Tetrabutylammonium decatungstate) as a direct HAT photocatalyst, they successfully obtained diverse chiral AEAs with high yields and excellent diastereoselectivity under mild conditions (Scheme 112). Excited TBADT underwent a hydrogen atom transfer towards the corresponding methylaryl substrates, followed by sulfinyl imine addition and subsequent back-hydrogen atom transfer from the photocatalyst. Thus, the chiral AEA was synthetized with both good yield and ee.

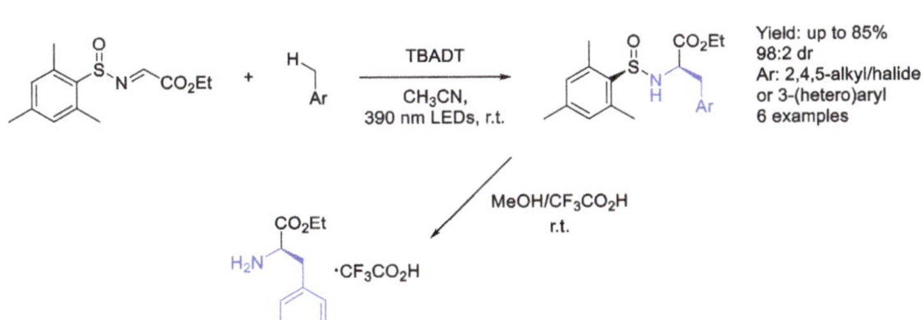

Scheme 112. Wolframium-catalyzed asymmetric synthesis of an AEA by Masson et al. in 2024.

3.6. *Iron*

Also in 2024, Chang et al. [147] reported a visible-light-promoted enantioselective a-amidation of aldehydes by organo-iron catalysis towards the synthesis of chiral 2-arylethylacetamides (Scheme 113). Interestingly, mechanistic studies revealed that in situ-generated [Fe(II)Cl_3^-] via visible-light-promoted LMCT effectively activated dioxazolones, affording an iron-acylnitrenoid radical that was subsequently inserted into chiral enamine intermediates. Thus, good yields and ee were achieved.

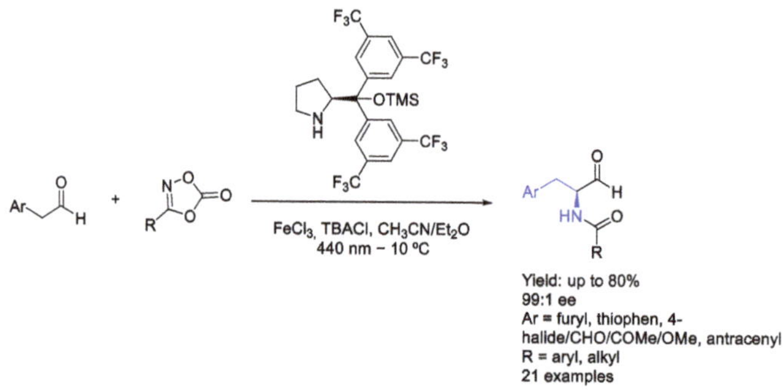

Scheme 113. Chiral 2-arylethylacetamides by Chang et al. in 2024.

4. Conclusions

Because of their very well-known bioactive properties, 2-arylethylamine motif compounds are of great interest in both synthetic and medicinal chemistry. As small nitrogen-containing compounds, they are able to pass the blood–brain barrier; thus, they are used in treatments related to Central Neural System. The human body is chiral and, thus, chirality introduction into synthetic methodologies is a mandatory approach in the pharmaceutical industry at present. This asymmetric synthesis can present several challenges when faced in a laboratory. Herein, we present the most recent advances in asymmetric synthesis of 2-arylethylamines using transition metal catalysis as the main tool to introduce chirality in either C-1, C-2, or in both carbons of the aforementioned motif. The main methodology when it comes to chiral synthesis is presented and discussed, considering conditions, challenges, yields, and ee/dr. Also, when considered, interesting and novel catalytic cycles are presented and discussed.

Conventional metal catalysis supported with a chiral organo complex is the main core of the review, organized by metal abundancy in the literature of recent years. While copper and palladium-based catalysis are the richest when it comes to different substrates and targets, rhodium and cobalt have mainly been used for asymmetric hydrogenation or hydroamination of different substrates. On the other hand, less abundant metals such as ruthenium, iridium, or nickel are not focused on one type of methodology, also considering that iridium is mainly used in photocatalysis. Photocatalysis is presented as separate section covering the main advances in recent years, showing that iridium is the most abundant transition metal used towards the synthesis of chiral AEAs.

Derivatization of the aforementioned compounds has proved to be very important in the pharmaceutical industry, as shown in the presented review.

Author Contributions: Conceptualization, A.M., Á.G.-G., C.T.N., N.G.L., D.D. and N.M.G.; investigation, A.M. and Á.G.-G.; data curation, A.M. and Á.G.-G.; writing—original draft preparation, A.M.; writing—review and editing, A.M., Á.G.-G., C.T.N., N.G.L., D.D. and N.M.G. All authors have read and agreed to the published version of the manuscript.

Funding: The authors gratefully acknowledge the financial support of this work provided by Ministerio de Ciencia, Innovación y Universidades: (PID2020-118303GB-I00 and MCIN/AEI/10.13039/501100011033), Junta de Castilla y León (SA076P20) and Universidad de Salamanca (Programme I, GIR PRONABIO-LACT).

Institutional Review Board Statement: Not applicable.

Informed Consent Statement: Not applicable.

Data Availability Statement: No new data were created or analyzed in this study. Data sharing is not applicable to this article.

Conflicts of Interest: The authors declare no conflicts of interest.

References

1. Sulzer, D.; Sonders, M.S.; Poulsen, N.W.; Galli, A. Mechanisms of Neurotransmitter Release by Amphetamines: A Review. *Prog. Neurobiol.* **2005**, *75*, 406–433. [CrossRef]
2. Bentley, K.W. β-Phenylethylamines and the Isoquinoline Alkaloids. *Nat. Prod. Rep.* **2006**, *23*, 444–463. [CrossRef]
3. Chinta, S.J.; Andersen, J.K. Dopaminergic Neurons. *Int. J. Biochem. Cell Biol.* **2005**, *37*, 942–946. [CrossRef] [PubMed]
4. Nieto, C.T.; Cascón, A.M.; Arroyo, L.B.; Martin, D.D.; Garrido, N.M. 2-Phenethylamines in Medicinal Chemistry: A Review. *Molecules* **2022**, *28*, 855. [CrossRef] [PubMed]
5. Nieto, C.; Manchado, A.; García-González, Á.; Díez, D.; Garrido, N.M. 2-Heteroarylethylamines in Medicinal Chemistry: A Review of 2-Phenethylamine Satellite Chemical Space. *Beilstein J. Org. Chem.* **2024**, *20*, 1880–1893. [CrossRef] [PubMed]
6. Manchado, A.; García-González, Á.; Nieto, C.T.; Díez, D.; Garrido, N.M. Asymmetric Synthesis of 2-Arylethylamines: A Metal-Free Review of the New Millennium. *Molecules* **2024**, *29*, 5729. [CrossRef]
7. Abdine, R.A.A.; Hedouin, G.; Colobert, F.; Wencel-Delord, J. Metal-Catalyzed Asymmetric Hydrogenation of C=N Bonds. *ACS Catal.* **2021**, *11*, 215–247. [CrossRef]
8. Cabré, A.; Verdaguer, X.; Riera, A. Recent Advances in the Enantioselective Synthesis of Chiral Amines via Transition Metal-Catalyzed Asymmetric Hydrogenation. *Chem. Rev.* **2022**, *122*, 269–339. [CrossRef]
9. Mathew, S.; Renn, D.; Rueping, M. Advances in One-Pot Chiral Amine Synthesis Enabled by Amine Transaminase Cascades: Pushing the Boundaries of Complexity. *ACS Catal.* **2023**, *13*, 5584–5598. [CrossRef]
10. Pozhydaiev, V.; Muller, C.; Moran, J.; Lebœuf, D. Catalytic Synthesis of β-(Hetero)Arylethylamines: Modern Strategies and Advances. *Angew. Chem.-Int. Ed.* **2023**, *62*, e202309289. [CrossRef]
11. Kwon, Y.; Wang, Q. Recent Advances in 1,2-Amino(Hetero)Arylation of Alkenes. *Chem. Asian J.* **2022**, *17*, e202200215. [CrossRef]
12. Sun, J.; Zheng, G.; Zhang, G.; Li, Y.; Zhang, Q. Recent Advances in Enantioselective Construction of C-N Bonds Involving Radical Intermediates. *Org. Chem. Front.* **2025**, *12*, 1671–1694. [CrossRef]
13. Ju, X.; Lee, M.; Leung, J.C.; Lee, J. Synthesis and Functionalization of Aziridines: A Perspective View from Pharmaceutical Industries. *Eur. J. Org. Chem.* **2025**, e202401414. [CrossRef]
14. Yin, Y.; Zhao, X.; Jiang, Z. Asymmetric Photocatalytic Synthesis of Enantioenriched Azaarene Derivatives. *Chin. J. Org. Chem.* **2022**, *42*, 1609–1625. [CrossRef]
15. Zeng, W.; Chemler, S.R. Copper(II)-Catalyzed Enantioselective Intramolecular Carboamination of Alkenes. *J. Am. Chem. Soc.* **2007**, *129*, 12948–12949. [CrossRef]
16. Hajra, S.; Sinha, D. Catalytic Enantioselective Aziridoarylation of Aryl Cinnamyl Ethers toward Synthesis of Trans-3-Amino-4-Arylchromans. *J. Org. Chem.* **2011**, *76*, 7334–7340. [CrossRef] [PubMed]
17. Sakae, R.; Matsuda, N.; Hirano, K.; Satoh, T.; Miura, M. Highly Stereoselective Synthesis of (Borylmethyl)Cyclopropylamines by Copper-Catalyzed Aminoboration of Methylenecyclopropanes. *Org. Lett.* **2014**, *16*, 1228–1231. [CrossRef] [PubMed]
18. Kato, K.; Hirano, K.; Miura, M. Synthesis of Β-Boryl-α-Aminosilanes by Copper-Catalyzed Aminoboration of Vinylsilanes. *Angew. Chem. Int. Ed.* **2016**, *128*, 14612–14616. [CrossRef]
19. Parra, A.; Amenós, L.; Guisán-Ceinos, M.; López, A.; García Ruano, J.L.; Tortosa, M. Copper-Catalyzed Diastereo- and Enantioselective Desymmetrization of Cyclopropenes: Synthesis of Cyclopropylboronates. *J. Am. Chem. Soc.* **2014**, *136*, 15833–15836. [CrossRef]
20. Yang, Y.; Perry, I.B.; Buchwald, S.L. Copper-Catalyzed Enantioselective Addition of Styrene-Derived Nucleophiles to Imines Enabled by Ligand-Controlled Chemoselective Hydrocupration. *J. Am. Chem. Soc.* **2016**, *138*, 9787–9790. [CrossRef]
21. Ge, C.; Liu, R.R.; Gao, J.R.; Jia, Y.X. Cu(I)-Catalyzed Enantioselective Friedel-Crafts Alkylation of Indoles with 2-Aryl-N-Sulfonylaziridines as Alkylating Agents. *Org. Lett.* **2016**, *18*, 3122–3125. [CrossRef] [PubMed]
22. Chen, Z.; Huo, Y.; An, P.; Wang, X.; Song, C.; Ma, Y. Paracyclophane-Based N-Heterocyclic Carbene as Efficient Catalyst or as Ligand for Copper Catalyst for Asymmetric α-Silylation of: N -Tosylaldimines. *Org. Chem. Front.* **2016**, *3*, 1725–1737. [CrossRef]
23. Pascual-Escudero, A.; de Cózar, A.; Cossío, F.P.; Adrio, J.; Carretero, J.C. Alkenyl Arenes as Dipolarophiles in Catalytic Asymmetric 1,3-Dipolar Cycloaddition Reactions of Azomethine Ylides. *Angew. Chem.-Int. Ed.* **2016**, *55*, 15334–15338. [CrossRef]
24. Peng, F.; Chen, Y.; Chen, C.Y.; Dormer, P.G.; Kassim, A.; McLaughlin, M.; Reamer, R.A.; Sherer, E.C.; Song, Z.J.; Tan, L.; et al. Asymmetric Formal Synthesis of the Long-Acting DPP-4 Inhibitor Omarigliptin. *J. Org. Chem.* **2017**, *82*, 9023–9029. [CrossRef]
25. Wang, D.; Wang, F.; Chen, P.; Lin, Z.; Liu, G. Enantioselective Copper-Catalyzed Intermolecular Amino- and Azidocyanation of Alkenes in a Radical Process. *Angew. Chem.* **2017**, *129*, 2086–2090. [CrossRef]

26. Wang, D.; Wu, L.; Wang, F.; Wan, X.; Chen, P.; Lin, Z.; Liu, G. Asymmetric Copper-Catalyzed Intermolecular Aminoarylation of Styrenes: Efficient Access to Optical 2,2-Diarylethylamines. *J. Am. Chem. Soc.* **2017**, *139*, 6811–6814. [CrossRef] [PubMed]
27. Hu, Z.; Fu, L.; Chen, P.; Cao, W.; Liu, G. Enantioselective Intermolecular Aminoalkynylation of Styrenes via Copper-Catalyzed Radical Relay. *Org. Lett.* **2021**, *23*, 129–134. [CrossRef]
28. Yang, P.J.; Qi, L.; Liu, Z.; Yang, G.; Chai, Z. Lewis Acid Catalyzed Dynamic Kinetic Asymmetric Transformation of Racemic N-Sulfonylaziridines. *J. Am. Chem. Soc.* **2018**, *140*, 17211–17217. [CrossRef]
29. Li, Z.; Zhang, M.; Zhang, Y.; Liu, S.; Zhao, J.; Zhang, Q. Multicomponent Cyclopropane Synthesis Enabled by Cu-Catalyzed Cyclopropene Carbometalation with Organoboron Reagent: Enantioselective Modular Access to Polysubstituted 2-Arylcyclopropylamines. *Org. Lett.* **2019**, *21*, 5432–5437. [CrossRef]
30. Chang, X.; Yang, Y.; Shen, C.; Xue, K.S.; Wang, Z.F.; Cong, H.; Tao, H.Y.; Chung, L.W.; Wang, C.J. β-Substituted Alkenyl Heteroarenes as Dipolarophiles in the Cu(I)-Catalyzed Asymmetric 1,3-Dipolar Cycloaddition of Azomethine Ylides Empowered by a Dual Activation Strategy: Stereoselectivity and Mechanistic Insight. *J. Am. Chem. Soc.* **2021**, *143*, 3519–3535. [CrossRef]
31. Nishino, S.; Miura, M.; Hirano, K. An Umpolung-Enabled Copper-Catalysed Regioselective Hydroamination Approach to α-Amino Acids. *Chem. Sci.* **2021**, *12*, 11525–11537. [CrossRef] [PubMed]
32. Nishino, S.; Nishii, Y.; Hirano, K. Anti-Selective Synthesis of β-Boryl-α-Amino Acid Derivatives by Cu-Catalysed Borylamination of α,β-Unsaturated Esters. *Chem. Sci.* **2022**, *13*, 14387–14394. [CrossRef] [PubMed]
33. Nakamura, S.; Nishino, S.; Hirano, K. Synthesis of α-Aminophosphonates by Umpolung-Enabled Cu-Catalyzed Regioselective Hydroamination. *J. Org. Chem.* **2023**, *88*, 1270–1281. [CrossRef]
34. Ren, K.; Yuan, R.; Gui, Y.Y.; Chen, X.W.; Min, S.Y.; Wang, B.Q.; Yu, D.G. Cu-Catalyzed Reductive Aminomethylation of 1,3-Dienes with N,O-Acetals: Facile Construction of β-Chiral Amines with Quaternary Stereocenters. *Org. Chem. Front.* **2022**, *10*, 467–472. [CrossRef]
35. Dai, L.; Chen, Y.-Y.; Xiao, L.-J.; Zhou, Q.-L. Intermolecular Enantioselective Benzylic C(Sp3)-H Amination via Cationic Copper Catalysis. *Angew. Chem.* **2023**, *62*, e202304427. [CrossRef] [PubMed]
36. Chen, J.J.; Zhang, J.Y.; Fang, J.H.; Du, X.Y.; Xia, H.D.; Cheng, B.; Li, N.; Yu, Z.L.; Bian, J.Q.; Wang, F.L.; et al. Copper-Catalyzed Enantioconvergent Radical C(Sp3)-N Cross-Coupling of Activated Racemic Alkyl Halides with (Hetero)Aromatic Amines under Ambient Conditions. *J. Am. Chem. Soc.* **2023**, *145*, 14686–14696. [CrossRef]
37. Klake, R.K.; Sieber, J.D. Aminoalcohol Synthesis through Nonprecious Metal Catalysis: Enantioselective Cu-Catalyzed Reductive Coupling of Aldehydes and Allenamides. *Org. Lett.* **2023**, *25*, 4730–4734. [CrossRef] [PubMed]
38. Geng, R.; Peng, L.; Guo, C. Dynamic Kinetic Stereodivergent Transformations of Propargylic Ammonium Salts via Dual Nickel and Copper Catalysis. *Chin. Chem. Lett.* **2024**, *35*, 109433. [CrossRef]
39. Zhu, J.; Rahim, F.; Zhou, P.; Zhang, A.; Malcolmson, S.J. Copper-Catalyzed Diastereo-, Enantio-, and (Z)-Selective Aminoallylation of Ketones through Reductive Couplings of Azatrienes for the Synthesis of Allylic 1,2-Amino Tertiary Alcohols. *J. Am. Chem. Soc.* **2024**, *146*, 20270–20278. [CrossRef]
40. Lee, M.; Sulwey, D.; Rotella, M.E.; Kim, W.S.; Ju, X.; Jiang, Q.; Kozlowski, M.C.; Lee, J. Mechanistic Insights of Copper Catalyzed Trifluoromethyl Aziridine Opening: Regioselective and Stereospecific Aryl Grignard Addition. *Org. Lett.* **2024**, *26*, 2713–2717. [CrossRef]
41. Huang, N.; Luo, J.; Liao, L.; Zhao, X. Catalytic Enantioselective Aminative Difunctionalization of Alkenes. *J. Am. Chem. Soc.* **2024**, *146*, 7029–7038. [CrossRef] [PubMed]
42. Wallace, D.J.; Campos, K.R.; Shultz, C.S.; Klapars, A.; Zewge, D.; Crump, B.R.; Phenix, B.D.; McWilliams, J.C.; Krska, S.; Sun, Y.; et al. New Efficient Asymmetric Synthesis of Taranabant, a CB1R Inverse Agonist for the Treatment of Obesity. *Org. Process Res. Dev.* **2009**, *13*, 84–90. [CrossRef]
43. Valdez, S.C.; Leighton, J.L. Tandem Asymmetric Aza-Darzens/Ring-Opening Reactions: Dual Functionality from the Silane Lewis Acid. *J. Am. Chem. Soc.* **2009**, *131*, 14638–14639. [CrossRef]
44. Praquin, C.F.B.; De Koning, P.D.; Peach, P.J.; Howard, R.M.; Spencer, S.L. Development of an Asymmetric Hydrogenation Route to (S)- N -Boc-2,6-Dimethyltyrosine. *Org. Process Res. Dev.* **2011**, *15*, 1124–1129. [CrossRef]
45. Fox, M.E.; Jackson, M.; Meek, G.; Willets, M. Large-Scale Synthesis of a Substituted d -Phenylalanine Using Asymmetric Hydrogenation. *Org. Process Res. Dev.* **2011**, *15*, 1163–1171. [CrossRef]
46. Anderson, N.A.; Fallon, B.J.; Valverde, E.; MacDonald, S.J.F.; Pritchard, J.M.; Suckling, C.J.; Watson, A.J.B. Asymmetric Rhodium-Catalysed Addition of Arylboronic Acids to Acyclic Unsaturated Esters Containing a Basic γ-Amino Group. *Synlett* **2012**, *23*, 2817–2821. [CrossRef]
47. Appell, R.B.; Boulton, L.T.; Daugs, E.D.; Hansen, M.; Hanson, C.H.; Heinrich, J.; Kronig, C.; Lloyd, R.C.; Louks, D.; Nitz, M.; et al. The Large-Scale Synthesis of (S)-N-Boc-Bis(4-Fluorophenyl)Alanine. *Org. Process Res. Dev.* **2013**, *17*, 69–76. [CrossRef]
48. Villa-Marcos, B.; Xiao, J. Metal and Organo-Catalysed Asymmetric Hydroaminomethylation of Styrenes. *Cuihua Xuebao/Chin. J. Catal.* **2015**, *36*, 106–112. [CrossRef]

49. Chen, C.; Jin, S.; Zhang, Z.; Wei, B.; Wang, H.; Zhang, K.; Lv, H.; Dong, X.Q.; Zhang, X. Rhodium/Yanphos-Catalyzed Asymmetric Interrupted Intramolecular Hydroaminomethylation of Trans-1,2-Disubstituted Alkenes. *J. Am. Chem. Soc.* **2016**, *138*, 9017–9020. [CrossRef]
50. Lin, T.Y.; Wu, H.H.; Feng, J.J.; Zhang, J. Transfer of Chirality in the Rhodium-Catalyzed Chemoselective and Regioselective Allylic Alkylation of Hydroxyarenes with Vinyl Aziridines. *Org. Lett.* **2017**, *19*, 2897–2900. [CrossRef]
51. Meng, J.; Li, X.H.; Han, Z.Y. Enantioselective Hydroaminomethylation of Olefins Enabled by Rh/Brønsted Acid Relay Catalysis. *Org. Lett.* **2017**, *19*, 1076–1079. [CrossRef]
52. Lang, Q.; Gu, G.; Cheng, Y.; Yin, Q.; Zhang, X. Highly Enantioselective Synthesis of Chiral γ-Lactams by Rh-Catalyzed Asymmetric Hydrogenation. *ACS Catal.* **2018**, *8*, 4824–4828. [CrossRef]
53. Bai, X.Y.; Zhang, W.W.; Li, Q.; Li, B.J. Highly Enantioselective Synthesis of Propargyl Amides through Rh-Catalyzed Asymmetric Hydroalkynylation of Enamides: Scope, Mechanism, and Origin of Selectivity. *J. Am. Chem. Soc.* **2018**, *140*, 506–514. [CrossRef] [PubMed]
54. Farr, C.M.B.; Kazerouni, A.M.; Park, B.; Poff, C.D.; Won, J.; Sharp, K.R.; Baik, M.H.; Blakey, S.B. Designing a Planar Chiral Rhodium Indenyl Catalyst for Regio- And Enantioselective Allylic C-H Amidation. *J. Am. Chem. Soc.* **2020**, *142*, 13996–14004. [CrossRef] [PubMed]
55. Mi, R.; Zhang, X.; Wang, J.; Chen, H.; Lan, Y.; Wang, F.; Li, X. Rhodium-Catalyzed Regio-, Diastereo-, and Enantioselective Three-Component Carboamination of Dienes via C-H Activation. *ACS Catal.* **2021**, *11*, 6692–6697. [CrossRef]
56. Wu, L.; Xu, H.; Gao, H.; Li, L.; Chen, W.; Zhou, Z.; Yi, W. Chiral Allylic Amine Synthesis Enabled by the Enantioselective CpXRh(III)-Catalyzed Carboaminations of 1,3-Dienes. *ACS Catal.* **2021**, *11*, 2279–2287. [CrossRef]
57. Yang, X.; Hong, K.; Zhang, S.; Zhang, Z.; Zhou, S.; Huang, J.; Xu, X.; Hu, W. Asymmetric Three-Component Reaction of Two Diazo Compounds and Hyrdroxylamine Derivatives for the Access to Chiral α-Alkoxy-β-Amino-Carboxylates. *ACS Catal.* **2022**, *12*, 12302–12309. [CrossRef]
58. Kliemann, M.N.; Teeuwen, S.; Weike, C.; Franciò, G.; Leitner, W. Rhodium-Catalyzed Asymmetric Hydrohydrazonemethylation of Styrenes: Access to Chiral Hydrazones, Hydrazides, Hydrazines and Amines. *Adv. Synth. Catal.* **2022**, *364*, 4006–4012. [CrossRef]
59. Lamartina, C.W.; Chartier, C.A.; Lee, S.; Shah, N.H.; Rovis, T. Modular Synthesis of Unnatural Peptides via Rh(III)-Catalyzed Diastereoselective Three-Component Carboamidation Reaction. *J. Am. Chem. Soc.* **2023**, *145*, 1129–1135. [CrossRef]
60. Lamartina, C.W.; Chartier, C.A.; Hirano, J.M.; Shah, N.H.; Rovis, T. Crafting Unnatural Peptide Macrocycles via Rh(III)-Catalyzed Carboamidation. *J. Am. Chem. Soc.* **2024**, *146*, 20868–20877. [CrossRef]
61. Mi, R.; Ding, Z.; Yu, S.; Crabtree, R.H.; Li, X. O-Allylhydroxyamine: A Bifunctional Olefin for Construction of Axially and Centrally Chiral Amino Alcohols via Asymmetric Carboamidation. *J. Am. Chem. Soc.* **2023**, *145*, 8150–8162. [CrossRef] [PubMed]
62. Huang, J.Z.; Ying, V.Y.; Seyedsayamdost, M.R. Synthesis of Non-canonical Tryptophan Variants via Rh-catalyzed C–H Functionalization of Anilines. *Angew. Chem. Int. Ed.* **2025**, *64*, e202414998. [CrossRef] [PubMed]
63. Jiao, B.; Wang, F.; Lv, H. Rhodium-Catalyzed Asymmetric Hydrogenation of Tetrasubstituted α,β-Unsaturated Amides: Efficient Access to Chiral β-Amino Amides†. *Chin. J. Chem.* **2024**, *42*, 2641–2646. [CrossRef]
64. Jia, X.; Hao, G.L.; Feng, M.; Jiang, H.; Wang, S.G.; Huang, L. Rh(III)-Catalyzed Diastereo- and Enantioselective Regiodivergent (Hetero)Arylamidation of (Homo)Allylic Sulfides. *J. Am. Chem. Soc.* **2024**, *146*, 9768–9778. [CrossRef]
65. Goddard, C.M.L.; Massah, A.R.; Jackson, R.F.W. Improved Synthesis of Phenylethylamine Derivatives by Negishi Cross-Coupling Reactions. *Tetrahedron* **2010**, *66*, 9175–9181. [CrossRef]
66. Trost, B.M.; Osipov, M.; Dong, G. Palladium-Catalyzed Dynamic Kinetic Asymmetric Transformations of Vinyl Aziridines with Nitrogen Heterocycles: Rapid Access to Biologically Active Pyrroles and Indoles. *J. Am. Chem. Soc.* **2010**, *132*, 15800–15807. [CrossRef]
67. Mai, D.N.; Wolfe, J.P. Asymmetric Palladium-Catalyzed Carboamination Reactions for the Synthesis of Enantiomerically Enriched 2-(Arylmethyl)- and 2-(Alkenylmethyl)Pyrrolidines. *J. Am. Chem. Soc.* **2010**, *132*, 12157–12159. [CrossRef]
68. White, D.R.; Hutt, J.T.; Wolfe, J.P. Asymmetric Pd-Catalyzed Alkene Carboamination Reactions for the Synthesis of 2-Aminoindane Derivatives. *J. Am. Chem. Soc.* **2015**, *137*, 11246–11249. [CrossRef]
69. Duda, M.L.; Michael, F.E. Palladium-Catalyzed Cross-Coupling of n -Sulfonylaziridines with Boronic Acids. *J. Am. Chem. Soc.* **2013**, *135*, 18347–18349. [CrossRef]
70. Takeda, Y.; Ikeda, Y.; Kuroda, A.; Tanaka, S.; Minakata, S. Pd/NHC-Catalyzed Enantiospecific and Regioselective Suzuki-Miyaura Arylation of 2-Arylaziridines: Synthesis of Enantioenriched 2-Arylphenethylamine Derivatives. *J. Am. Chem. Soc.* **2014**, *136*, 8544–8547. [CrossRef]
71. Takeda, Y.; Kuroda, A.; Sameera, W.M.C.; Morokuma, K.; Minakata, S. Palladium-Catalyzed Regioselective and Stereo-Invertive Ring-Opening Borylation of 2-Arylaziridines with Bis(Pinacolato)Diboron: Experimental and Computational Studies. *Chem. Sci.* **2016**, *7*, 6141–6152. [CrossRef]

72. Takeda, Y.; Shibuta, K.; Aoki, S.; Tohnai, N.; Minakata, S. Catalyst-Controlled Regiodivergent Ring-Opening C(Sp3)-Si Bond-Forming Reactions of 2-Arylaziridines with Silylborane Enabled by Synergistic Palladium/Copper Dual Catalysis. *Chem. Sci.* **2019**, *10*, 8642–8647. [CrossRef] [PubMed]
73. Takeda, Y.; Matsuno, T.; Sharma, A.K.; Sameera, W.M.C.; Minakata, S. Asymmetric Synthesis of B2-Aryl Amino Acids through Pd-Catalyzed Enantiospecific and Regioselective Ring-Opening Suzuki–Miyaura Arylation of Aziridine-2-Carboxylates. *Chem.-A Eur. J.* **2019**, *25*, 10226–10231. [CrossRef] [PubMed]
74. Chan, K.S.L.; Fu, H.Y.; Yu, J.Q. Palladium(II)-Catalyzed Highly Enantioselective C-H Arylation of Cyclopropylmethylamines. *J. Am. Chem. Soc.* **2015**, *137*, 2042–2046. [CrossRef] [PubMed]
75. Zhou, K.; Zhu, Y.; Fan, W.; Chen, Y.; Xu, X.; Zhang, J.; Zhao, Y. Late-Stage Functionalization of Aromatic Acids with Aliphatic Aziridines: Direct Approach to Form β-Branched Arylethylamine Backbones. *ACS Catal.* **2019**, *9*, 6738–6743. [CrossRef]
76. Ahmad, A.; Dutta, H.S.; Kumar, M.; Raziullah; Gangwar, M.K.; Koley, D. Directing Group Guided Site-Selective Diversification of Indoles by Aziridine: Synthesis of β-Indolylethylamines. *Org. Lett.* **2022**, *24*, 2783–2787. [CrossRef]
77. Liu, J.H.; Zhou, Q.; Lin, Y.; Wu, Z.L.; Cai, T.; Wen, W.; Huang, Y.M.; Guo, Q.X. Modular Chiral-Aldehyde/Palladium Catalysis Enables Atom-Economical α-Allylation of N-Unprotected Amino Acid Esters with 1,3-Dienes and Allenes. *ACS Catal.* **2023**, *13*, 6013–6022. [CrossRef]
78. Wang, C.; Xi, Y.; Xia, T.; Qu, J.; Chen, Y. Pd(0)-Catalyzed Diastereoselective and Enantioselective Intermolecular Heck-Miyaura Borylation of Internal Enamides for the β-Aminoboronate Ester Synthesis. *ACS Catal.* **2024**, *14*, 418–425. [CrossRef]
79. Hansen, K.B.; Hsiao, Y.; Xu, F.; Rivera, N.; Clausen, A.; Kubryk, M.; Krska, S.; Rosner, T.; Simmons, B.; Balsells, J.; et al. Highly Efficient Asymmetric Synthesis of Sitagliptin. *J. Am. Chem. Soc.* **2009**, *131*, 8798–8804. [CrossRef]
80. Steinhuebel, D.; Sun, Y.; Matsumura, K.; Sayo, N.; Saito, T. Direct Asymmetric Reductive Amination. *J. Am. Chem. Soc.* **2009**, *131*, 11316–11317. [CrossRef]
81. Steinhuebel, D.P.; Krska, S.W.; Alorati, A.; Baxter, J.M.; Belyk, K.; Bishop, B.; Palucki, M.; Sun, Y.; Davies, I.W. Asymmetric Hydrogenation of Protected Allylic Amines. *Org. Lett.* **2010**, *12*, 4201–4203. [CrossRef] [PubMed]
82. Otsuka, M.; Yokoyama, H.; Endo, K.; Shibata, T. Ru-Catalyzed β-Selective and Enantioselective Addition of Amines to Styrenes Initiated by Direct Arene-Exchange. *Org. Biomol. Chem.* **2012**, *10*, 3815–3818. [CrossRef]
83. Komiyama, M.; Itoh, T.; Takeyasu, T. Scalable Ruthenium-Catalyzed Asymmetric Synthesis of a Key Intermediate for the B2-Adrenergic Receptor Agonist. *Org. Process Res. Dev.* **2015**, *19*, 315–319. [CrossRef]
84. Wysocki, J.; Schlepphorst, C.; Glorius, F. Asymmetric Homogeneous Hydrogenation of 2-Pyridones. *Synlett* **2015**, *26*, 1557–1562. [CrossRef]
85. Beliaev, A. Development of the Asymmetric Hydrogenation Step for Multikilogram Production of Etamicastat. *Org. Process Res. Dev.* **2016**, *20*, 724–732. [CrossRef]
86. Gallardo-Donaire, J.; Hermsen, M.; Wysocki, J.; Ernst, M.; Rominger, F.; Trapp, O.; Hashmi, A.S.K.; Schäfer, A.; Comba, P.; Schaub, T. Direct Asymmetric Ruthenium-Catalyzed Reductive Amination of Alkyl-Aryl Ketones with Ammonia and Hydrogen. *J. Am. Chem. Soc.* **2018**, *140*, 355–361. [CrossRef]
87. Hu, L.; Zhang, Y.; Zhang, Q.; Yin, Q.; Zhang, X. Ruthenium-Catalyzed Direct Asymmetric Reductive Amination of Diaryl and Sterically Hindered Ketones with Ammonium Salts and H_2. *Angew. Chem.* **2020**, *132*, 5359–5363. [CrossRef]
88. Brandes, D.S.; Sirvent, A.; Mercado, B.Q.; Ellman, J.A. Three-Component 1,2-Carboamidation of Bridged Bicyclic Alkenes via RhIII-Catalyzed Addition of C-H Bonds and Amidating Reagents. *Org. Lett.* **2021**, *23*, 2836–2840. [CrossRef]
89. Hu, L.; Wang, Y.Z.; Xu, L.; Yin, Q.; Zhang, X. Highly Enantioselective Synthesis of N-Unprotected Unnatural α-Amino Acid Derivatives by Ruthenium-Catalyzed Direct Asymmetric Reductive Amination. *Angew. Chem.-Int. Ed.* **2022**, *61*, e202202552. [CrossRef]
90. Nie, X.; Ritter, C.W.; Hemming, M.; Ivlev, S.I.; Xie, X.; Chen, S.; Meggers, E. Nitrene-Mediated Enantioselective Intramolecular Olefin Oxyamination to Access Chiral γ-Aminomethyl-γ-Lactones. *Angew. Chem.-Int. Ed.* **2023**, *62*, e202314398. [CrossRef]
91. Yi, Z.Q.; Deng, B.W.; Chen, F.; He, Y.M.; Fan, Q.H. Asymmetric Hydrogenation of Dibenzo-Fused Azepines with Chiral Cationic Ruthenium Diamine Catalysts. *New J. Chem.* **2023**, *47*, 11492–11497. [CrossRef]
92. Wang, Y.Z.; Liu, S.D.; Cheng, L.; Liu, L.; Li, C.J. Asymmetric Addition of Hydrazones as Alkyl Carbanion Equivalents with Aryl Imines in Water. *Org. Chem. Front.* **2023**, *10*, 3021–3026. [CrossRef]
93. Ding, Z.; Luo, Y.; Yuan, Q.; Wang, G.; Yu, Z.; Zhao, M.; Liu, D.; Zhang, W. Ru-Catalyzed Asymmetric Hydrogenation of α,β-Unsaturated γ-Lactams. *J. Am. Chem. Soc.* **2024**, *146*, 25312–25320. [CrossRef]
94. Huang, C.Y.; Doyle, A.G. Nickel-Catalyzed Negishi Alkylations of Styrenyl Aziridines. *J. Am. Chem. Soc.* **2012**, *134*, 9541–9544. [CrossRef] [PubMed]
95. Huang, C.Y.; Doyle, A.G. Electron-Deficient Olefin Ligands Enable Generation of Quaternary Carbons by Ni-Catalyzed Cross-Coupling. *J. Am. Chem. Soc.* **2015**, *137*, 5638–5641. [CrossRef]
96. Woods, B.P.; Orlandi, M.; Huang, C.Y.; Sigman, M.S.; Doyle, A.G. Nickel-Catalyzed Enantioselective Reductive Cross-Coupling of Styrenyl Aziridines. *J. Am. Chem. Soc.* **2017**, *139*, 5688–5691. [CrossRef] [PubMed]

97. He, J.; Xue, Y.; Han, B.; Zhang, C.; Wang, Y.; Zhu, S. Nickel-Catalyzed Asymmetric Reductive 1,2-Carboamination of Unactivated Alkenes. *Angew. Chem.* **2020**, *132*, 2348–2352. [CrossRef]
98. Kang, T.; Kim, N.; Cheng, P.T.; Zhang, H.; Foo, K.; Engle, K.M. Nickel-Catalyzed 1,2-Carboamination of Alkenyl Alcohols. *J. Am. Chem. Soc.* **2021**, *143*, 13962–13970. [CrossRef]
99. Hu, X.; Cheng-Sánchez, I.; Cuesta-Galisteo, S.; Nevado, C. Nickel-Catalyzed Enantioselective Electrochemical Reductive Cross-Coupling of Aryl Aziridines with Alkenyl Bromides. *J. Am. Chem. Soc.* **2023**, *145*, 6270–6279. [CrossRef]
100. Wang, Y.Z.; Wang, Z.H.; Eshel, I.L.; Sun, B.; Liu, D.; Gu, Y.C.; Milo, A.; Mei, T.S. Nickel/Biimidazole-Catalyzed Electrochemical Enantioselective Reductive Cross-Coupling of Aryl Aziridines with Aryl Iodides. *Nat. Commun.* **2023**, *14*, 2322. [CrossRef]
101. Lyu, X.; Seo, C.; Faber, T.; Kim, D.; Seo, S.; Chang, S. Intramolecular Hydroamidation of Alkenes Enabling Asymmetric Synthesis of β-Lactams via Unprecedented Mechanism of NiH Catalysis. *Nat. Cat.* **2023**, *6*, 784–795. [CrossRef]
102. Wang, C.Y.; Huang, Y.L.; Xu, W.C.; Gao, Q.; Liu, P.; Bi, Y.X.; Liu, G.K.; Wang, X.S. Nickel-Catalyzed Asymmetric Decarboxyarylation with NHP Esters of α-Amino Acid to Chiral Benzylamines. *Org. Lett.* **2023**, *25*, 6964–6968. [CrossRef] [PubMed]
103. Samanta, S.; Biswas, P.; O'Bannon, B.C.; Powers, D.C. β-Phenethylamine Synthesis: N-Pyridinium Aziridines as Latent Dual Electrophiles. *Angew. Chem.-Int. Ed.* **2024**, *63*, e202406335. [CrossRef]
104. Bender, T.; Fürstner, A. Enantioselective Synthesis of Vic-Aminoalcohol Derivatives by Nickel-Catalyzed Reductive Coupling of Aldehydes with Protected Amino-Pentadienoates. *J. Am. Chem. Soc.* **2024**, *146*, 33295–33301. [CrossRef] [PubMed]
105. Lan, Y.; Han, Q.; Liao, P.; Chen, R.; Fan, F.; Zhao, X.; Liu, W. Nickel-Catalyzed Enantioselective C(Sp3)-C(Sp3) Cross-Electrophile Coupling of N-Sulfonyl Styrenyl Aziridines with Alkyl Bromides. *J. Am. Chem. Soc.* **2024**, *146*, 25426–25432. [CrossRef] [PubMed]
106. Lyu, X.; Jeon, E.; Seo, C.; Kim, D.; Chang, S. Nickel-Catalyzed Asymmetric Homobenzylic Hydroamidation of Aryl Alkenes to Access Chiral β-Arylamides. *J. Am. Chem. Soc.* **2025**, *147*, 8928–8938. [CrossRef]
107. Gao, P.; Foster, D.; Sipos, G.; Skelton, B.W.; Sobolev, A.N.; Dorta, R. Chiral NHC-Iridium Complexes and Their Performance in Enantioselective Intramolecular Hydroamination and Ring-Opening Amination Reactions. *Organometallics* **2020**, *39*, 556–573. [CrossRef]
108. Jouffroy, M.; Nguyen, T.M.; Cordier, M.; Blot, M.; Roisnel, T.; Gramage-Doria, R. Iridium-Catalyzed Direct Reductive Amination of Ketones and Secondary Amines: Breaking the Aliphatic Wall. *Chemistry* **2022**, *28*, e202201078. [CrossRef]
109. Ma, S.; Xi, Y.; Fan, H.; Roediger, S.; Hartwig, J.F. Enantioselective Hydroamination of Unactivated Terminal Alkenes. *Chem* **2022**, *8*, 532–542. [CrossRef]
110. Zhao, W.; Li, B.J. Directing Group Repositioning Strategy Enabled Site- and Enantioselective Addition of Heteroaromatic C-H Bonds to Acyclic Internal Alkenes. *J. Am. Chem. Soc.* **2023**, *145*, 6861–6870. [CrossRef]
111. Yajima, T.; Katayama, A.; Ito, T.; Kawada, T.; Yabushita, K.; Yasuda, T.; Ohta, T.; Katayama, T.; Utsumi, N.; Kayaki, Y.; et al. Asymmetric Reductive Amination of α-Keto Acids Using Ir-Based Hydrogen Transfer Catalysts: An Access to Unprotected Unnatural α-Amino Acids. *Org. Lett.* **2024**, *26*, 1426–1431. [CrossRef] [PubMed]
112. Liu, J.; Robinson, D.; Li, B.; Liu, S.-Y. Synthesis of Chiral δ-Aminoboronic Esters by Enantioselec- Tive Hydrogenation of 1,2-Azaborines. *Chem. Rxiv* **2024**. [CrossRef]
113. Chen, M.H.; Hsieh, J.C.; Lee, Y.H.; Cheng, C.H. Controlled Synthesis of Enantioselective 1-Aminoindenes via Cobalt-Catalyzed [3 + 2] Annulation Reaction. *ACS Catal.* **2018**, *8*, 9364–9369. [CrossRef]
114. Wang, C.; Ge, S. Versatile Cobalt-Catalyzed Enantioselective Entry to Boryl-Functionalized All-Carbon Quaternary Stereogenic Centers. *J. Am. Chem. Soc.* **2018**, *140*, 10687–10690. [CrossRef] [PubMed]
115. Ozols, K.; Onodera, S.; Woźniak, Ł.; Cramer, N. Cobalt(III)-Catalyzed Enantioselective Intermolecular Carboamination by C–H Functionalization. *Angew. Chem.-Int. Ed.* **2021**, *60*, 655–659. [CrossRef]
116. Li, A.; Song, X.; Ren, Q.; Bao, P.; Long, X.; Huang, F.; Yuan, L.; Zhou, J.S.; Qin, X. Cobalt-Catalyzed Asymmetric Deuteration of α-Amidoacrylates for Stereoselective Synthesis of α,β-Dideuterated α-Amino Acids. *Angew. Chem.-Int. Ed.* **2023**, *62*, e202301091. [CrossRef]
117. Yang, H.; Hu, Y.; Zou, Y.; Zhang, Z.; Zhang, W. Cobalt-Catalyzed Efficient Asymmetric Hydrogenation of α-Primary Amino Ketones. *JACS Au* **2023**, *3*, 2981–2986. [CrossRef]
118. Mendelsohn, L.N.; MacNeil, C.S.; Esposito, M.R.; Pabst, T.P.; Leahy, D.K.; Davies, I.W.; Chirik, P.J. Asymmetric Hydrogenation of Indazole-Containing Enamides Relevant to the Synthesis of Zavegepant Using Neutral and Cationic Cobalt Precatalysts. *Org. Lett.* **2024**, *26*, 2718–2723. [CrossRef]
119. Cui, T.; Ye, C.X.; Thelemann, J.; Jenisch, D.; Meggers, E. Enantioselective and Enantioconvergent Iron-Catalyzed C(Sp3)-H Aminations to Chiral 2-Imidazolidinones. *Chin. J. Chem.* **2023**, *41*, 2065–2070. [CrossRef]
120. Holst, H.M.; Floreancig, J.T.; Ritts, C.B.; Race, N.J. Aziridine Opening via a Phenonium Ion Enables Synthesis of Complex Phenethylamine Derivatives. *Org. Lett.* **2022**, *24*, 501–505. [CrossRef]
121. Ghorai, M.K.; Tiwari, D.P.; Jain, N. Lewis Acid Catalyzed SN2-Type Ring Opening of N -Activated Aziridines with Electron-Rich Arenes/Heteroarenes. *J. Org. Chem.* **2013**, *78*, 7121–7130. [CrossRef] [PubMed]

122. Trost, B.M.; Hung, C.I.J.; Gnanamani, E. Tuning the Reactivity of Ketones through Unsaturation: Construction of Cyclic and Acyclic Quaternary Stereocenters via Zn-ProPhenol Catalyzed Mannich Reactions. *ACS Catal.* **2019**, *9*, 1549–1557. [CrossRef]
123. Mohamadpour, F.; Amani, A.M. Photocatalytic Systems: Reactions, Mechanism, and Applications. *RSC Adv.* **2024**, *14*, 20609–20645. [CrossRef]
124. Marzo, L.; Pagire, S.K.; Reiser, O.; König, B. Photokatalyse Mit Sichtbarem Licht: Welche Bedeutung Hat Sie Für Die Organische Synthese? *Angew. Chem.* **2018**, *130*, 10188–10228. [CrossRef]
125. König, B. Photocatalysis in Organic Synthesis—Past, Present, and Future. *Eur. J. Org. Chem.* **2017**, *2017*, 1979–1981. [CrossRef]
126. David, A.N.; David, W.C. MacMillan Merging Photoredox Catalysis with Organocatalysis: The Direct Alkylation of Aldehydes. *Science (1979)* **2008**, *322*, 77–80. [CrossRef]
127. Pitre, S.P.; Overman, L.E. Strategic Use of Visible-Light Photoredox Catalysis in Natural Product Synthesis. *Chem. Rev.* **2022**, *122*, 1717–1751. [CrossRef]
128. Großkopf, J.; Kratz, T.; Rigotti, T.; Bach, T. Enantioselective Photochemical Reactions Enabled by Triplet Energy Transfer. *Chem. Rev.* **2022**, *122*, 1626–1653. [CrossRef] [PubMed]
129. Weinstain, R.; Slanina, T.; Kand, D.; Klán, P. Visible-to-NIR-Light Activated Release: From Small Molecules to Nanomaterials. *Chem. Rev.* **2020**, *120*, 13135–13272. [CrossRef]
130. Proctor, R.S.J.; Davis, H.J.; Phipps, R.J. Catalytic Enantioselective Minisci-Type Addition to Heteroarenes. *Science* **2018**, *360*, 419–422. [CrossRef]
131. Monos, T.M.; Mcatee, R.C.; Stephenson, C.R.J. Arylsulfonylacetamides as Bifunctional Reagents for Alkene Aminoarylation. *Science* **2018**, *361*, 1369–1373. [CrossRef]
132. Zheng, S.; Gutiérrez-Bonet, Á.; Molander, G.A. Merging Photoredox PCET with Ni-Catalyzed Cross-Coupling: Cascade Amidoarylation of Unactivated Olefins. *Chem* **2019**, *5*, 339–352. [CrossRef] [PubMed]
133. Whalley, D.M.; Duong, H.A.; Greaney, M.F. A Visible Light-Mediated, Decarboxylative, Desulfonylative Smiles Rearrangement for General Arylethylamine Syntheses. *Chem. Commun.* **2020**, *56*, 11493–11496. [CrossRef]
134. Noten, E.A.; McAtee, R.C.; Stephenson, C.R.J. Catalytic Intramolecular Aminoarylation of Unactivated Alkenes with Aryl Sulfonamides. *Chem. Sci.* **2022**, *13*, 6942–6949. [CrossRef] [PubMed]
135. Noten, E.A.; Ng, C.H.; Wolesensky, R.M.; Stephenson, C.R.J. A General Alkene Aminoarylation Enabled by N-Centered Radical Reactivity of Sulfinamides. *Nat. Chem.* **2024**, *16*, 599–606. [CrossRef] [PubMed]
136. Shi, C.; Guo, L.; Gao, H.; Luo, M.; Yang, C.; Xia, W. Highly Diastereoselective Synthesis of γ-Lactams Enabled by Photoinduced Deaminative [3 + 2] Annulation Reaction. *Org. Lett.* **2022**, *24*, 4365–4370. [CrossRef]
137. Hervieu, C.; Kirillova, M.S.; Hu, Y.; Cuesta-Galisteo, S.; Merino, E.; Nevado, C. Chiral Arylsulfinylamides as Reagents for Visible Light-Mediated Asymmetric Alkene Aminoarylations. *Nat. Chem.* **2024**, *16*, 607–614. [CrossRef]
138. Hu, H.; Shi, Z.; Guo, X.; Zhang, F.H.; Wang, Z. A Radical Approach for Asymmetric α-C-H Addition of N-Sulfonyl Benzylamines to Aldehydes. *J. Am. Chem. Soc.* **2024**, *146*, 5316–5323. [CrossRef]
139. Sun, H.; Yang, C.; Lin, R.; Xia, W. Regioselective Ring-Opening Nucleophilic Addition of Aziridines through Photoredox Catalyst. *Adv. Synth. Catal.* **2014**, *356*, 2775–2780. [CrossRef]
140. Tu, J.L.; Tang, W.; Liu, F.; Liu, F. Photoredox-Neutral Alkene Aminoarylation for the Synthesis of 1,4,5,6-Tetrahydropyridazines. *Org. Chem. Front.* **2021**, *8*, 3712–3717. [CrossRef]
141. Shu, X.; Huan, L.; Huang, Q.; Huo, H. Direct Enantioselective C(Sp3)-H Acylation for the Synthesis of α-Amino Ketones. *J. Am. Chem. Soc.* **2020**, *142*, 19058–19064. [CrossRef] [PubMed]
142. Steiman, T.J.; Liu, J.; Mengiste, A.; Doyle, A.G. Synthesis of β-Phenethylamines via Ni/Photoredox Cross-Electrophile Coupling of Aliphatic Aziridines and Aryl Iodides. *J. Am. Chem. Soc.* **2020**, *142*, 7598–7605. [CrossRef] [PubMed]
143. Jiang, H.; Yu, X.; Daniliuc, C.G.; Studer, A. Three-Component Aminoarylation of Electron-Rich Alkenes by Merging Photoredox with Nickel Catalysis. *Angew. Chem.-Int. Ed.* **2021**, *60*, 14399–14404. [CrossRef]
144. Qian, S.; Lazarus, T.M.; Nicewicz, D.A. Enantioselective Amino- and Oxycyanation of Alkenes via Organic Photoredox and Copper Catalysis. *J. Am. Chem. Soc.* **2023**, *145*, 18247–18252. [CrossRef]
145. Yang, F.; Chi, L.; Ye, Z.; Gong, L. Photoinduced Regiodivergent and Enantioselective Cross-Coupling of Glycine Derivatives with Hydrocarbon Feedstocks. *J. Am. Chem. Soc.* **2025**, *147*, 1767–1780. [CrossRef] [PubMed]
146. Leone, M.; Milton, J.P.; Gryko, D.; Neuville, L.; Masson, G. TBADT-Mediated Photocatalytic Stereoselective Radical Alkylation of Chiral N-Sulfinyl Imines: Towards Efficient Synthesis of Diverse Chiral Amines. *Chem.-A Eur. J.* **2024**, *30*, e202400363. [CrossRef]
147. Hore, S.; Jeong, J.; Kim, D.; Chang, S. Visible-Light-Promoted Enantioselective α-Amidation of Aldehydes by Harnessing Organo-Iron Dual Catalysis. *J. Am. Chem. Soc.* **2024**, *146*, 22172–22179. [CrossRef]

molecules

Article

PyBox–La(OTf)$_3$-Catalyzed Enantioselective Diels–Alder Cycloadditions of 2-Alkenoylpyridines with Cyclopentadiene

Hao Wei [1], Yujie Zhang [1], Sanlin Jin [1], Ying Yu [2,*], Ning Chen [1], Jiaxi Xu [1] and Zhanhui Yang [1,*]

1 Department of Organic Chemistry, College of Chemistry, Beijing University of Chemical Technology, Beijing 100029, China; w18439551277@126.com (H.W.); 2022201038@buct.edu.cn (Y.Z.); 2022201048@buct.edu.cn (S.J.); 2011500030@buct.edu.cn (N.C.); 2007500055@buct.edu.cn (J.X.)

2 China United Test & Certification Co., Ltd., Beijing 100088, China

* Correspondence: yhb_5158@163.com (Y.Y.); zhyang@mail.buct.edu.cn (Z.Y.)

Abstract: The PyBox–La(OTf)$_3$-catalyzed enantioselective Diels–Alder cycloaddition of 2-alk-2-enoylpyridines with cyclopentadiene is realized, producing enantiopure disubstituted norbornenes, which possess four contiguous stereocenters and are biologically relevant structures in up to 92:8 dr and 99:1 er.

Keywords: PyBox; rare earth; asymmetric catalysis; Diels–Alder cycloaddition; norbornene derivatives

1. Introduction

Enantioenriched norbornene scaffolds with multiple stereocenters are prevalent motifs in many pharmaceutically relevant and naturally occurring compounds (Figure 1) [1–6]. Therefore, the construction of such scaffolds constitutes an intriguing topic in the synthetic community. The catalytic asymmetric Diels–Alder cycloadditions of auxiliary-substituted alkenes with cyclopentadienes provide a straightforward and atom-economical access to the norbornene scaffolds, and are anticipated to produce good levels of chirality transfer and diastereocontrol by virtue of interactions between auxiliary groups and chiral catalysts [7]. In this regard, alkenoyl pyridines, which can chelate their pyridinyl nitrogen and carbonyl oxygen atoms to the metal center of a chiral catalyst, are demonstrated as robust dienophiles to undergo asymmetric Diels–Alder cycloadditions with cyclopentadienes, even in aqueous media (Scheme 1). Documented catalysts and ligands that are applicable to this useful type of reaction include Engberts's chiral amino acid–Cu(II) system (74% ee) [8], Pedro's chiral bis(oxazoline)–Cu(II) system (19% ee) [9], Lin and Feng's *N,N′*-dioxide–Ni(II) system (up to 95:5 dr and 96% ee) [10] and Ollevier's chiral bipyridine diol–Fe(III) system (up to 93:7 dr and 84% ee) [11]. In addition, numerous chiral artificial metalloenzymes and DNA-modified catalysts were also developed [12–17]. In spite of the above advances, new alternative catalytic systems are still in demand for effecting high levels of diastereo- and enantiocontrol, a purpose that was hitherto only satisfied by Lin and Feng's *N,N′*-dioxide–Ni(II) catalysts [10].

The past decades have witnessed fruitful advancements on rare earth-catalyzed asymmetric Diels–Alder cycloadditions [18], which encouraged us to provide a rare-earth protocol for enantiocontrol. The key factor that underlies this protocol is the choice of a chiral ligand. Our previous research with didentate pyridyl imidazolines [19–22] led us to consider similar chirally modified nitrogen ligands. Considering the special coordination numbers of rare earth salts, we designed and synthesized several tridentate chiral nitrogen ligands and evaluated their chirality transfer ability in the rare earth-catalyzed Diels–Alder cycloadditions between 2-alk-2-enoylpyridines and cyclopentadiene. We eventually found that the combination of chiral pyridine-2,6-bisoxazolines and La(OTf)$_3$ displays the best results in terms of diastereo- and enantiocontrol, producing enantioenriched norbornenes in good yields and enantiopurities (Scheme 1b).

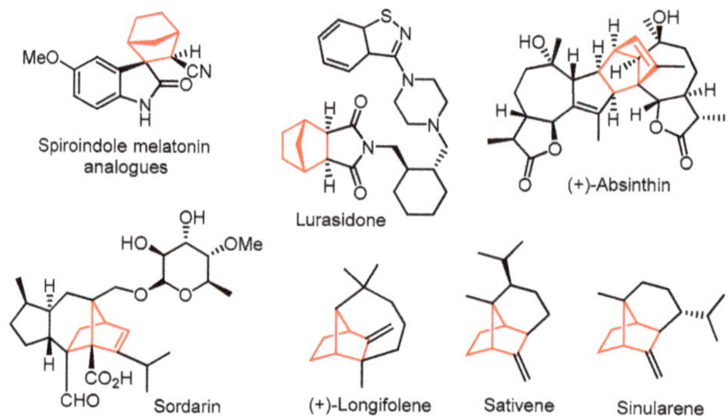

Figure 1. Representative bioactive compunds.

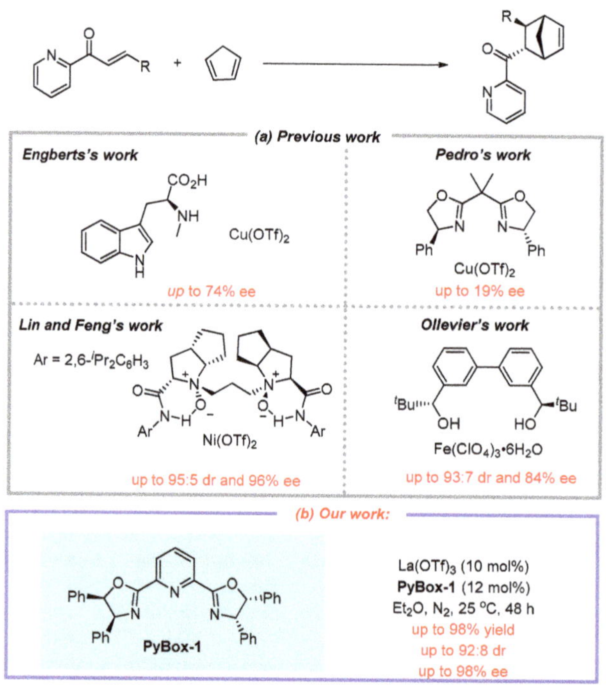

Scheme 1. Previous work (**a**) and our work (**b**) on the asymmetric Diels–Alder cycloadditions of 2-alk-2-enoylpyridines with cyclopentadiene.

2. Results and Discussion

We first optimized the reaction conditions using (*E*)-styrenoyl pyridine (**1a**) and cyclopentadiene (**2A**) as the model substrates. The selected attempts on ligand screening are listed in Table 1 (for more details, see Tables S1 and S2 in the Supplementary Materials). Pyridine-2,6-bisimidazoline ligands **PyBim-1** and **PyBim-2** produced **3aA** in 84:16 and 75:25 er, respectively [23–25]. In contrast, **3aA** was produced in 94:6 and 68:32 er values when the respective pyridine-2,6-bisoxazoline ligands **PyBox-1** and **PyBox-2** were used. The reaction with 1,10-phenanthroline-2-oxazoline (**PhenOx**) delivered **3aA** as a nearly racemic mixture. Apparently, **PyBox-1** produced the best result in terms of enantioselectivity. It deserves mention that similar ligands were also employed by Desimoni's group in 2002 in the Diels–Alder *exo*-cycloadditions between alkenoyl-1,3-oxazolidin-2-ones and cyclopentadiene [26]. With this ligand, other lanthanum(III) salts such as $LaCl_3$ and $La(BF_4)_3$ were evaluated, but nearly no enantioselectivity was observed. Thus, from the conditions listed in Table 1, entry 3 was selected as the optimal one.

Table 1. Selected optimization of the reaction conditions.

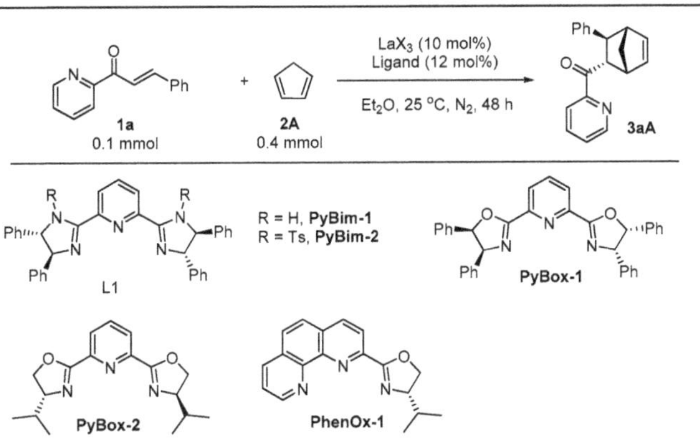

Entry	Catalyst	Ligand	Yield [a] (%)	Er [b]	Dr [b]
1	$La(OTf)_3$	**PyBim-1**	54	84:16	51:49
2	$La(OTf)_3$	**PyBim-2**	67	85.5:14.5	79.5:20.5
3	$La(OTf)_3$	**PyBox-1**	91	94:6	88:12
4	$La(OTf)_3$	**PyBox-2**	95	68:32	89:11
5	$La(OTf)_3$	**PhenOx-1**	87	58:42	86.5:13.5
6	$LaCl_3$	**PyBox-1**	48	51.5:48.5	84:16
7	$La(BF_4)_3$	**PyBox-1**	71	51:49	89:11

[a] Isolated yields on column chromatography. [b] Enantioselective and diastereoselective ratios obtained by chiral HPLC. dr = *endo:exo* ratios.

With the optimal conditions at hand, a number of dienophiles, that is, 2-alk-2-enoylpyridines **1a–r**, were reacted with cyclopentadiene. The results regarding enantio- and diastereoselectivities (*endo*/*exo*) are summarized in Figure 2. 2-Alk-2-enoylpyridines bearing electronically diverse aryl substituents were first screened. In addition to (*E*)-styrenoylpyridine (**1a**), the halogen-substituted dienophiles **1b–1h** all productively gave desired products **3bA–3hA** in 80–98 yields, 64:36–89:11 drs, and most importantly, 87:13–95.5:4.5 ers. The highest enantioselectivity (95.5:4.5 er) was observed in the reaction producing **3cA**. Strongly electron-withdrawing cyano-substituted dienophile **1i**, as well as those with electron-donating alkyl and methoxy substituents, produced desired products **3iA–3lA** in 87:13–93:7 ers with varying diastereoselectivities and yields. Dienophiles with heteroaryls and fused aryls were also amenable, and the corresponding products were procured in 81:19 to 90:10 er. The alkyl-substituted 2-alk-2-enoylpyridines **1p–1r** were also tested under the optimal conditions. Dienophiles **1p** and **1q** with more steric hindrance delivered products **3pA** and **3qA** in excellent 99:1 and 95:5 ers, respectively, whilst dienophile **1r** with less steric cyclopropyl only afforded moderate enantioselectivity of 77:23. Acyclic dienes were also examined. 2,3-Dimethylbuta-1,3-diene (**2B**) afforded **3aB** at a 45% yield with 82:18 er and >20:1 dr. However, other attempted dienes, including 2-methylbuta-1,3-diene (**2C**), cyclohexa-1,3-diene (**2D**) and furan (**2E**), did not react with **1a**.

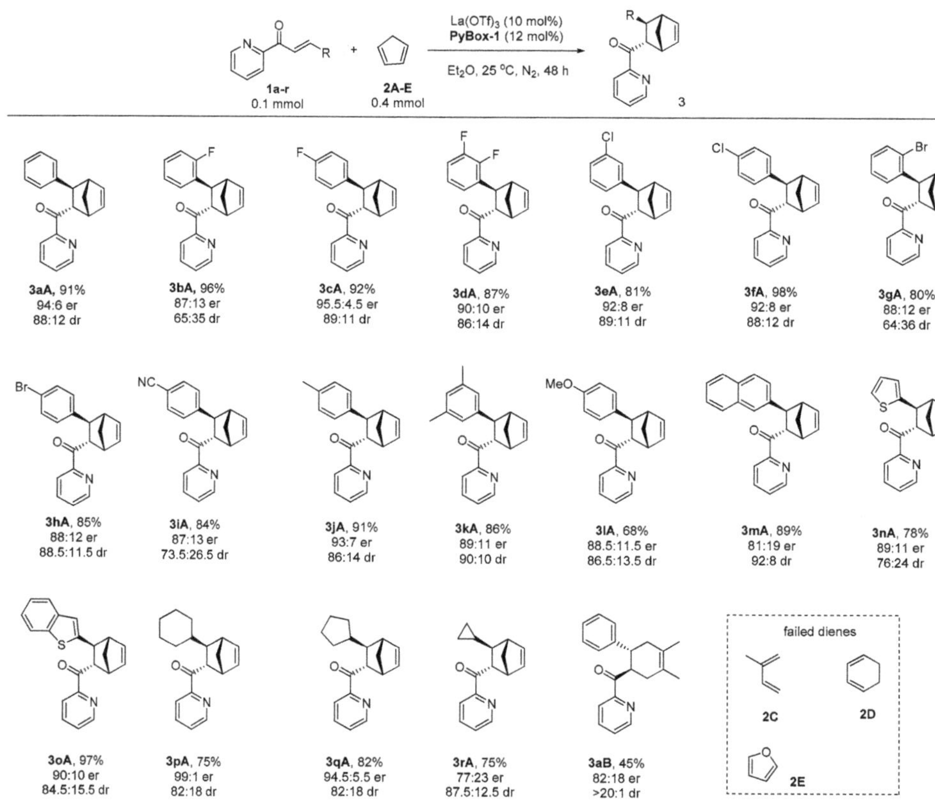

Figure 2. Scope of substrates.

Mechanistic studies were performed (Scheme 2). Mixing **PyBox-1** with La(OTf)$_3$ afforded a stable complex **4** (Scheme 2a). The complex can even be prepared in a decagram scale at a 91% yield. In the presence of 10 mol% of **4**, the cycloaddition of **1a** and **2A** produced **3aA** at a 91% yield, 91.5:8.5 er and 89:11 dr, a result almost identical with that obtained with separately added **PyBox-1** and La(OTf)$_3$. In the control experiment (Scheme 2b), the reaction of *E*-chalcone (**1s**) with cyclopentadiene (**2A**) did not yield the desired product 3sA, highlighting the essence of the pyridine auxiliary and nitrogen coordination to the metal center. Based on the above results, a plausible enantiocontrol model is proposed (Scheme 2c). Ligand exchange between **A** and substrate **1a** afforded chiral complex **B**. The C$_2$-asymmetry created a chiral pocket that allowed cyclopentadiene to approach the dienophile from the southwest upper face of **B**. A π-stacking interaction between the pyridinyl ring and the upward phenyl ring in proximity to the nitrogen atom was proposed to stabilize the transition state **TS** [27]. Once product **3a** was formed, it was still coordinated with the rare earth metal. Ligand exchange between **C** and **1a** ultimately produced **3aA** and opened a new catalytic cycle.

The **PyBox**-La(OTf)$_3$-catalyzed cycloaddition is featured with good scalability (Scheme 3a). In the gram-scale reaction, **3aA** was obtained in 93.5:6.5 er, 86:14 dr with a 98% yield. The obtained difunctionalized norbornene **3aA** possesses two reactive chemical handles to undergo further manipulations (Scheme 3b). Epoxidation of the endocyclic C=C bond yielded epoxide **5**, with two new stereocenters generated at a 73% yield, leaving the electron-rich pyridine nitrogen intact. The facial selectivity was assigned according to the empirical rule proposed by Brown and coworkers [27]. Grignard addition to the carboxyl group led to a tertiary alcohol **6** at a 92%

yield, whereas the $NaBH_4$ reduction produced a secondary alcohol **7** at a 93% yield. However, the ozonolysis of **3aA** led to complex mixtures with all the starting material consumed.

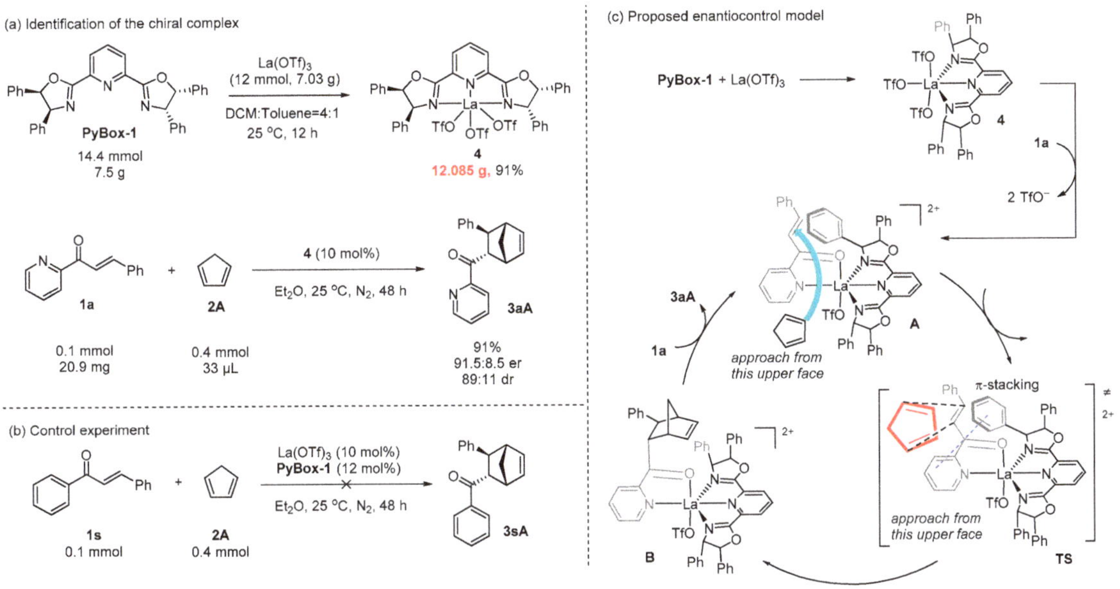

Scheme 2. Proposed mechanism.

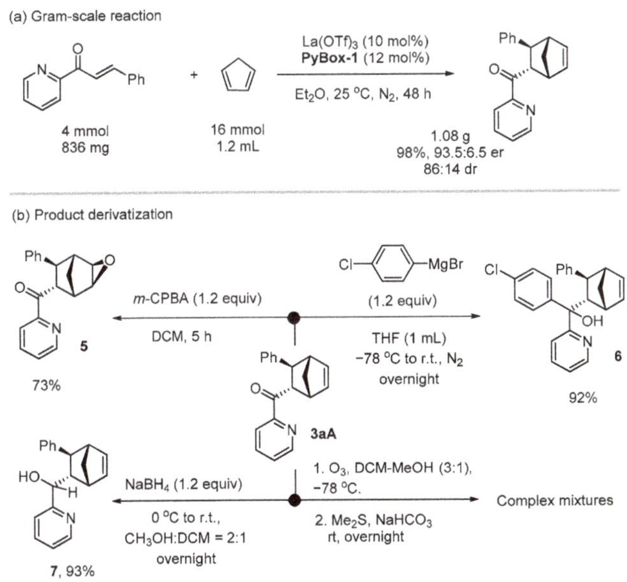

Scheme 3. Gram-scale reaction and product derivatization.

3. Materials and Methods

3.1. Materials and Instruments

Unless otherwise noted, all materials were purchased from commercial suppliers. Dichloromethane (DCM) and dichloroethane (DCE) were refluxed over CaH_2; tetrahydrofuran (THF) and toluene (PhMe) were refluxed over lithium aluminum hydride. The solvents were freshly distilled prior to use. Column chromatography was performed on silica gel (normal phase, 200–300 mesh) from Anhui Liangchen Silicon Material Co., Ltd. (Lu'An, China), with petroleum ether (PE, bp. 60–90 °C) and ethyl acetate (EA) as eluent. Reactions were monitored using thin-layer chromatography (TLC) on GF_{254} silica gel plates (0.2 mm) from Anhui Liangchen Silicon Material Co., Ltd. The plates were visualized via UV light using other TLC stains (1% potassium permanganate in water; 10 g of iodine absorbed on 30 g of silica gel). ^{1}H and ^{13}C NMR spectra were recorded on a Bruker 400 MHz spectrometer, usually in $CDCl_3$ as an internal standard, and the chemical shifts (δ) were reported in parts per million (ppm). Multiplicities are indicated as s (singlet), d (doublet), t (triplet), q (quartet), dd (double doublet), m (multiplet) and b (broad). Coupling constants (J) are reported in Hertz (Hz). HRMS measurements were carried out on an Agilent LC/MSD TOF mass spectrometer. The enantiomeric excesses were determined via HPLC analysis using Agilent Technologies 1260 Infinity equipment, and the employed chiral stationary phase column are specified in the individual experiment by comparing the enantiomeric samples with the appropriate racemic mixtures.

Substrate **1a–r** were prepared according to Caggiano's study [28]. The nuclear magnetic spectra of **1a** to **1c**, **1e** to **1p** and **1r** are in agreement with the predecessors. **PyBim-2** [23], and **PhenOx-1** [29] were prepared according to published procedures.

3.2. General Procedure for Reduction of 2-Alkenoylpyridines

To a solution of ketone (5 mmol) and aldehyde (5.5 mmol) in ethanol (15mL), aqueous sodium hydroxide solution (5 mL, 2.5 M) was added dropwise at 0 °C. The reaction mixture was further stirred at room temperature until the completion of reaction (detected by TLC). Then, the reaction mixture was filtered and washed with ethanol–water solution (1/1, *v*/*v*) and dried. The precipitate was recrystallized in methanol to obtain pure unsaturated ketone products.

3.2.1. (*E*)-3-(2,3-Difluorophenyl)-1-(pyridin-2-yl)prop-2-en-1-one (**1d**)

Yellow solid, 1.018 g, yield 83%, R_f = 0.5 (PE/EA = 5:1, *v/v*). **^{1}H NMR** (400 MHz, $CDCl_3$) δ 8.75 (ddd, J = 4.8, 1.7, 0.9 Hz, 1H), 8.38 (d, J = 16.3 Hz, 1H), 8.19 (dt, J = 7.8, 1.1 Hz, 1H), 8.04 (d, J = 16.3 Hz, 1H), 7.89 (td, J = 7.7, 1.7 Hz, 1H), 7.52 (dddd, J = 15.1, 7.5, 4.5, 1.5 Hz, 2H), 7.23–7.09 (m, 2H). **^{13}C NMR** (101 MHz, $CDCl_3$) δ 189.2, 153.8, 148.9, 137.0, 135.5, 127.1, 125.5 (d, J = 8.7 Hz), 124.3 (d, J = 5.6 Hz), 124.2 (t, J = 5.8, 11.4 Hz), 123.8, 123.0, 118.7 (d, J = 17.3 Hz), 118.6. **^{19}F NMR** (376 MHz, $CDCl_3$) δ −137.72, −139.62. **HRMS** (ESI): m/z $[M+H]^+$ calculated for $C_{14}H_{10}F_2NO^+$ 246.0725, found 246.0732.

3.2.2. (*E*)-3-Cyclopentyl-1-(pyridin-2-yl)prop-2-en-1-one (**1q**)

Colorless oil, 532.6 mg, yield 53%, R_f = 0.65 (PE/EA = 5:1, *v/v*). **^{1}H NMR** (400 MHz, $CDCl_3$) δ 8.70 (ddd, J = 4.8, 1.7, 0.9 Hz, 1H), 8.12 (dt, J = 7.9, 1.1 Hz, 1H), 7.84 (td, J = 7.7, 1.7 Hz, 1H), 7.59 (dd, J = 15.6, 1.1 Hz, 1H), 7.45 (ddd, J = 7.5, 4.8, 1.3 Hz, 1H), 7.22 (dd, J = 15.6, 8.2 Hz, 1H), 2.77 (h, J = 8.1, 7.6 Hz, 1H), 1.96–1.87 (m, 2H), 1.75–1.48 (m, 6H). **^{13}C NMR** (101 MHz, $CDCl_3$) δ 189.6, 154.6, 154.2, 148.7, 136.8, 126.6, 122.8, 122.4, 43.5, 32.6, 25.4. **HRMS** (ESI): m/z $[M+H]^+$ calculated for $C_{13}H_{16}NO^+$ 202.1226, found 202.1234.

3.3. General Procedure for Reduction of **PyBim-1** and **PyBox**

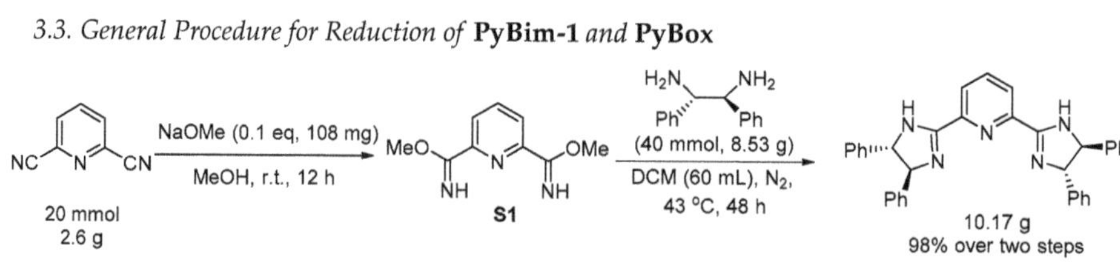

This was prepared according to Beller's procedure, with slight modification [24]. The ^{1}H and ^{13}C NMR spectra are in agreement with those reported [24]. ***Procedure***: To an oven-dried round-bottom flask equipped with a magnetic stirring bar was added 2,6-pyridinedicarbonitrile (2.6 g, 20 mmol), sodium methoxide (108 mg, 2 mmol), and methanol (40 mL). The mixture was stirred at ambient temperature. After 12 h, the solution was transferred to an oven-dried flask and then quenched with acetic acid (0.24 mL). The solvent was evaporated under reduced pressure to obtain product **S1** (>99% yield).

To an oven-dried round-bottom flask equipped with a magnetic stirring bar was added **S1** and (1*S*, 2*S*)-1,2-diphenylethane-1,2-diamine (8.53 g, 40 mmol). The flask was sealed immediately with a rubber stopper and protected with a nitrogen balloon by evacuation-backfill operations (repeated three times). Dry dichloromethane (DCM, 60 mL) was injected to the flask via a syringe, and the reaction mixture was kept at 43 °C for 48 h. The solvent was evaporated under reduced pressure, and the crude mixture was subjected to column chromatography (PE:EA:Et_3N = 20:10:1) on silica gel to afford the corresponding crude product. Final purification by recrystallization from ethyl acetate and petroleum ether afforded **PyBim-1** in 98% yield (10.17 g) as a white solid.

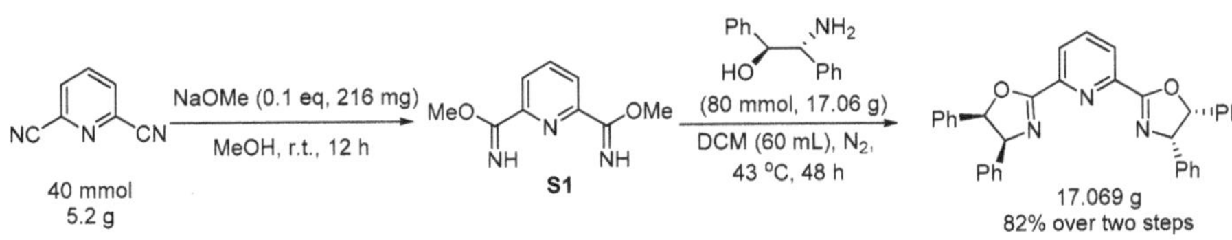

This was prepared according to Fokin's procedure, with slight modification [30]. The ^{1}H and ^{13}C NMR spectrum and data are in agreement with those reported. ***Procedure***: To an oven-dried round-bottom flask equipped with a magnetic stirring bar was added 2,6-pyridinedicarbonitrile (5.2 g, 40 mmol), sodium methoxide (216 mg, 4 mmol), and methanol (80 mL). The reaction mixture was stirred at ambient temperature. After 12 h, the solution was transferred to an oven-dried flask and quenched with acetic acid (0.48 mL); the solvent was evaporated under reduced pressure to obtain product **S1** (>99% yield). To an oven-dried round-bottom flask equipped with a magnetic stirring bar was added **S1** and (1*S*, 2*R*)-2-amino-1,2-diphenylethan-1-ol (17.06 g, 80 mmol). The flask was sealed immediately with a rubber stopper and protected with a nitrogen balloon by evacuation-backfill operations (repeated three times). Dry DCM (120 mL) was injected to the flask via a syringe, and the reaction mixtures was kept at 43 °C. After 48 h, the solvent was evaporated, and the remaining mixture was solidified by MeOH and washed with MeOH and EA to obtain **PyBox-1** (17.069 g, 82%) as a white solid.

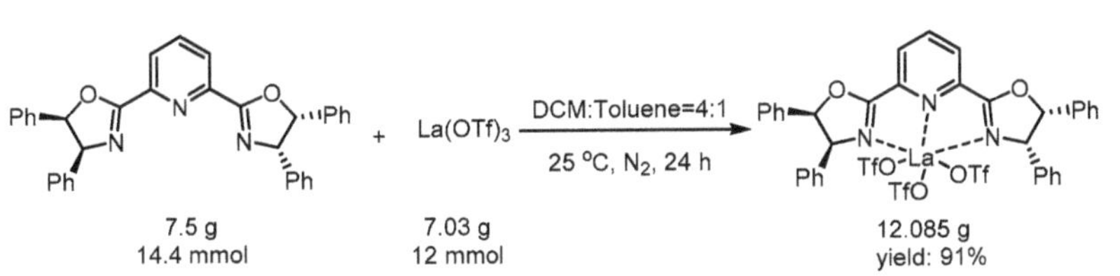

Procedure: To an oven-dried round-bottom flask equipped with a magnetic stirring bar was successively added **PyBox-1** (7.5 g, 14.4 mmol) and La(OTf)$_3$ (7.03 g, 12 mmol). The flask was sealed immediately with a rubber stopper and protected with a nitrogen balloon by evacuation-backfill operations (repeated three times). A mixture of DCM and toluene (v/v = 4:1) (120 mL) was injected to the tube via a syringe, and the mixture was stirred for about 24 h. The solvent was evaporated under reduced pressure. The crude mixture was washed with Et_2O to obtain the product (12.085 g, 91%) as a white solid.

3.4. General Procedure for Reduction of Chiral Product 3

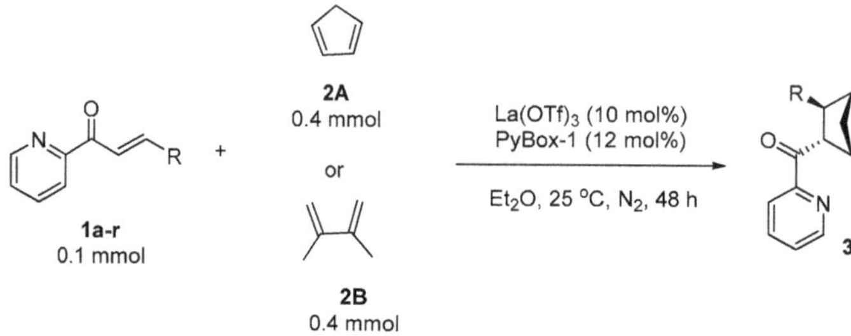

General Procedure: To an oven-dried reaction tube equipped with a magnetic stirring bar was added 2-alkenoyl pyridines **1** (0.1 mmol, 1.0 equiv), $La(OTf)_3$ (5.9 mg, 0.01 mmol) and ligand **PyBox-1** (6.3 mg, 0.012 mmol). The tube was sealed immediately with a rubber stopper and protected with a nitrogen balloon by evacuation-backfill operations (repeated three times). Dry Et_2O (1 mL) was injected to the tube via a syringe. The resultant mixture was stirred for about 1 h, followed by addition of cyclopentadiene (**2A**) (33 μL, 0.4 mmol, 4.0 equiv) or 2,3-dimethylbuta-2,3-diene (**2B**) (90 μL, 0.4 mmol, 4.0 equiv) via a microsyringe. The mixture was stirred at 25 °C for 48 h. The solvent was evaporated under reduced pressure, and the crude mixture was subjected to column chromatography on silica gel to afford the corresponding products.

3.4.1. ((1*R*,2*S*,3*S*,4*S*)-3-Phenylbicyclo[2.2.1]hept-5-en-2-yl)(pyridin-2-yl)methanone (**3aA**)

Colorless oil, 25 mg, yield 91%, R_f = 0.5 (PE/EA = 10:1, *v*/*v*). **^{1}H NMR** (400 MHz, $CDCl_3$) δ 8.76–8.60 (d, *J* = 4.4 Hz, 1H), 8.00 (d, *J* = 7.8 Hz, 1H), 7.79 (td, *J* = 7.7, 1.8 Hz, 1H), 7.42 (ddd, *J* = 7.6, 4.8, 1.3 Hz, 1H), 7.36–7.06 (m, 6H), 6.49 (dd, *J* = 5.6, 3.2 Hz, 1H), 5.82 (dd, *J* = 5.6, 2.8 Hz, 1H), 4.53 (dd, *J* = 5.2, 3.4 Hz, 1H), 3.54 (s, 1H), 3.46 (d, *J* = 4.8 Hz, 1H), 3.09 (s, 1H), 2.07 (d, *J* = 8.4 Hz, 1H), 1.61 (dd, *J* = 8.4, 2.0 Hz, 1H). **^{13}C NMR** (101 MHz, $CDCl_3$) δ 201.0, 153.5, 148.8, 144.6, 139.3, 136.7, 132.8, 128.3, 127.6, 126.8, 125.7, 122.1, 54.2, 49.3, 48.7, 48.2, 45.5.

HPLC (Daicel Chiralpak OD-H, ^{n}hexane/^{i}PrOH = 97:3, 0.65 mL/min, T = 15 °C): t_{R1} (major) = 17.172 min, t_{R2} (minor) = 12.070 min; dr: 89:11; er: 94:6.

$[\alpha]^{20}_D$ = +135.7 (c = 1.0, $CHCl_3$).

3.4.2. ((1*R*,2*S*,3*S*,4*S*)-3-(2-Fluorophenyl)bicyclo[2.2.1]hept-5-en-2-yl)(pyridin-2-yl)methanone (**3bA**)

Colorless oil, 28 mg, yield 96%, R_f = 0.55 (PE/EA = 10:1, *v*/*v*). **^{1}H NMR** (400 MHz, $CDCl_3$) δ 8.66 (ddd, *J* = 4.8, 1.8, 0.9 Hz, 1H), 8.05 (dt, *J* = 7.9, 1.1 Hz, 1H), 7.82 (td, *J* = 7.7, 1.8 Hz, 1H), 7.48–7.40 (m, 2H), 7.19–7.08 (m, 2H), 6.94 (ddd, *J* = 10.5, 7.9, 1.5 Hz, 1H), 6.48 (dd, *J* = 5.7, 3.2 Hz, 1H), 5.88 (dd, *J* = 5.6, 2.8 Hz, 1H), 4.53 (dd, *J* = 5.1, 3.5 Hz, 1H), 3.53 (d, *J* = 5.0 Hz, 1H), 3.49 (s, 1H), 3.19 (d, *J* = 2.1 Hz, 1H), 1.97 (d, *J* = 8.5 Hz, 1H), 1.62 (dd, *J* = 8.5, 1.8 Hz, 1H). **^{13}C NMR** (101 MHz, $CDCl_3$) δ 200.2,160.2 (d, *J* = 246 Hz), 153.5, 148.8, 138.3, 136.8, 133.3, 131.8, 127.2 (d, *J* = 9.3 Hz), 126.8, 123.8 (d, *J* = 2.8 Hz), 122.2, 115.2 (d, *J* = 22.7 Hz), 52.5, 48.5, 48.1, 47.6, 39.2. **^{19}F NMR** (376 MHz, $CDCl_3$) δ −139.01. **HRMS** (ESI): *m*/*z* $[M+H]^+$ calculated for $C_{19}H_{17}FNO^+$ 294.1289, found 294.1288.

HPLC (Daicel Chiralpak OD-H, ^{n}hexane/^{i}PrOH = 97:3, 0.65 mL/min, T = 15 °C): t_{R1} (major) = 14.393 min, t_{R2} (minor) = 13.080 min; dr: 65:35; er: 87:13.

$[\alpha]^{20}_D$ = +81.4 (c = 1.7, $CHCl_3$).

3.4.3. ((1*R*,2*S*,3*S*,4*S*)-3-(4-Fluorophenyl)bicyclo[2.2.1]hept-5-en-2-yl)(pyridin-2-yl)methanone (**3cA**)

Colorless oil, 27 mg, yield 92%, R_f = 0.55 (PE/EA = 10:1, *v*/*v*). **^{1}H NMR** (400 MHz, $CDCl_3$) δ 8.68 (d, *J* = 4.6 Hz, 1H), 8.00 (d, *J* = 7.8 Hz, 1H), 7.82 (td, *J* = 7.7, 1.8 Hz, 1H), 7.46 (ddd, *J* = 7.6, 4.8, 1.3 Hz, 1H), 7.30–7.21 (m, 2H), 7.01–6.85 (m, 2H), 6.48 (dd, *J* = 5.7,

3.2 Hz, 1H), 5.83 (dd, J = 5.6, 2.8 Hz, 1H), 4.46 (dd, J = 5.2, 3.4 Hz, 1H), 3.54 (s, 1H), 3.42 (d, J = 4.7 Hz, 1H), 3.07–3.01 (dd, J = 1.2 Hz, 1H), 2.03 (d, J = 8.5 Hz, 1H), 1.62 (dd, J = 8.5, 1.8 Hz, 1H). **^{13}C NMR** (101 MHz, $CDCl_3$) δ 201.0, 160.2, 153.5, 148.9, 140.2, 139.3, 136.8, 132.9, 129.0, 128.9 (d, J = 7.8 Hz), 126.9, 122.2, 115.1 (d, J = 21.0 Hz), 54.5, 49.4, 48.7, 48.1, 44.9. **^{19}F NMR** (376 MHz, $CDCl_3$) δ −117.84.

HPLC (Daicel Chiralpak OD-H, ^{n}hexane/^{i}PrOH = 97:3, 0.65 mL/min, T = 15 °C): t_{R1} (major) = 14.534 min, t_{R2} (minor) = 11.045 min; dr: 89:11; er:95.5:4.5.

$[\alpha]^{20}_D$ = +127.6 (c = 2.3, $CHCl_3$).

3.4.4. ((1*R*,2*S*,3*S*,4*S*)-3-(2,3-Difluorophenyl)bicyclo[2.2.1]hept-5-en-2-yl)(pyridin-2-yl)methanone (**3dA**)

Colorless oil, 27 mg, yield 87%, R_f = 0.5 (PE/EA = 10:1, v/v). **^{1}H NMR** (400 MHz, $CDCl_3$) δ 8.66 (ddd, J = 4.8, 1.8, 0.9 Hz, 1H), 8.05 (dt, J = 7.9, 1.1 Hz, 1H), 7.83 (td, J = 7.7, 1.8 Hz, 1H), 7.46 (ddd, J = 7.6, 4.8, 1.3 Hz, 1H), 7.17 (ddd, J = 7.0, 4.6, 2.0 Hz, 1H), 7.04–6.93 (m, 2H), 6.47 (dd, J = 5.6, 3.2 Hz, 1H), 5.89 (dd, J = 5.6, 2.8 Hz, 1H), 4.51 (dd, J = 5.1, 3.5 Hz, 1H), 3.55 (d, J = 5.2 Hz, 1H), 3.51 (s, 1H), 3.17 (d, J = 1.7 Hz, 1H), 1.95 (d, J = 8.6 Hz, 1H), 1.63 (dd, J = 8.6, 1.8 Hz, 1H). **^{13}C NMR** (101 MHz, $CDCl_3$) δ 199.8, 153.3, 148.8, 138.2, 136.9, 134.5 (d, J = 10.7 Hz), 133.5, 127.0, 123.6 (t, J = 5.8, 11.9 Hz), 122.3, 121.9 (d, J = 2.7 Hz), 114.6 (d, J = 5.0 Hz), 52.6, 48.5, 48.2, 47.8, 39.1. **^{19}F NMR** (376 MHz, $CDCl_3$) δ −111.33, 116.03. **HRMS** (ESI): m/z $[M+H]^+$ calculated for $C_{19}H_{16}F_2NO^+$ 312.1194, found 312.1199.

HPLC (Daicel Chiralpak OD-H, ^{n}hexane/^{i}PrOH = 97:3, 0.65 mL/min, T = 15 °C): t_{R1} (major) = 16.212 min, t_{R2} (minor) = 11.443 min; dr: 86:14; er: 90:10.

$[\alpha]^{20}_D$ = +93.5 (c = 2.1, $CHCl_3$).

3.4.5. ((1*S*,2*R*,3*R*,4*R*)-3-(3-Chlorophenyl)bicyclo[2.2.1]hept-5-en-2-yl)(pyridin-2-yl) methanone (**3eA**)

Colorless oil, 25 mg, yield 81%, R_f = 0.5 (PE/EA = 10:1, v/v). **^{1}H NMR** (400 MHz, $CDCl_3$) δ 8.68 (d, J = 4.3 Hz, 1H), 8.01 (d, J = 7.9 Hz, 1H), 7.83 (td, J = 7.7, 1.8 Hz, 1H), 7.46 (ddd, J = 7.6, 4.7, 1.3 Hz, 1H), 7.30–7.07 (m, 4H), 6.47 (dd, J = 5.7, 3.2 Hz, 1H), 5.83 (dd, J = 5.6, 2.8 Hz, 1H), 4.46 (dd, J = 5.2, 3.4 Hz, 1H), 3.56 (s, 1H), 3.42 (d, J = 4.2 Hz, 1H), 3.07 (d, J = 1.8 Hz, 1H), 2.03 (d, J = 8.5 Hz, 1H), 1.63 (dd, J = 8.6, 1.8 Hz, 1H). **^{13}C NMR** (101 MHz, $CDCl_3$) δ 200.7, 153.4, 148.9, 146.8, 139.2, 136.9, 134.2, 133.0, 129.6, 127.6, 127.0, 126.0, 122.2, 54.4, 49.1, 48.7, 48.2, 45.3.

HPLC (Daicel Chiralpak AD-H, ^{n}hexane/^{i}PrOH = 99.5:0.5, 0.5 mL/min, T = 12 °C): t_{R1} (major) = 36.171 min, t_{R2} (minor) = 38.433 min; dr: 89:11; er: 92:8.

$[\alpha]^{20}_D$ = +112.8 (c = 1.6, $CHCl_3$).

3.4.6. ((1*R*,2*S*,3*S*,4*S*)-3-(4-Chlorophenyl)bicyclo[2.2.1]hept-5-en-2-yl)(pyridin-2-yl) methanone (**3fA**)

Colorless oil, 30 mg, yield 98%, R_f = 0.5 (PE/EA = 10:1, v/v). **^{1}H NMR** (400 MHz, $CDCl_3$) δ 8.70 (d, J = 4.8 Hz, 1H), 8.03 (d, J = 7.9 Hz, 1H), 7.85 (td, J = 7.7, 1.8 Hz, 1H), 7.48 (ddd, J = 7.5, 4.7, 1.3 Hz, 1H), 7.26 (m, 4H), 6.50 (dd, J = 5.7, 3.2 Hz, 1H), 5.85 (dd, J = 5.6, 2.8 Hz, 1H), 4.48 (dd, J = 5.2, 3.4 Hz, 1H), 3.57 (s, 1H), 3.43 (dd, J = 5.2, 1.8 Hz, 1H), 3.07 (d, J = 1.7 Hz, 1H), 2.03 (d, J = 8.5 Hz, 1H), 1.64 (dd, J = 8.5, 1.8 Hz, 1H). **^{13}C NMR** (101 MHz, $CDCl_3$) δ 200.9, 153.4, 148.9, 143.1, 139.2, 136.8, 133.0, 131.5, 128.9, 128.4, 127.0, 122.2, 54.4, 49.2, 48.7, 48.1, 45.0.

HPLC (Daicel Chiralpak OD-H, ^{n}hexane/^{i}PrOH = 97:3, 0.65 mL/min, T = 15 °C): t_{R1} (major) = 14.238 min, t_{R2} (minor) = 10.897 min; dr: 88:12; er: 92:8.

$[\alpha]^{20}_D$ = +104.4 (c = 1.4, $CHCl_3$).

3.4.7. ((1*R*,2*S*,3*S*,4*S*)-3-(2-Bromophenyl)bicyclo[2.2.1]hept-5-en-2-yl)(pyridin-2-yl) methanone (**3gA**)

Colorless oil, 28 mg, yield 80%, R_f = 0.5 (PE/EA = 10:1, v/v). **^{1}H NMR** (400 MHz, $CDCl_3$) δ 8.66 (d, J = 4.9 Hz, 1H), 8.02 (d, J = 7.9 Hz, 1H), 7.81 (td, J = 7.7, 1.8 Hz, 1H), 7.53 (ddd, J = 11.1, 8.0, 1.6 Hz, 2H), 7.44 (ddd, J = 7.6, 4.8, 1.3 Hz, 1H), 7.32–7.25 (m, 1H), 7.05

(td, J = 7.7, 1.7 Hz, 1H), 6.53 (dd, J = 5.6, 3.2 Hz, 1H), 5.90 (dd, J = 5.6, 2.8 Hz, 1H), 4.68 (dd, J = 5.1, 3.5 Hz, 1H), 3.60 (d, J = 3.5 Hz, 1H), 3.48 (s, 1H), 3.06 (dd, J = 3.3, 1.7 Hz, 1H), 1.99 (d, J = 8.5 Hz, 1H), 1.59 (dd, J = 8.5, 1.8 Hz, 1H). **^{13}C NMR** (101 MHz, $CDCl_3$) δ 200.3, 153.6 148.8, 143.4, 138.5, 136.8, 133.6, 133.1, 127.9, 127.4, 127.2, 126.8, 126.3, 122.2, 51.2, 49.9, 48.4, 47.7, 46.3.

HPLC (Daicel Chiralpak OD-H, ^{n}hexane/^{i}PrOH = 99:1, 0.65 mL/min, T = 15 °C): t_{R1} (major) = 18.929 min, t_{R2} (minor) = 17.916 min; dr: 64:36; er: 88:12.

$[\alpha]^{20}_D$ = +61.9 (c = 2.3, $CHCl_3$).

3.4.8. ((1*R*,2*S*,3*S*,4*S*)-3-(4-Bromophenyl)bicyclo[2.2.1]hept-5-en-2-yl)(pyridin-2-yl) methanone (**3hA**)

Colorless oil, 30 mg, yield 85%, R_f = 0.5 (PE/EA = 10:1, *v*/*v*). **^{1}H NMR** (400 MHz, $CDCl_3$) δ 8.67 (ddd, J = 4.8, 1.8, 0.9 Hz, 1H), 8.00 (dt, J = 7.9, 1.1 Hz, 1H), 7.82 (td, J = 7.7, 1.8 Hz, 1H), 7.46 (ddd, J = 7.5, 4.7, 1.3 Hz, 1H), 7.41–7.35 (m, 2H), 7.21–7.15 (m, 2H), 6.47 (dd, J = 5.6, 3.2 Hz, 1H), 5.83 (dd, J = 5.7, 2.8 Hz, 1H), 4.46 (dd, J = 5.2, 3.4 Hz, 1H), 3.55 (s, 1H), 3.39 (d, J = 4.6 Hz, 1H), 3.04 (d, J = 2.2 Hz, 1H), 2.01 (d, J = 8.6 Hz, 1H), 1.62 (dd, J = 8.6, 1.8 Hz, 1H). **^{13}C NMR** (101 MHz, $CDCl_3$) δ 200.8, 153.4, 148.8, 143.7, 139.2, 136.9, 133.0, 131.3, 129.4, 127.0, 122.2, 119.5, 54.4, 49.2, 48.7, 48.1, 45.1.

HPLC (Daicel Chiralpak OD-H, ^{n}hexane/^{i}PrOH = 97:3, 0.65 mL/min, T = 15 °C): t_{R1} (major) = 15.277 min, t_{R2} (minor) = 11.426 min; dr: 88.5:11.5; er: 88:12.

$[\alpha]^{20}_D$ = +90.2 (c = 2.3, $CHCl_3$).

3.4.9. 4-((1*S*,2*S*,3*S*,4*R*)-3-Picolinoylbicyclo[2.2.1]hept-5-en-2-yl)benzonitrile (**3iA**)

Colorless oil, 25 mg, yield 84%, R_f = 0.6 (PE/EA = 10:1, *v*/*v*). **^{1}H NMR** (400 MHz, $CDCl_3$) δ 8.67 (d, J = 4.5 Hz, 1H), 8.01 (d, J = 7.6 Hz 1H), 7.88–7.82 (m, 1H), 7.60–7.54 (m, 2H), 7.48 (ddd, J = 7.6, 4.0, 2.8 Hz, 1H), 7.40 (d, J = 8.3 Hz, 2H), 6.48 (dd, J = 5.7, 3.2 Hz, 1H), 5.86 (dd, J = 5.6, 2.8 Hz, 1H), 4.45 (dd, J = 5.3, 3.4 Hz, 1H), 3.59 (s, 1H), 3.49 (d, J = 4.8 Hz, 1H), 3.10 (dd, J = 3.3, 1.6 Hz, 1H), 1.99 (d, J = 8.6 Hz, 1H), 1.65 (dd, J = 8.6, 1.8 Hz, 1H). **^{13}C NMR** (101 MHz, $CDCl_3$) δ 200.4, 153.2, 150.5, 148.9, 139.0, 136.9, 133.3, 132.2, 128.4, 127.1, 122.3, 119.0, 109.6, 54.5, 48.8, 48.7, 48.2, 45.8.

HPLC (Daicel Chiralpak OD-H, ^{n}hexane/^{i}PrOH = 97:3, 0.65 mL/min, T = 13 °C): t_{R1} (major) = 34.765 min, t_{R2} (minor) = 40.230 min; dr: 73.5:26.5; er:87:13.

$[\alpha]^{20}_D$ = +79.6 (c = 2.0, $CHCl_3$).

3.4.10. Pyridin-2-yl((1*R*,2*S*,3*S*,4*S*)-3-(p-tolyl)bicyclo[2.2.1]hept-5-en-2-yl)methanone (**3jA**)

Colorless oil, 26 mg, yield 91%, R_f = 0.6 (PE/EA = 10:1, *v*/*v*). **^{1}H NMR** (400 MHz, $CDCl_3$) δ 8.67 (d, J = 4.0 Hz, 1H), 7.99 (d, J = 7.8 Hz, 1H), 7.80 (td, J = 7.7, 1.8 Hz, 1H), 7.44 (ddd, J = 7.6, 4.8, 1.3 Hz, 1H), 7.21 (d, J = 8.0 Hz, 2H), 7.08 (d, J = 7.9 Hz, 2H), 6.48 (dd, J = 5.6, 3.2 Hz, 1H), 5.81 (dd, J = 5.6, 2.8 Hz, 1H), 4.52 (dd, J = 5.2, 3.4 Hz, 1H), 3.53 (s, 1H), 3.41 (d, J = 4.8 Hz, 1H), 3.05 (s, 1H), 2.30 (s, 3H), 2.06 (d, J = 8.4 Hz, 1H), 1.59 (dd, J = 8.5, 1.8 Hz, 1H). **^{13}C NMR** (101 MHz, $CDCl_3$) δ 201.2, 153.6, 148.8, 141.5, 139.4, 136.8, 135.2, 132.7, 129.0, 127.5, 126.8, 122.1, 54.1, 49.6, 48.7, 48.2, 45.2, 20.9.

HPLC (Daicel Chiralpak OD-H, ^{n}hexane/^{i}PrOH = 97:3, 0.65 mL/min, T = 12 °C): t_{R1} (major) = 16.212 min, t_{R2} (minor) = 11.443 min; dr: 86:14; er: 93:7.

$[\alpha]^{20}_D$ = +111.6 (c = 2.3, $CHCl_3$).

3.4.11. ((1*R*,2*S*,3*S*,4*S*)-3-(3,5-Dimethylphenyl)bicyclo[2.2.1]hept-5-en-2-yl)(pyridin-2-yl)methanone (**3kA**)

Colorless oil, 24 mg, yield 86%, R_f = 0.6 (PE/EA = 10:1, *v*/*v*). **^{1}H NMR** (400 MHz, $CDCl_3$) δ 8.69 (d, J = 4.6 Hz, 1H), 8.00 (d, J = 7.8 Hz, 1H), 7.81 (td, J = 7.7, 1.7 Hz, 1H), 7.44 (ddd, J = 7.5, 4.7, 1.2 Hz, 1H), 6.94 (s, 2H), 6.81 (s, 1H), 6.48 (dd, J = 5.7, 3.2 Hz, 1H), 5.80 (dd, J = 5.6, 2.8 Hz, 1H), 4.49 (dd, J = 5.2, 3.4 Hz, 1H), 3.54 (s, 1H), 3.38 (d, J = 4.9 Hz, 1H), 3.05 (s, 1H), 2.27 (s, 6H), 2.09 (d, J = 8.4 Hz, 1H), 1.60 (dd, J = 8.5, 1.8 Hz, 1H). **^{13}C NMR** (101 MHz, $CDCl_3$) δ 201.1, 153.6, 148.8, 144.5, 139.5, 137.8, 136.8, 132.7, 127.5, 126.8, 125.5,

122.2, 54.2, 49.5, 48.7, 48.3, 45.3, 21.4. **HRMS** (ESI): m/z $[M+H]^+$ calculated for $C_{21}H_{22}NO^+$ 304.1696, found 304.1692.

HPLC (Daicel Chiralpak OD-H, ^{n}hexane/^{i}PrOH = 97:3, 0.65 mL/min, T = 11 °C): t_{R1} (major) = 25.518 min, t_{R2} (minor) = 9.663 min; dr: 90:10; er: 89:11.

$[\alpha]^{20}_D$ = +111.8 (c = 1.9, $CHCl_3$).

3.4.12. ((1*R*,2*S*,3*S*,4*S*)-3-(4-Methoxyphenyl)bicyclo[2.2.1]hept-5-en-2-yl)(pyridin-2-yl) methanone (**3lA**)

Colorless oil, 21 mg, yield 68%, R_f = 0.4 (PE/EA = 10:1, *v*/*v*). **^{1}H NMR** (400 MHz, $CDCl_3$) δ 8.68 (ddd, *J* = 4.8, 1.7, 0.9 Hz, 1H), 8.00 (dt, *J* = 7.8, 1.1 Hz, 1H), 7.81 (td, *J* = 7.7, 1.8 Hz, 1H), 7.44 (ddd, *J* = 7.6, 4.8, 1.3 Hz, 1H), 7.26–7.22 (m, 2H), 6.86–6.80 (m, 2H), 6.48 (dd, *J* = 5.7, 3.2 Hz, 1H), 5.81 (dd, *J* = 5.7, 2.8 Hz, 1H), 4.49 (dd, *J* = 5.2, 3.4 Hz, 1H), 3.77 (s, 3H), 3.53 (s, 1H), 3.38 (d, *J* = 4.4 Hz, 1H), 3.05–3.00 (m, 1H), 2.08–2.03 (m, 1H), 1.60 (dd, *J* = 8.5, 1.8 Hz, 1H). **^{13}C NMR** (101 MHz, $CDCl_3$) δ 201.2, 157.7, 153.6, 148.8, 139.4, 136.8, 132.7, 128.5, 126.8, 122.1, 113.7, 55.2, 54.2, 49.6, 48.7, 48.1, 44.9.

HPLC (Daicel Chiralpak OD-H, ^{n}hexane/^{i}PrOH = 99:1, 0.65 mL/min, T = 12 °C): t_{R1} (major) = 40.987 min, t_{R2} (minor) = 22.953 min; dr: 90:10; er: 89:11.

$[\alpha]^{20}_D$ = +91.5 (c = 2.2, $CHCl_3$).

3.4.13. ((1*R*,2*S*,3*S*,4*S*)-3-(Naphthalen-2-yl)bicyclo[2.2.1]hept-5-en-2-yl)(pyridin-2-yl) methanone (**3mA**)

Colorless oil, 29 mg, yield 89%, R_f = 0.65 (PE/EA = 10:1, *v*/*v*). **^{1}H NMR** (400 MHz, $CDCl_3$) δ 8.66 (d, *J* = 4.5 Hz, 1H), 8.16–8.11 (m, 1H), 8.08 (d, *J* = 8.0 Hz, 1H), 7.83 (td, *J* = 7.5, 1.7 Hz, 2H), 7.73 (d, *J* = 8.2 Hz, 1H), 7.66 (d, *J* = 7.2 Hz, 1H), 7.50–7.41 (m, 4H), 6.66 (dd, *J* = 5.7, 3.2 Hz, 1H), 5.96 (dd, *J* = 5.6, 2.8 Hz, 1H), 4.73 (dd, *J* = 5.1, 3.5 Hz, 1H), 4.10 (d, *J* = 4.9 Hz, 1H), 3.57 (s, 1H), 3.24 (s, 1H), 2.23 (d, *J* = 8.4 Hz, 1H), 1.72 (dd, *J* = 8.5, 1.8 Hz, 1H). **^{13}C NMR** (101 MHz, $CDCl_3$) δ 201.2, 153.5, 148.8, 140.7, 139.0, 136.8, 133.8, 133.4, 132.7, 128.6, 126.9, 126.6, 125.9, 125.4, 125.3, 124.3, 123.0, 122.2, 52.7, 49.9, 48.8, 48.5, 41.6. **HRMS** (ESI): m/z $[M+H]^+$ calculated for $C_{23}H_{20}NO^+$ 326.1539, found 326.1543.

HPLC (Daicel Chiralpak OD-H, ^{n}hexane/^{i}PrOH = 98:2, 0.65 mL/min, T = 15 °C): t_{R1} (major) = 25.574 min, t_{R2} (minor) = 12.468 min; dr: 92:2; er: 81:19.

$[\alpha]^{20}_D$ = +92.6 (c = 1.2, $CHCl_3$).

3.4.14. Pyridin-2-yl((1*R*,2*S*,3*S*,4*S*)-3-(thiophen-2-yl)bicyclo[2.2.1]hept-5-en-2-yl) methanone (**3nA**)

Colorless oil, 22 mg, yield 78%, R_f = 0.4 (PE/EA = 10:1, *v*/*v*). **^{1}H NMR** (400 MHz, $CDCl_3$) δ 8.70 (ddd, *J* = 4.8, 1.7, 0.9 Hz, 1H), 8.00 (dt, *J* = 7.8, 1.1 Hz, 1H), 7.82 (td, *J* = 7.7, 1.8 Hz, 1H), 7.46 (ddd, *J* = 7.6, 4.8, 1.3 Hz, 1H), 7.14–7.08 (m, 1H), 6.93–6.88 (m, 2H), 6.45 (dd, *J* = 5.7, 3.2 Hz, 1H), 5.80 (dd, *J* = 5.6, 2.8 Hz, 1H), 4.57 (dd, *J* = 4.9, 3.4 Hz, 1H), 3.63 (dd, *J* = 4.9, 1.7 Hz, 1H), 3.55 (s, 1H), 3.06 (s, 1H), 2.11 (d, *J* = 8.7 Hz, 1H), 1.64 (dd, *J* = 8.7, 1.8 Hz, 1H). **^{13}C NMR** (101 MHz, $CDCl_3$) δ 200.4, 153.4, 148.9, 138.6, 136.8, 132.7, 126.9, 126.6, 123.7, 122.9, 122.2, 55.7, 51.5, 48.6, 48.5, 41.6.

HPLC (Daicel Chiralpak OD-H, ^{n}hexane/^{i}PrOH = 99:1, 0.65 mL/min, T = 17 °C): t_{R1} (major) = 20.617 min, t_{R2} (minor) = 16.158 min; dr: 76:24; er: 89:11.

$[\alpha]^{20}_D$ = +125.6 (c = 1.4, $CHCl_3$).

3.4.15. ((1*R*,2*S*,3*S*,4*S*)-3-(Benzo[b]thiophen-2-yl)bicyclo[2.2.1]hept-5-en-2-yl)(pyridin-2-yl)methanone (**3oA**)

Colorless oil, 32 mg, yield 97%, R_f = 0.4 (PE/EA = 10:1, *v*/*v*). **^{1}H NMR** (400 MHz, $CDCl_3$) δ 8.70 (d, *J* = 4.6 Hz, 1H), 8.01 (d, *J* = 7.8 Hz, 1H), 7.81 (td, *J* = 7.7, 1.8 Hz, 1H), 7.76–7.71 (m, 1H), 7.64 (dd, *J* = 7.6, 1.4 Hz, 1H), 7.45 (ddd, *J* = 7.6, 4.8, 1.3 Hz, 1H), 7.29–7.21 (m, 2H), 7.13 (s, 1H), 6.48 (dd, *J* = 5.7, 3.2 Hz, 1H), 5.84 (dd, *J* = 5.7, 2.8 Hz, 1H), 4.67 (dd, *J* = 5.0, 3.4 Hz, 1H), 3.69 (d, *J* = 5.1 Hz, 1H), 3.59 (s, 1H), 3.18 (s, 1H), 2.14 (d, *J* = 8.7 Hz, 1H), 1.67 (dd, *J* = 8.7, 1.8 Hz, 1H). **^{13}C NMR** (101 MHz, $CDCl_3$) δ 200.2, 153.3, 149.8, 148.9, 140.0,

139.0, 138.6, 136.8, 132.9, 127.0, 124.1, 123.5, 122.8, 122.2, 122.0, 120.0, 55.1, 51.0, 48.7, 48.6, 42.3. **HRMS** (ESI): *m/z* [M+H]$^+$ calculated for $C_{21}H_{18}NOS^+$ 332.1104, found 332.1110.

HPLC (Daicel Chiralpak OD-H, ^{n}hexane/^{i}PrOH = 97:3, 0.65 mL/min, T = 11 °C): t_{R1} (major) = 28.571 min, t_{R2} (minor) = 14.914 min; dr: 84.5:15.5; er: 90:10.

[α]^{20}$_D$ = +109.6 (c = 2.3, $CHCl_3$).

3.4.16. ((1*R*,2*S*,3*S*,4*S*)-3-Cyclohexylbicyclo[2.2.1]hept-5-en-2-yl)(pyridin-2-yl) methanone (**3pA**)

Colorless oil, 21 mg, yield 78%, R_f = 0.6 (PE/EA = 10:1, *v*/*v*). **^{1}H NMR** (400 MHz, $CDCl_3$) δ 8.75–8.71 (m, 1H), 7.98 (dt, *J* = 7.8, 1.2 Hz, 1H), 7.83 (dd, *J* = 7.7, 1.8 Hz, 1H), 7.46 (ddd, *J* = 7.6, 4.8, 1.4 Hz, 1H), 6.33 (dd, *J* = 5.7, 3.3 Hz, 1H), 5.68 (dd, *J* = 5.6, 2.8 Hz, 1H), 4.12 (dd, *J* = 5.0, 3.4 Hz, 1H), 3.30 (s, 1H), 2.91 (d, *J* = 2.3 Hz, 1H), 2.02–1.94 (m, 1H), 1.80 (ddd, *J* = 10.3, 5.0, 1.7 Hz, 1H), 1.77–1.71 (m, 2H), 1.63–1.56 (m, 3H), 1.47–1.43 (m, 1H), 1.24–1.07 (m, 6H). **^{13}C NMR** (101 MHz, $CDCl_3$) δ 201.8, 153.6, 148.8, 139.3, 136.8, 131.7, 126.7, 122.2, 51.4, 48.5, 48.1, 47.8, 44.6, 42.3, 32.9, 32.5, 26.6, 26.5, 26.3.

HPLC (Daicel Chiralpak AD-H, ^{n}hexane/^{i}PrOH = 99:1, 0.8 mL/min, T = 12 °C): t_{R1} (major) = 10.062 min, t_{R2} (minor) = 8.462 min; dr: 82:18; er: 99:1.

[α]^{20}$_D$ = +70.2 (c = 2.8, $CHCl_3$).

3.4.17. ((1*R*,2*S*,3*S*,4*S*)-3-Cyclopentylbicyclo[2.2.1]hept-5-en-2-yl)(pyridin-2-yl) methanone (**3qA**)

Colorless oil, 22 mg, yield 82%, R_f = 0.6 (PE/EA = 10:1, *v*/*v*). **^{1}H NMR** (400 MHz, $CDCl_3$) δ 8.73 (d, *J* = 4.7 Hz, 1H), 7.99 (d, *J* = 7.8 Hz, 1H), 7.83 (td, *J* = 7.7, 1.7 Hz, 1H), 7.47 (ddd, *J* = 7.6, 4.8, 1.3 Hz, 1H), 6.36 (dd, *J* = 5.7, 3.2 Hz, 1H), 5.68 (dd, *J* = 5.7, 2.8 Hz, 1H), 4.14–4.09 (m, 1H), 3.33 (s, 1H), 2.80–2.73 (m, 1H), 1.97–1.39 (m, 12H), 1.33–1.00 (m, 3H). **^{13}C NMR** (101 MHz, $CDCl_3$) δ 201.9, 153.8, 148.8, 139.2, 136.8, 131.4, 126.7, 122.1, 52.3, 48.4, 48.2, 47.5, 47.1, 45.6, 32.4, 32.2, 25.3, 25.0. **HRMS** (ESI): *m/z* [M+H]$^+$ calculated for $C_{18}H_{22}NO^+$ 268.1696, found 268.1700.

HPLC (Daicel Chiralpak AD-H, ^{n}hexane/^{i}PrOH = 99:1, 0.5 mL/min, T = 16 °C): t_{R1} (major) = 13.391 min, t_{R2} (minor) = 14.348 min; dr: 82:18; er: 94.5:5.5.

[α]^{20}$_D$ = +99.4 (c = 3.7, $CHCl_3$).

3.4.18. ((1*R*,2*S*,3*S*,4*S*)-3-Cyclopropylbicyclo[2.2.1]hept-5-en-2-yl)(pyridin-2-yl) methanone (**3rA**)

Colorless oil, 18 mg, yield 75%, R_f = 0.55 (PE/EA = 10:1, *v*/*v*). **^{1}H NMR** (400 MHz, $CDCl_3$) δ 8.72 (dt, *J* = 4.8, 1.2 Hz, 1H), 7.95 (d, *J* = 7.8 Hz, 1H), 7.81 (td, *J* = 7.7, 1.8 Hz, 1H), 7.46 (ddd, *J* = 7.5, 4.8, 1.3 Hz, 1H), 6.29 (dd, *J* = 5.7, 3.2 Hz, 1H), 5.66 (dd, *J* = 5.7, 2.8 Hz, 1H), 4.21–4.16 (m, 1H), 3.37 (s, 1H), 2.80 (s, 1H), 1.91 (d, *J* = 8.4 Hz, 1H), 1.53 (dd, *J* = 8.5, 1.9 Hz, 1H), 1.41 (dd, *J* = 4.8, 2.5 Hz, 1H), 0.76 (dddd, *J* = 12.8, 9.8, 8.0, 4.9 Hz, 1H), 0.49–0.36 (m, 2H), 0.22 (dq, *J* = 9.3, 4.8 Hz, 1H), 0.06–0.00 (m, 1H). **^{13}C NMR** (101 MHz, $CDCl_3$) δ 201.7, 153.8, 148.8, 139.0, 136.8, 131.8, 126.7, 122.1, 53.5, 48.6, 48.1, 48.0, 47.2, 15.9, 5.0, 4.4. **HRMS** (ESI): *m/z* [M+H]$^+$ calculated for $C_{16}H_{18}NO^+$ 240.1383, found 240.1389.

HPLC (Daicel Chiralpak AD-H, ^{n}hexane/^{i}PrOH = 99.5:0.5, 0.5 mL/min, T = 18 °C): t_{R1} (major) = 15.321 min, t_{R2} (minor) = 16.065 min; dr: 87.5:12.5; er: 77:23.

[α]^{20}$_D$ = +46.2 (c = 3.8, $CHCl_3$).

3.4.19. ((1*S*,2*S*)-4,5-Dimethyl-1,2,3,6-tetrahydro-[1,1′-biphenyl]-2-yl)(pyridin-2-yl) methanone (**3aB**)

Colorless oil, 13 mg, yield 45%, R_f = 0.75 (PE/EA = 5:1, *v*/*v*). **^{1}H NMR** (400 MHz, $CDCl_3$) δ 8.66 (dt, *J* = 4.9, 1.3 Hz, 1H), 7.78–7.73 (m, 1H), 7.68 (td, *J* = 7.6, 1.7 Hz, 1H), 7.38 (ddd, *J* = 7.4, 4.8, 1.4 Hz, 1H), 7.23–7.17 (m, 2H), 7.11 (dd, *J* = 8.4, 6.8 Hz, 2H), 7.06–6.99 (m, 1H), 4.74 (ddd, *J* = 11.6, 10.2, 5.7 Hz, 1H), 3.26 (ddd, *J* = 11.5, 9.9, 6.5 Hz, 1H), 2.36–2.20 (m, 4H), 1.67 (s, 6H). **^{13}C NMR** (101 MHz, $CDCl_3$) δ 204.7, 153.3, 148.7, 144.9, 136.7, 128.1, 127.5, 126.8, 125.9, 125.5, 124.2, 122.0, 45.3, 43.1, 41.2, 36.4, 18.7. **HRMS** (ESI): *m/z* [M+H]$^+$ calculated for $C_{20}H_{22}NO^+$ 292.1696, found 292.1697.

HPLC (Daicel Chiralpak OD-H, ^{n}hexane/^{i}PrOH = 98:2, 1.0 mL/min, T = 14 °C): t_{R1} (major) = 7.349 min, t_{R2} (minor) = 6.717 min; dr: >20:1; er: 82:18.

$[\alpha]^{20}_D$ = −21.5 (c = 0.2, $CHCl_3$).

3.5. Synthesis and Characterization of Product **5**

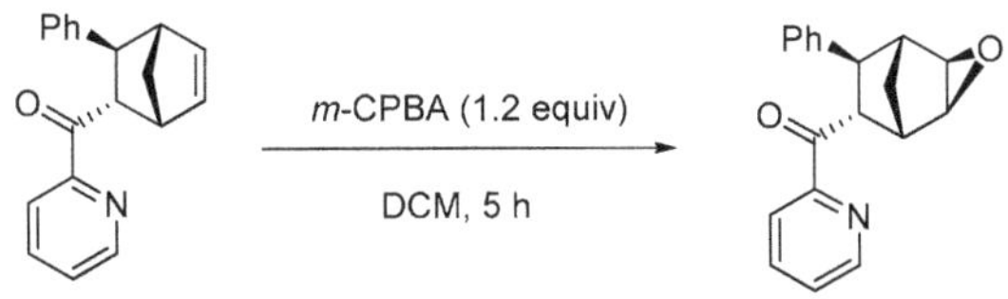

3aA **5**, 73%

Procedure: To an oven-dried reaction tube equipped with a magnetic stirring bar was added **3aA** (55 mg, 0.2 mmol), *m*-CPBA (53 mg, 0.24 mmol) and DCM (2 mL). The reaction was stirred at ambient temperature for 5 h. After completion of the reaction, as monitored using TLC, the mixture was washed with $NaHCO_3$ solution and dried over $MgSO_4$. After removal of the solvent in vacuo, the reaction mixture was purified via column chromatography on silica gel with PE and EA as eluent to afford product **5**.

((1*R*,2*R*,4*S*,5*S*,6*R*,7*S*)-7-Phenyl-3-oxatricyclo[3.2.1.02,4]octan-6-yl)(pyridin-2-yl) methanone (**5**)

Colorless oil, 42 mg, yield 73%, R_f = 0.3 (PE/EA = 10:1, *v*/*v*). **^{1}H NMR** (400 MHz, $CDCl_3$) δ 8.69 (dt, *J* = 4.8, 1.2 Hz, 1H), 8.10 (d, *J* = 7.8 Hz, 1H), 7.88 (td, *J* = 7.7, 1.7 Hz, 1H), 7.50 (ddd, *J* = 7.6, 4.8, 1.3 Hz, 1H), 7.32–7.27 (m, 4H), 7.24–7.16 (m, 1H), 4.47 (dd, *J* = 5.6, 3.8 Hz, 1H), 3.70 (d, *J* = 5.1 Hz, 1H), 3.47 (dd, *J* = 3.7, 1.4 Hz, 1H), 3.28–3.24 (m, 1H), 3.02 (d, *J* = 3.3 Hz, 1H), 2.85 (d, *J* = 1.7 Hz, 1H), 1.66–1.57 (s, 2H). **^{13}C NMR** (101 MHz, $CDCl_3$) δ 200.6, 153.2, 149.0, 143.9, 137.0, 128.5, 128.2, 128.1, 126.1, 122.2, 57.4, 51.8, 50.0, 43.3, 43.2, 42.8, 26.7. **HRMS** (ESI): *m/z* $[M+H]^+$ calculated for $C_{19}H_{18}NO_2^+$ 292.1332, found 292.1337.

3.6. Synthesis and Characterization of Product **6**

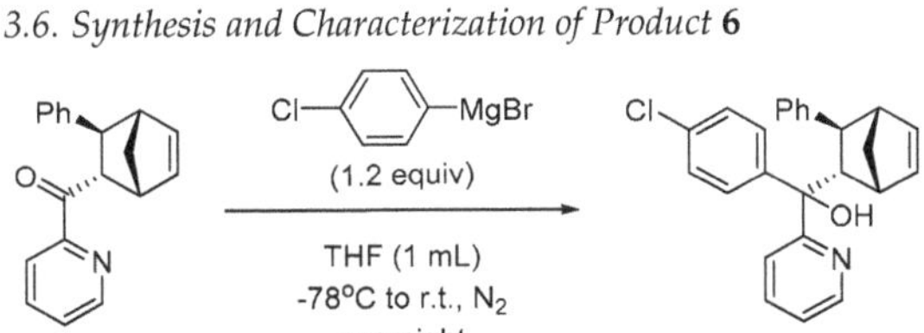

3aA **6**, 92%

To an oven-dried reaction tube equipped with a magnetic stirring bar was added **3aA** (55 mg, 0.2 mmol). The tube was sealed immediately with a rubber stopper and protected with a nitrogen balloon by evacuation-backfill operations repeated three times. Dry THF (1 mL) and (4-chlorophenyl) magnesium bromide (58 µL, 1.2 equiv) was injected to the tube via a syringe at −78 °C. The mixture was stirred at 25 °C overnight. Water (3.0 mL) was added dropwise, and the mixture was dried over sodium sulfate and filtered. After removal of the solvent in vacuo, the reaction mixture was purified using column chromatography on silica gel with PE and EA as eluent to afford product **6**.

(4-Chlorophenyl)((1*R*,2*S*,3*S*,4*S*)-3-phenylbicyclo[2.2.1]hept-5-en-2-yl)(pyridin-2-yl) methanol (**6**)

Colorless oil, 91 mg, yield 92%, R_f = 0.35 (PE/EA = 5:1, *v*/*v*). It is a mixture of three diastereoisomers; only the characteristic signals in ^{1}H NMR are provided. **^{1}H NMR** (400 MHz, $CDCl_3$) **^{1}H NMR** (400 MHz, chloroform-d) δ 8.50 (d, *J* = 4.5 Hz, 1H), 8.34 (d, *J* = 4.6 Hz, 1H), 8.24 (d, *J* = 4.7 Hz, 1H). **HRMS** (ESI): *m/z* $[M+H]^+$ calculated for $C_{25}H_{23}ClNO^+$ 388.1463, found 388.1465.

3.7. Synthesis and Characterization of Product 7

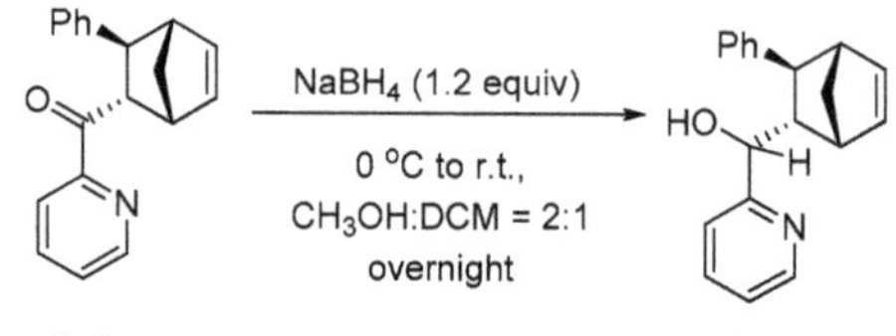

Procedure: To an oven-dried reaction tube equipped with a magnetic stirring bar was added **3aA** (55 mg, 0.2 mmol) and $NaBH_4$ (53 mg, 0.24 mmol) in CH_3OH-DCM (v/v = 2:1) (3 mL) at 0 °C. The reaction was stirred at ambient temperature overnight. Water (3.0 mL) was added dropwise, and the mixture was dried over sodium sulfate and filtered. After removal of the solvent in vacuo, the reaction mixture was purified using column chromatography on silica gel with PE and EA as eluent to afford product **7**.

(S)-((1*R*,2*S*,3*S*,4*S*)-3-Phenylbicyclo[2.2.1]hept-5-en-2-yl)(pyridin-2-yl)methanol (**7**)

Colorless oil, 51.5 mg, yield 93%, R_f = 0.5 (PE/EA = 2:1, v/v). **^{1}H NMR** (400 MHz, $CDCl_3$) δ 8.38 (d, *J* = 4.6 Hz, 1H), 7.43 (td, *J* = 7.7, 1.8 Hz, 1H), 7.14–6.97 (m, 5H), 6.86 (dd, *J* = 6.8, 1.9 Hz, 2H), 6.44 (dd, *J* = 5.7, 3.1 Hz, 1H), 6.36 (dd, *J* = 5.6, 2.9 Hz, 1H), 4.26 (d, *J* = 9.3 Hz, 1H), 3.61 (s, 1H), 3.22 (s, 1H), 2.76 (s, 1H), 2.58 (ddd, *J* = 9.0, 5.4, 3.3 Hz, 1H), 2.35 (d, *J* = 5.0 Hz, 1H), 1.74 (d, *J* = 8.7 Hz, 1H), 1.52 (dd, *J* = 8.7, 1.8 Hz, 1H). **^{13}C NMR** (101 MHz, $CDCl_3$) δ 161.8, 148.6, 144.1, 138.4, 136.3, 135.0, 127.9, 127.4, 125.5, 122.4, 121.4, 54.2, 50.6, 47.7, 46.4, 44.8. **HRMS** (ESI): *m/z* $[M+H]^+$ calculated for $C_{19}H_{20}NO^+$ 278.1539, found 278.1535.

4. Conclusions

We have identified a chiral PyBox–La$(OTf)_3$ complex for catalyzing enantioselective Diels–Alder cycloadditions of 2-alk-2-enoylpyridines with cyclopentadiene. The asymmetric reactions proceeded efficiently, displaying good levels of diastereo- and enantiocontrol (up to 92:8 dr and 99:1 er). Enantiopure disubstituted norbornenes, which possess four contiguous stereocenters and are biologically relevant structures, are produced conveniently in this way. Further manipulations of these structures were also demonstrated, yielding the more densely functionalized norbornene derivatives. We hope our catalytic protocol could benefit the synthetic and medicinal chemists who are associated with enantioenriched norbornenes.

Supplementary Materials: The following supporting information can be downloaded at: https://www.mdpi.com/article/10.3390/molecules29132978/s1, Further results of reaction condition optimization, copies of ^{1}H NMR, ^{13}C NMR, ^{19}F NMR and HPLC spectra of products and are included in the supporting information.

Author Contributions: Conceptualization, Z.Y.; Methodology, Z.Y.; Validation, H.W.; Formal analysis, H.W.; Investigation, H.W., Y.Z. and S.J.; Resources, Y.Y., J.X. and Z.Y.; Data curation, H.W.; Writing—original draft, Z.Y.; Writing—review & editing, H.W., N.C. and J.X.; Visualization, H.W.; Supervision, J.X. and Z.Y.; Project administration, Z.Y.; Funding acquisition, Y.Y., J.X. and Z.Y. All authors have read and agreed to the published version of the manuscript.

Funding: This research was supported by the National Key Research and Development Program of China (no. 2022YFF0709803).

Institutional Review Board Statement: Not applicable.

Informed Consent Statement: Not applicable.

Data Availability Statement: Data are contained within the article.

Conflicts of Interest: Author Ying Yu was employed by the company China United Test & Certification Co., Ltd. The remaining authors declare that the research was conducted in the absence of any commercial or financial relationships that could be construed as a potential conflict of interest.

References

1. Zhang, W.; Luo, S.; Fang, F.; Chen, Q.; Hu, H.; Jia, X.; Zhai, H. Total Synthesis of Absinthin. *J. Am. Chem. Soc.* **2005**, *127*, 18–19. [CrossRef] [PubMed]
2. Schule, A.; Liang, H.; Vors, J.P.; Ciufolini, M.A. Synthetic studies toward sordarin: Building blocks for the terpenoid core and for analogues thereof. *J. Org. Chem.* **2009**, *74*, 1587–1597. [CrossRef] [PubMed]
3. Antczak, K.; Kingston, J.F.; Fallis, A.G.; Hanson, A.W. A general intramolecular Diels–Alder approach to tricyclic sesquiterpenes: Stereoselective total syntheses of (±)-sinularene and (±)-5-epi-sinularene. *Can. J. Chem.* **1987**, *65*, 114–123. [CrossRef]
4. Bo, L.; Fallis, A.G. Direct total synthesis of (+)-longifolene via an intramolecular Diels-Alder strategy. *J. Am. Chem. Soc.* **2002**, *112*, 4609–4610. [CrossRef]
5. Lozinskaya, N.A.; Volkova, M.S.; Seliverstov, M.Y.; Temnov, V.V.; Sosonyuk, S.E.; Proskurnina, M.V.; Zefirov, N.S. Syntheses of spiroindole melatonin analogues via 2-(indolin-3-ylidene)acetonitrile cycloadditions. *Mendeleev Commun.* **2014**, *24*, 260–261. [CrossRef]
6. Wang, J.; Li, Z.; Xu, Y.; Hu, W.; Zheng, G.; Zheng, L.; Ren, T. Synthesis and Tribological Behavior of Bridged Bicyclic Polymers as Lubricants. *Ind. Eng. Chem. Res.* **2020**, *59*, 20730–20739. [CrossRef]
7. Jiang, X.; Wang, R. Recent Developments in Catalytic Asymmetric Inverse-Electron-Demand Diels–Alder Reaction. *Chem. Rev.* **2013**, *113*, 5515–5546. [CrossRef] [PubMed]
8. Otto, S.; Boccaletti, G.; Engberts, J.B.F.N. A Chiral Lewis-Acid-Catalyzed Diels—Alder Reaction. Water-Enhanced Enantioselectivity. *J. Am. Chem. Soc.* **1998**, *120*, 4238–4239. [CrossRef]
9. Barroso, S.; Blay, G.; Pedro, J.R. 2-alkenoyl pyridine N-oxides, highly efficient dienophiles for the enantioselective Cu(II)-bis(oxazoline) catalyzed Diels-Alder reaction. *Org. Lett.* **2007**, *9*, 1983–1986. [CrossRef]
10. Lu, Y.; Zhou, Y.; Lin, L.; Zheng, H.; Fu, K.; Liu, X.; Feng, X. N,N′-Dioxide/nickel(ii)-catalyzed asymmetric Diels–Alder reaction of cyclopentadiene with 2,3-dioxopyrrolidines and 2-alkenoyl pyridines. *Chem. Commun.* **2016**, *52*, 8255–8258. [CrossRef]
11. Meng, D.; Li, D.; Ollevier, T. Recyclable iron(ii) caffeine-derived ionic salt catalyst in the Diels-Alder reaction of cyclopentadiene and alpha,beta-unsaturated N-acyl-oxazolidinones in dimethyl carbonate. *RSC Adv.* **2019**, *9*, 21956–21963. [CrossRef] [PubMed]
12. Bos, J.; Fusetti, F.; Driessen, A.J.; Roelfes, G. Enantioselective artificial metalloenzymes by creation of a novel active site at the protein dimer interface. *Angew. Chem. Int. Ed.* **2012**, *51*, 7472–74725. [CrossRef] [PubMed]
13. Filice, M.; Romero, O.; Gutierrez-Fernandez, J.; de Las Rivas, B.; Hermoso, J.A.; Palomo, J.M. Synthesis of a heterogeneous artificial metallolipase with chimeric catalytic activity. *Chem. Commun.* **2015**, *51*, 9324–9327. [CrossRef] [PubMed]
14. Jiang, J.; Meng, Y.; Zhang, L.; Liu, M. Self-Assembled Single-Walled Metal-Helical Nanotube (M-HN): Creation of Efficient Supramolecular Catalysts for Asymmetric Reaction. *J. Am. Chem. Soc.* **2016**, *138*, 15629–15635. [CrossRef] [PubMed]
15. Li, Y.; Wang, C.; Hao, J.; Cheng, M.; Jia, G.; Li, C. Higher-order human telomeric G-quadruplex DNA metalloenzyme catalyzed Diels-Alder reaction: An unexpected inversion of enantioselectivity modulated by K^+ and NH_4^+ ions. *Chem. Commun.* **2015**, *51*, 13174–13177. [CrossRef] [PubMed]
16. Marek, J.J.; Hennecke, U. Why DNA Is a More Effective Scaffold than RNA in Nucleic Acid-Based Asymmetric Catalysis-Supramolecular Control of Cooperative Effects. *Chemistry* **2017**, *23*, 6009–6013. [CrossRef] [PubMed]
17. Park, S.; Okamura, I.; Sakashita, S.; Yum, J.H.; Acharya, C.; Gao, L.; Sugiyama, H. Development of DNA Metalloenzymes Using a Rational Design Approach and Application in the Asymmetric Diels–Alder Reaction. *ACS Catal.* **2015**, *5*, 4708–4712. [CrossRef]
18. Fukuzawa, S.-I.; Metoki, K.; Esumi, S.-I. Asymmetric Diels–Alder reactions in supercritical carbon dioxide catalyzed by rare earth complexes. *Tetrahedron* **2003**, *59*, 10445–10452. [CrossRef]
19. Liu, C.; Chen, Y.; Yang, Z. Iridium-Catalyzed Stereoselective Transfer Hydrogenation of 1,5-Benzodiazepines. *J. Org. Chem.* **2022**, *87*, 12001–12018. [CrossRef]
20. Yang, S.; Tang, W.; Yang, Z.; Xu, J. Iridium-Catalyzed Highly Efficient and Site-Selective Deoxygenation of Alcohols. *ACS Catal.* **2018**, *8*, 9320–9326. [CrossRef]
21. Wang, J.; Wang, T.; Du, H.; Chen, N.; Xu, J.; Yang, Z. Accessing para-Alkylphenols via Iridium-Catalyzed Site-Specific Deoxygenation of Alcohols. *J. Org. Chem.* **2023**, *88*, 12572–12584. [CrossRef] [PubMed]
22. Wang, T.; Miao, R.; Luo, R.; Xu, J.; Yang, Z. Furan-2-yl Anions as γ-Oxo/Hydroxyl Acyl Anion Equivalents Enabled by Iridium-Catalyzed Chemoselective Reduction. *Org. Lett.* **2023**, *25*, 4705–4710. [CrossRef] [PubMed]
23. Bhor, S.; Anilkumar, G.; Tse, M.K.; Klawonn, M.; Döbler, C.; Bitterlich, B.; Grotevendt, A.; Beller, M. Synthesis of a New Chiral N,N,N-Tridentate Pyridinebisimidazoline Ligand Library and Its Application in Ru-Catalyzed Asymmetric Epoxidation. *Org. Lett.* **2005**, *7*, 3393–3396. [CrossRef] [PubMed]
24. Anilkumar, G.; Bhor, S.; Tse, M.K.; Klawonn, M.; Bitterlich, B.; Beller, M. Synthesis of a novel class of chiral N,N,N-tridentate pyridinebisimidazoline ligands and their application in Ru-catalyzed asymmetric epoxidations. *Tetrahedron Asymmetry* **2005**, *16*, 3536–3561. [CrossRef]
25. Mendoza, S.D.; Rombola, M.; Tao, Y.; Zuend, S.J.; Götz, R.; McLaughlin, M.J.; Reisman, S.E. Expanding the Chiral Monoterpene Pool: Enantioselective Diels–Alder Reactions of α-Acyloxy Enones. *Org. Lett.* **2022**, *24*, 3802–3806. [CrossRef] [PubMed]
26. Desimoni, G.; Faita, G.; Guala, M.; Pratelli, C. An efficient catalyst for highly enantioselective exo-Diels–Alder reaction between alkenoyl-1,3-oxazolidin-2-ones and cyclopentadiene. *Tetrahedron* **2002**, *58*, 2929–2935. [CrossRef]

27. Brown, H.C.; Kawakami, J.H.; Ikegami, S. Additions to bicyclic olefins. III. Stereochemistry of the epoxidation of norbornene, 7,7-dimethylnorbornene, and related bicyclic olefins. Steric effects in the 7,7-dimethylnorbornyl system. *J. Am. Chem. Soc.* **1970**, *92*, 6914–6917. [CrossRef]
28. Ciupa, A.; Mahon, M.F.; De Bank, P.A.; Caggiano, L. Simple pyrazoline and pyrazole "turn on" fluorescent sensors selective for Cd^{2+} and Zn^{2+} in MeCN. *Org. Biomol. Chem.* **2012**, *10*, 8753–8757. [CrossRef]
29. Nguyen, H.; Gagné, M.R. Enantioselective Cascade Cyclization/Protodemetalation of Polyenes with $N3Pt^{2+}$ Catalysts. *ACS Catal.* **2014**, *4*, 855–859. [CrossRef]
30. Meng, J.-C.; Fokin, V.V.; Finn, M.G. Kinetic resolution by copper-catalyzed azide–alkyne cycloaddition. *Tetrahedron Lett.* **2005**, *46*, 4543–4546. [CrossRef]

molecules

MDPI

Article

Enantioselective Synthesis of the Sex Pheromone of *Sitodiplosis mosellana* (Géhin) and Its Stereoisomers

Jianan Wang [1], Xiaoyang Li [1], Yun Zhou [2], Qinghua Bian [1] and Jiangchun Zhong [1,*]

[1] Department of Applied Chemistry, China Agricultural University, Beijing 100193, China; bianqinghua@cau.edu.cn (Q.B.)

[2] Institute of Industrial Crops, Shandong Academy of Agricultural Sciences, Jinan 250100, China; zysass2021@163.com

* Correspondence: zhong@cau.edu.cn; Tel.: +86-010-6273-1356

Abstract: (2*S*,7*S*)-2,7-Nonanediyl dibutyrate is the sex pheromone of *Sitodiplosis mosellana* (Géhin). In this study, this sex pheromone and its three stereoisomers were prepared. Central to this strategy was the ring opening of chiral epoxide with an alkynyllithium and the hydrogenation of the triple bond. Moreover, this approach consisted of six steps, and the total yields were 59–64%.

Keywords: orange wheat blossom midge; chiral source; sex pheromone; asymmetric synthesis

1. Introduction

The orange wheat blossom midge, *Sitodiplosis mosellana* (Géhin) (Diptera: Cecidomyiidae), is a notorious pest worldwide [1,2], which has caused significant yield losses in wheat (*Triticum aestivum* Linnaeus) [3,4]. The adult females often oviposit on wheat heads, and the larvae feed on the developing seeds [5]. This leads to shriveling and presprouting damage, resulting in a reduction in both the yield and quality of wheat harvested [6,7]. The current integrated pest management programs for *S. mosellana* mainly rely on pesticides [8,9], the conservation of natural enemies [10] and resistant cultivars [11,12].

Pheromone-based pest control is an alternative strategy with the advantages of environmental friendliness, high efficiency and rarely inducing pest resistance [13,14]. In 2000, Gries identified that the sex pheromone of *S. mosellana* was (2*S*,7*S*)-2,7-nonanediyl dibutyrate ((2*S*,7*S*)-1) (Figure 1) via coupled gas chromatographic-electroantennographic detection (GC-EAD) with GC–mass spectrometry (MS) and trap experiments in wheat fields [15]. Furthermore, Gries prepared (2*S*,7*S*)-**1** using CuI, which catalyzed the reaction of *S*-propylene oxide with Grignard reagent, and through the hydrolytic kinetic resolution of epoxide using Jacobsen's catalyst [15]. The pheromone could be utilized for monitoring and trapping the orange wheat blossom midge [16,17], which has attracted significant interest from chemists. Previous syntheses have employed various approaches, including chiral sources of (*S*)-but-3-yn-2-ol [18,19] and (*S*)-5-hexen-2-ol [20], and the esterification of racemic nonanediyl dibutyrate using *Pseudomonas cepacia* Burkh lipase [21]. To further research the biological activities of the sex pheromone, herein, new and efficient synthesis of the sex pheromone of *S. mosellana* and its stereoisomers (Figure 1) is achieved.

 https://doi.org/10.3390/molecules30030671

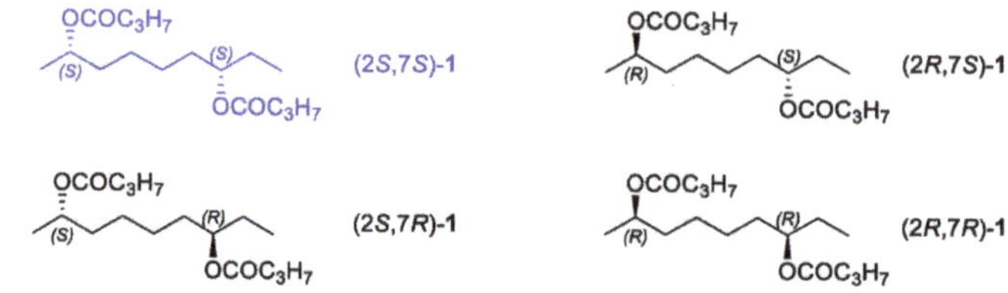

Figure 1. Sex pheromone of *S. mosellana* and its stereoisomers.

2. Results and Discussion

2.1. Retrosynthetic Analysis

The retrosynthetic analysis of the sex pheromone of *S. mosellana* (2*S*,7*S*)-**1** is shown in Scheme 1. The target pheromone (2*S*,7*S*)-**1** was obtained through the acylation of alkynyl diol (2*S*,7*S*)-**8** with butyric anhydride (**10**). The chiral secondary hydroxy at C2 of (2*S*,7*S*)-**8** was envisioned to be constructed through the addition of (*S*)-2-methyloxirane ((*S*)-**6**) with chiral TBDPS ether (*S*)-**5** and *n*-BuLi. The other stereocenter of (2*S*,7*S*)-**8** was obtained from the chiral source (*S*)-2-ethyloxirane ((*S*)-**2**). Following a similar procedure for sex pheromone (2*S*,7*S*)-**1**, its stereoisomers (2*R*,7*S*)-**1**, (2*S*,7*R*)-**1**, (2*R*,7*R*)-**1** were prepared.

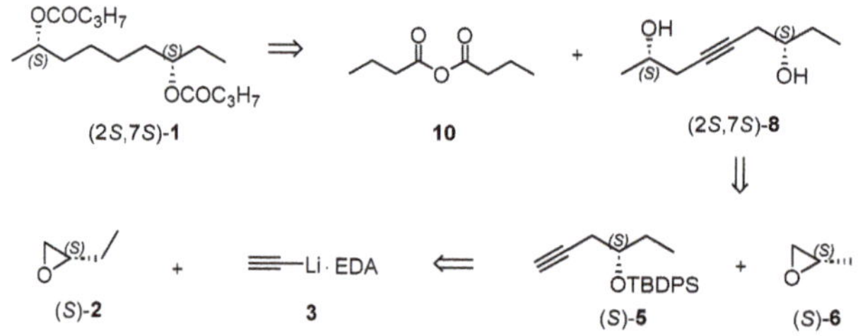

Scheme 1. Retrosynthetic analysis of sex pheromone of *S. mosellana* (2*S*,7*S*)-**1**.

2.2. Synthesis of Chiral TBDPS Ethers

Based on the retrosynthetic analysis of the sex pheromone (2*S*,7*S*)-**1**, our synthesis started from the preparation of chiral TBDPS ethers (*S*)- and (*R*)-**5** (Scheme 2). The reaction of (*S*)-2-ethyloxirane ((*S*)-**2**) with lithium acetylide ethylenediamine complex (**3**) in DMSO yielded (*S*)-hex-5-yn-3-ol ((*S*)-**4**) (87% yield) [22,23]. The specific rotation of chiral alcohol (*S*)-**4** was identical to that in the literature data [24]. The secondary hydroxy of (*S*)-**4** was then protected by a treatment with TBDPSCl and imidazole to provide chiral TBDPS ether (*S*)-**5** in a 95% yield [25,26]. Similarly, (*R*)-tert-butyl(hex-5-yn-3-yloxy)diphenylsilane ((*R*)-**5**) was prepared via the ring opening of (*R*)-2-ethyloxirane ((*R*)-**2**) with lithium acetylide ethylenediamine complex (**3**) and the protection of hydroxy of (*R*)-**4** with TBDPSCl.

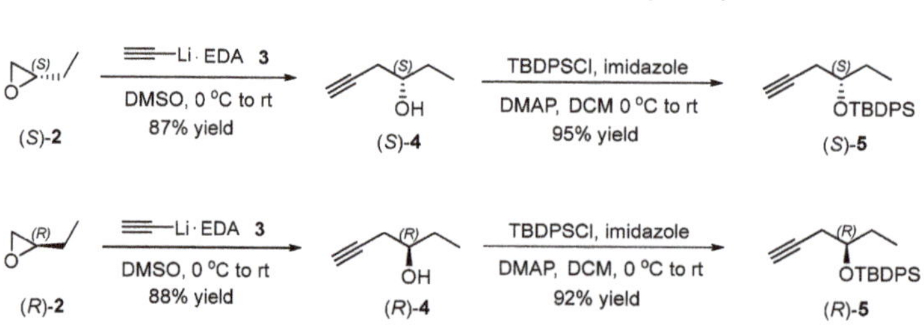

Scheme 2. Synthesis of chiral TBDPS ethers (*S*)- and (*R*)-**5**.

2.3. Synthesis of Sex Pheromone of S. mosellana

With the key chiral block (*S*)- **5** in hand, we next prepared the sex pheromone of *S. mosellana* (Scheme 3). In the presence of boron trifluoride-diethyl etherate, the addition of (*S*)-2-methyloxirane ((*S*)-**6**) with alkynyllithium, prepared in situ from the chiral TBDPS ether (*S*)-**5** and *n*-BuLi, provided (2*S*,7*S*)-7-((tert-butyldiphenylsilyl)oxy)non-4-yn-2-ol ((2*S*,7*S*)-**7**) almost quantitatively [27,28]. The subsequent deprotection with TBAF gave (2*S*,7*S*)-non-4-yne-2,7-diol ((2*S*,7*S*)-**8**) in a 93% yield [29]. Finaly, the hydrogenation of the triple bond of chiral alkynyl diol (2*S*,7*S*)-**8** resulted in the formation of (2*S*,7*S*)-nonane-2,7-diol ((2*S*,7*S*)-**9**) [30,31], which was treated with butyric anhydride (**10**) to yield (2*S*,7*S*)-nonane-2,7-diyl dibutyrate ((2*S*,7*S*)-**1**) [32,33]. The NMR spectra, HRMS, and specific rotation of the sex pheromone (2*S*,7*S*)-**1** matched those in reference [20].

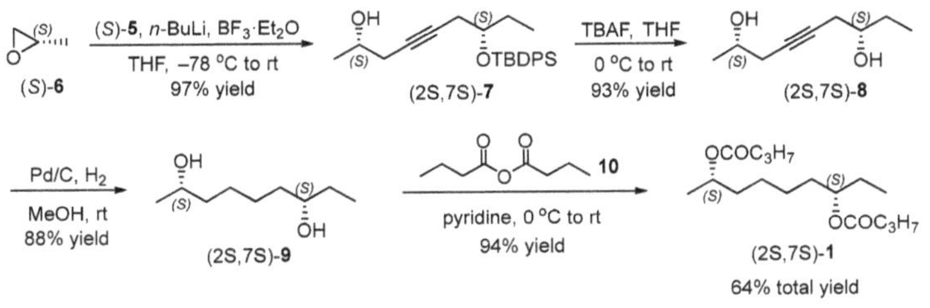

Scheme 3. Synthesis of sex pheromone of *S. mosellana*.

2.4. Synthesis of Stereoisomers of S. mosellana Sex Pheromone

Having achieved the synthesis of the sex pheromone of *S. mosellana*, we next investigated the preparation of its stereoisomers (Scheme 4). According to a similar procedure for the sex pheromone (2*S*,7*S*)-**1**, its stereoisomers (2*R*,7*S*)-**1**, (2*S*,7*R*)-**1** and (2*R*,7*R*)-**1** were synthesized from chiral TBDPS ether (*S*)-**5** and (*R*)-2-methyloxirane ((*R*)-**6**); chiral TBDPS ether (*R*)-**5** and (*S*)-2-methyloxirane ((*S*)-**6**); and chiral TBDPS ether (*R*)-**5** and (*R*)-2-methyloxirane ((*R*)-**6**), respectively. The structures of these stereoisomers were characterized by specific rotation, NMR spectra and HRMS (The NMR spectra are available in the Supplementary Materials).

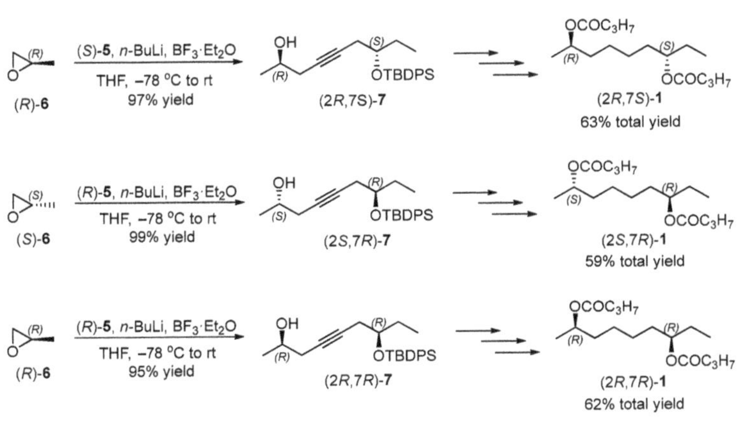

Scheme 4. Synthesis of stereoisomers of *S. mosellana* sex pheromone.

3. Materials and Methods

3.1. General Information

Unless otherwise stated, all reactions were performed under an argon atmosphere with a Schlenk line system. All commercial reagents and starting materials were used as received. The chiral reagents of (S)-2-ethyloxirane (268 RMB/100 mL), (R)-2-ethyloxirane (268 RMB/100 mL), (S)-2-methyloxirane (104 RMB/5 g) and (R)-2-methyloxirane (240 RMB/5 g) were purchased from Anhui Senrise Technology Co., Ltd., (Anhui, China). Dichloromethane, tetrahydrofuran, methanol and dimethyl sulfone were purchased from Beijing Ouhe Technology Co., Ltd., (Beijing, China), and purified according to standard procedures. ^{1}H NMR and ^{13}C NMR spectra were recorded on a Bruker Ascend™ 500 MHz spectrometer (Bruker Corporation, Billerica, MA, USA) at 500 MHz and 125 MHz, respectively. The chemical shifts were reported in ppm using TMS (0.00 ppm) and $CDCl_3$ (77.16 ppm) as internal standards. High-resolution mass spectra (HRMS) were obtained on a Waters LCT Premier™ (Waters Corporation, Milford, MA, USA) with an electrospray ionization (ESI) mass spectrometer. Optical rotations were measured on a Rudolph AUTOPOL-IV polarimeter (Rudolph Research Analytical, Flanders, NJ, USA).

3.2. Synthesis of (S)-5-hexyn-3-ol ((S)-**4**)

Under an argon atmosphere, (S)-2-ethyloxirane ((S)-**2**) (5.00 g, 69.35 mmol, >99% ee, 1.0 equiv.) was added slowly to a stirred suspension of lithium acetylide ethylenediamine complex (**3**) (15.96 g, 173.38 mmol, 2.5 equiv.) in DMSO (100 mL) at 0 °C. The reaction mixture was allowed to warm to room temperature and stirred continuously overnight. The reaction was then quenched by adding cold water (50 mL), followed by extraction with DCM (3 × 30 mL). The combined organic layers were washed with brine (30 mL), dried over anhydrous Na_2SO_4 and concentrated under reduced pressure to yield the crude product. The crude product was purified by silica gel column chromatography using an eluent of ethyl acetate and petroleum ether (1:5) to produce (S)-hex-5-yn-3-ol ((S)-**4**) (5.92 g, 87% yield) as a colorless oil. $[\alpha]_D^{25}$ = +2.78 (c 1.29, $CHCl_3$). Lit. [24] $[\alpha]_D$ = +3.00 (c 2.00, $CHCl_3$). ^{1}H NMR (500 MHz, $CDCl_3$) δ 3.65–3.60 (m, 1H), 2.39–2.34 (m, 1H), 2.28–2.23 (m, 1H), 2.04 (br s, 1H), 1.98 (t, J = 2.7 Hz, 1H), 1.57–1.45 (m, 2H), 0.89 (t, J = 7.5 Hz, 3H). ^{13}C NMR (125 MHz, $CDCl_3$) δ 81.0, 71.3, 70.8, 29.2, 27.0, 10.0. HRMS (ESI, m/z): calculated for $[M + Na]^+$ $C_6H_{10}ONa$ 121.0624, found: 121.0615.

3.3. Synthesis of (R)-5-hexyn-3-ol ((R)-**4**)

Using a similar procedure for chiral alcohol (S)-**4**, the addition of lithium acetylide ethylenediamine complex **3** (15.96 g, 173.38 mmol, 2.5 equiv.) to (R)-2-ethyloxirane ((R)-**2**) (5.00 g, 69.35 mmol, >99% ee, 1.0 equiv.) yielded (R)-5-Hexyn-3-ol ((R)-**4**) (5.98 g, 88% yield) as a colorless oil. $[\alpha]_D^{25}$ = −4.45 (c 0.99, $CHCl_3$). Lit. [24] $[\alpha]_D$ = −4.00 (c 2.00, $CHCl_3$). ^{1}H NMR (500 MHz, $CDCl_3$) δ 3.65–3.60 (m, 1H), 2.39–2.34 (m, 1H), 2.28–2.23 (m, 1H), 1.99–1.98 (m, 1H), 1.90–1.88 (m, 1H), 1.55–1.47 (m, 2H), 0.90 (t, J = 7.5 Hz, 3H). ^{13}C NMR (125 MHz, $CDCl_3$) δ 81.0, 71.3, 70.8, 29.2, 27.0, 10.0. HRMS (ESI, m/z): calculated for $[M + Na]^+$ $C_6H_{10}ONa$ 121.0624, found: 121.0612.

3.4. Synthesis of (S)-tert-butyl(hex-5-yn-3-yloxy)diphenylsilane ((S)-**5**)

Under an argon atmosphere, (S)-5-hexyn-3-ol ((S)-**4**) (8.54 g, 87.00 mmol, 1.0 equiv.) was added slowly to a solution of DMAP (2.13 g, 17.40 mmol, 0.2 equiv.) and imidazole (14.81 g, 217.50 mmol, 2.5 equiv.) in DCM (200 mL) at 0 °C. Subsequently, TBDPSCl (29.12 g, 104.4 mmol, 1.2 equiv.) was added. The mixture was allowed to warm to room temperature and stirred continuously overnight. The reaction was then quenched by adding water (50 mL), followed by a separation of the lower organic phase. The higher aqueous phase was extracted with DCM (3 × 30 mL). The combined organic layers were washed with brine (30 mL), dried over anhydrous Na_2SO_4 and concentrated under reduced pressure to yield the crude product. The crude product was purified by silica gel column chromatography using an eluent of ethyl acetate and petroleum ether (1:20) to produce (S)-tert-butyl(hex-5-yn-3-yloxy)diphenylsilane ((S)-**5**) (27.8 g, 95% yield) as a light yellow oil. $[\alpha]_D^{25}$ = −26.58 (c

4.91, $CHCl_3$). ^{1}H NMR (500 MHz, $CDCl_3$) δ 7.61–7.58 (m, 4H), 7.30–7.23 (m, 6H), 3.74–3.70 (m, 1H), 2.24–2.14 (m, 2H), 1.76 (t, J = 2.7 Hz, 1H), 1.60–1.47 (m, 2H), 0.97 (s, 9H), 0.74 (t, J = 7.5 Hz, 3H). ^{13}C NMR (125 MHz, $CDCl_3$) δ 136.03, 136.00, 134.4, 134.2, 129.8, 127.76, 127.69, 81.4, 72.6, 70.1, 28.7, 27.1, 26.0, 19.5, 9.1. HRMS (ESI, m/z): calculated for $[M + H]^+$ $C_{22}H_{29}OSi$ 337.1982, found: 339.1954.

3.5. Synthesis of (R)-tert-butyl(hex-5-yn-3-yloxy)diphenylsilane ((R)-**5**)

Using a similar procedure for chiral TBDPS ether (*S*)-**5**, the protection of (*R*)-5-hexyn-3-ol ((*R*)-**4**) (8.54 g, 87.00 mmol, 1.0 equiv.) with TBDPSCl (29.12 g, 104.40 mmol, 1.2 equiv.) yielded (*R*)-tert-butyl(hex-5-yn-3-yloxy)diphenylsilane ((*R*)-**5**) (26.9 g, 92% yield) as a light yellow oil. $[\alpha]_D^{25}$ = +25.77 (c 10.58, $CHCl_3$). ^{1}H NMR (500 MHz, $CDCl_3$) δ 7.62–7.58 (m, 4H), 7.33–7.25 (m, 6H), 3.75–3.71 (m, 1H), 2.24–2.15 (m, 2H), 1.79 (t, J = 2.7 Hz, 1H), 1.58–1.50 (m, 2H), 0.98 (s, 9H), 0.76 (t, J = 7.5 Hz, 3H). ^{13}C NMR (125 MHz, $CDCl_3$) δ 136.04, 136.01, 134.4, 134.2, 129.8, 127.69, 127.66, 127.64, 81.5, 72.6, 70.0, 28.7, 27.1, 26.0, 19.5, 9.1. HRMS (ESI, m/z): calculated for $[M + H]^+$ $C_{22}H_{29}OSi$ 337.1982, found: 339.1956.

3.6. Synthesis of (2S,7S)-7-((tert-butyldiphenylsilyl)oxy)non-4-yn-2-ol ((2S,7S)-**7**)

Under an argon atmosphere, *n*-BuLi (2.69 mL, 2.4 M in *n*-hexane, 6.45 mmol, 2.15 equiv.) was added dropwise to a stirred solution of (*S*)-tert-butyl(hex-5-yn-3-yloxy)diphenylsilane ((*S*)-**5**) (2.54 g, 7.50 mmol, 2.5 equiv.) in THF (15 mL) at −78 °C. The mixture was stirred for 1.5 h at the same temperature, and a solution of (*S*)-2-methyloxirane ((*S*)-**6**) (>99% ee, 174.3 mg, 3.00 mmol, 1.0 equiv.) in THF (1 mL) and boron trifluoride-diethyl etherate (852.0 mg, 6.00 mmol, 2.0 equiv.), which were pre-chilled at −78 °C, was then added. The reaction mixture was allowed to warm to room temperature and stirred continuously for 12 h. The reaction was quenched by adding saturated aqueous $NaHCO_3$ solution (5 mL), followed by separation of the organic phase. The aqueous phase was extracted with EtOAc (3 × 10 mL). The combined organic layers were washed with brine (10 mL), dried over anhydrous Na_2SO_4 and concentrated under reduced pressure to yield the crude product. The crude product was purified by silica gel column chromatography using an eluent of ethyl acetate and petroleum ether (1:5) to produce (2*S*,7*S*)-7-((tert-butyldiphenylsilyl)oxy)non-4-yn-2-ol ((2*S*,7*S*)-**7**) (1.15 g, 97% yield) as a colorless oil. $[\alpha]_D^{25}$ = −28.17 (c 1.53, $CHCl_3$). ^{1}H NMR (500 MHz, $CDCl_3$) δ 7.63–7.60 (m, 4H), 7.35–7.28 (m, 6H), 3.80–3.74 (m, 1H), 3.72–3.67 (m, 1H), 2.26–2.12 (m, 4H), 1.69 (br s, 1H), 1.57–1.49 (m, 2H), 1.13 (d, J = 6.2 Hz, 3H), 0.99 (s, 9H), 0.77 (t, J = 7.5 Hz, 3H). ^{13}C NMR (125 MHz, $CDCl_3$) δ 136.0, 134.5, 134.3, 129.73, 129.69, 127.7, 127.6, 80.2, 78.0, 73.0, 66.6, 29.6, 29.0, 27.1, 26.3, 22.3, 19.5, 9.2. HRMS (ESI, m/z): calculated for $[M + H]^+$ $C_{25}H_{35}O_2Si$ 395.2401, found: 395.2396.

3.7. Synthesis of (2R,7S)-7-((tert-butyldiphenylsilyl)oxy)non-4-yn-2-ol ((2R,7S)-**7**)

Using a similar procedure for chiral TBDPS ether alcohol (2*S*,7*S*)-**7**, the ring opening of (*R*)-2-methyloxirane (*R*)-**6** (>99% ee, 174.3 mg, 3.00 mmol, 1.0 equiv.) with (*S*)-tert-butyl(hex-5-yn-3-yloxy)diphenylsilane (*S*)-**5** (2.54 g, 7.50 mmol, 2.5 equiv.) and *n*-BuLi (2.69 mL, 2.4 M in *n*-hexane, 6.45 mmol, 2.15 equiv.) yielded (2*R*,7*S*)-7-((tert-butyldiphenylsilyl)oxy)non-4-yn-2-ol ((2*R*,7*S*)-**7**) (1.15 g, 97% yield) as a colorless oil. $[\alpha]_D^{25}$ = −30.66 (c 4.84, $CHCl_3$). ^{1}H NMR (500 MHz, $CDCl_3$) δ 7.70–7.67 (m, 4H), 7.43–7.35 (m, 6H), 3.87–3.81 (m, 1H), 3.80–3.75 (m, 1H), 2.33–2.20 (m, 4H), 1.99 (br s, 1H), 1.65–1.56 (m, 2H), 1.20 (d, J = 6.1 Hz, 3H), 1.06 (s, 9H), 0.84 (t, J = 7.5 Hz, 3H). ^{13}C NMR (125 MHz, $CDCl_3$) δ 136.0, 134.5, 134.2, 129.72, 129.68, 127.7, 127.6, 80.2, 78.0, 73.0, 66.6, 29.6, 28.9, 27.1, 26.3, 22.3, 19.5, 9.2. HRMS (ESI, m/z): calculated for $[M + H]^+$ $C_{25}H_{35}O_2Si$ 395.2401, found: 395.2401.

3.8. Synthesis of (2S,7R)-7-((tert-butyldiphenylsilyl)oxy)non-4-yn-2-ol ((2S,7R)-**7**)

Using a similar procedure for chiral TBDPS ether alcohol (2*S*,7*S*)-**7**, the ring opening of (*S*)-2-methyloxirane ((*S*)-**6**) (>99% ee, 174.3 mg, 3.00 mmol, 1.0 equiv.) with (*R*)-

tert-butyl(hex-5-yn-3-yloxy)diphenylsilane ((*R*)-**5**) (2.54 g, 7.50 mmol, 2.5 equiv.) and *n*-BuLi (2.69 mL, 2.4 M in *n*-hexane, 6.45 mmol, 2.15 equiv.) yielded (2*S*,7*R*)-7-((tert-butyldiphenylsilyl)oxy)non-4-yn-2-ol ((2*S*,7*R*)-**7**) (1.17 g, 99% yield) as a light yellow oil. $[\alpha]_D^{25}$ = +32.25 (c 3.28, $CHCl_3$). ^{1}H NMR (500 MHz, $CDCl_3$) δ 7.70–7.67 (m, 4H), 7.43–7.35 (m, 6H), 3.85–3.82 (m, 1H), 3.79–3.75 (m, 1H), 2.33–2.19 (m, 4H), 1.98 (br s, 1H), 1.64–1.57 (m, 2H), 1.20 (d, *J* = 6.1 Hz, 3H), 1.06 (s, 9H), 0.84 (t, *J* = 7.5 Hz, 3H). ^{13}C NMR (125 MHz, $CDCl_3$) δ 136.0, 134.5, 134.2, 129.72, 129.69, 127.7, 127.6, 80.2, 78.0, 73.0, 66.6, 29.6, 28.9, 27.1, 26.3, 22.3, 19.5, 9.2. HRMS (ESI, *m/z*): calculated for $[M + H]^+$ $C_{25}H_{35}O_2Si$ 395.2401, found: 395.2401.

3.9. Synthesis of (2R,7R)-7-((tert-butyldiphenylsilyl)oxy)non-4-yn-2-ol ((2R,7R)-**7**)

Using a similar procedure for chiral TBDPS ether alcohol (2*S*,7*S*)-**7**, the ring opening of (*R*)-2-methyloxirane ((*R*)-**6**) (>99% ee, 174.3 mg, 3.00 mmol, 1.0 equiv.) with (*R*)-tert-butyl(hex-5-yn-3-yloxy)diphenylsilane ((*R*)-**5**) (2.54 g, 7.50 mmol, 2.5 equiv.) and *n*-BuLi (2.69 mL, 2.4 M in *n*-hexane, 6.45 mmol, 2.15 equiv.) yielded (2*R*,7*R*)-7-((tert-butyldiphenylsilyl)oxy)non-4-yn-2-ol ((2*R*,7*R*)-**7**) (1.12 g, 95% yield) as a light yellow oil. $[\alpha]_D^{25}$ = +22.59 (c 3.40, $CHCl_3$). ^{1}H NMR (500 MHz, $CDCl_3$) δ 7.70–7.67 (m, 4H), 7.43–7.35 (m, 6H), 3.87–3.82 (m, 1H), 3.79–3.75 (m, 1H), 2.33–2.19 (m, 4H), 1.98 (br s, 1H), 1.65–1.56 (m, 2H), 1.20 (d, *J* = 6.2 Hz, 3H), 1.06 (s, 9H), 0.84 (t, *J* = 7.5 Hz, 3H). ^{13}C NMR (125 MHz, $CDCl_3$) δ 136.0, 134.5, 134.2, 129.73, 129.69, 127.7, 127.6, 80.1, 78.0, 73.0, 66.6, 29.6, 28.9, 27.1, 26.3, 22.3, 19.5, 9.2. HRMS (ESI, *m/z*): calculated for $[M + H]^+$ $C_{25}H_{35}O_2Si$ 395.2401, found: 395.2385.

3.10. Synthesis of (2S,7S)-non-4-yne-2,7-diol ((2S,7S)-**8**)

Under an argon atmosphere, TBAF (4 mL, 1 M in THF, 4.00 mmol, 2.0 equiv.) was added slowly to a stirred solution of (2*S*,7*S*)-7-((tert-butyldiphenylsilyl)oxy)non-4-yn-2-ol ((2*S*,7*S*)-**7**) (790.0 mg, 2.00 mmol, 1.0 equiv.) in THF (5 mL) at 0 °C. The reaction mixture was allowed to warm to room temperature and stirred continuously overnight. The reaction was then quenched by adding water (5 mL), followed by separation of the organic phase. The aqueous phase was extracted with EtOAc (3 × 8 mL). The combined organic layers were washed with brine (10 mL), dried over anhydrous Na_2SO_4 and concentrated under reduced pressure to yield the crude product. The crude product was purified by silica gel column chromatography using an eluent of ethyl acetate and petroleum ether (1:2) to produce (2*S*,7*S*)-non-4-yne-2,7-diol ((2*S*,7*S*)-**8**) (290.0 mg, 93% yield) as a colorless oil. $[\alpha]_D^{25}$ = +14.47 (c 2.24, $CHCl_3$). ^{1}H NMR (500 MHz, $CDCl_3$) δ 3.89–3.84 (m, 1H), 3.61–3.56 (m, 1H), 2.44 (br s, 2H), 2.38–2.30 (m, 2H), 2.26–2.21 (m, 2H), 1.53–1.45 (m, 2H), 1.18 (d, *J* = 6.3 Hz, 3H), 0.89 (t, *J* = 7.5 Hz, 3H). ^{13}C NMR (125 MHz, $CDCl_3$) δ 79.3, 79.2, 71.7, 66.6, 29.4, 29.3, 27.2, 22.4, 10.1. HRMS (ESI, *m/z*): calculated for $[M + H]^+$ $C_9H_{17}O_2$ 157.1223, found: 157.1225.

3.11. Synthesis of (2R,7S)-non-4-yne-2,7-diol ((2R,7S)-**8**)

Using a similar procedure for chiral alkynyl diol (2*S*,7*S*)-**8**, deprotection of (2*R*,7*S*)-7-((tert-butyldiphenylsilyl)oxy)non-4-yn-2-ol ((2*R*,7*S*)-**7**) (790.0 mg, 2.00 mmol, 1.0 equiv.) with TBAF (4 mL, 1 M in THF, 4.00 mmol, 2.0 equiv.) yielded (2*R*,7*S*)-non-4-yne-2,7-diol ((2*R*,7*S*)-**8**) (284 mg, 91% yield) as a colorless oil. $[\alpha]_D^{25}$ = −3.83 (c 1.25, $CHCl_3$). ^{1}H NMR (500 MHz, $CDCl_3$) δ 3.96–3.90 (m, 1H), 3.68–3.63 (m, 1H), 2.46–2.28 (m, 6H), 1.59–1.53 (m, 2H), 1.25 (d, *J* = 6.2 Hz, 3H), 0.96 (t, *J* = 7.5 Hz, 3H). ^{13}C NMR (125 MHz, $CDCl_3$) δ 79.3, 79.2, 71.7, 66.6, 29.4, 29.3, 27.3, 22.4, 10.1. HRMS (ESI, *m/z*): calculated for $[M + H]^+$ $C_9H_{17}O_2$ 157.1223, found: 157.1220.

*3.12. Synthesis of (2S,7R)-non-4-yne-2,7-diol ((2S,7R)-**8**)*

Using a similar procedure for chiral alkynyl diol (2*S*,7*S*)-**8**, deprotection of (2*S*,7*R*)-7-((tert-butyldiphenylsilyl)oxy)non-4-yn-2-ol ((2*S*,7*R*)-**7**) (790.0 mg, 2.00 mmol, 1.0 equiv.) with TBAF (4 mL, 1 M in THF, 4.00 mmol, 2.0 equiv.) yielded (2*S*,7*R*)-non-4-yne-2,7-diol ((2*S*,7*R*)-**8**) (264 mg, 86% yield) as a colorless oil. $[\alpha]_D^{25}$ = +3.18 (c 2.14, $CHCl_3$). ^1H NMR (500 MHz, $CDCl_3$) δ 3.96–3.90 (m, 1H), 3.67–3.64 (m, 1H), 2.62–2.28 (m, 6H), 1.59–1.54 (m, 2H), 1.25 (d, *J* = 6.2 Hz, 3H), 0.96 (t, *J* = 7.5 Hz, 3H). ^{13}C NMR (125 MHz, $CDCl_3$) δ 79.3, 79.2, 71.7, 66.6, 29.4, 29.2, 27.2, 22.4, 10.0. HRMS (ESI, *m*/*z*): calculated for $[M + H]^+$ $C_9H_{17}O_2$ 157.1223, found: 157.1221.

*3.13. Synthesis of (2R,7R)-non-4-yne-2,7-diol ((2R,7R)-**8**)*

Using a similar procedure for chiral alkynyl diol (2*S*,7*S*)-**8**, deprotection of (2*R*,7*R*)-7-((tert-butyldiphenylsilyl)oxy)non-4-yn-2-ol ((2*R*,7*R*)-**7**) (790.0 mg, 2.00 mmol, 1.0 equiv.) with TBAF (4 mL, 1 M in THF, 4.00 mmol, 2.0 equiv.) yielded (2*R*,7*R*)-non-4-yne-2,7-diol ((2*R*,7*R*)-**8**) (293 mg, 94% yield) as a colorless oil. $[\alpha]_D^{25}$ = −20.82 (c 1.13, $CHCl_3$). ^1H NMR (500 MHz, $CDCl_3$) δ 3.97–3.90 (m, 1H), 3.68–3.63 (m, 1H), 2.46–2.27 (m, 6H), 1.59–1.53 (m, 2H), 1.25 (d, *J* = 6.3 Hz, 3H), 0.96 (t, *J* = 7.5 Hz, 3H). ^{13}C NMR (125 MHz, $CDCl_3$) δ 79.4, 79.2, 71.7, 66.6, 29.4, 29.3, 27.3, 22.5, 10.1. HRMS (ESI, *m*/*z*): calculated for $[M + H]^+$ $C_9H_{17}O_2$ 157.1223, found: 157.1220.

*3.14. Synthesis of (2S,7S)-nonane-2,7-diol ((2S,7S)-**9**)*

Under a hydrogen atmosphere, (2*S*,7*S*)-non-4-yne-2,7-diol ((2*S*,7*S*)-**8**) (180.0 mg, 1.15 mmol, 1.0 equiv.) was add to a stirred solution of Pd/C (18.0 mg, 10%) in MeOH (5 mL) at room temperature. The reaction mixture was stirred continuously overnight under a hydrogen atmosphere, followed by being filtered through diatomaceous earth. The filtrate was concentrated under reduced pressure to yield the crude product. The crude product was purified by silica gel column chromatography using an eluent of ethyl acetate and petroleum ether (1:2) to produce (2*S*,7*S*)-nonane-2,7-diol ((2*S*,7*S*)-**9**) (162.0 mg, 88% yield) as a colorless oil. $[\alpha]_D^{25}$ = +14.77 (c 0.87, $CHCl_3$). ^1H NMR (500 MHz, $CDCl_3$) δ 3.77–3.71 (m, 1H), 3.49–3.44 (m, 1H), 1.58 (br s, 2H), 1.47–1.30 (m, 10H), 1.12 (d, *J* = 6.2 Hz, 3H), 0.87 (t, *J* = 7.5 Hz, 3H). ^{13}C NMR (125 MHz, $CDCl_3$) δ 73.3, 68.1, 39.4, 37.0, 30.3, 25.9, 25.7, 23.7, 10.0. HRMS (ESI, *m*/*z*): calculated for $[M + H]^+$ $C_9H_{21}O_2$ 161.1536, found: 161.1538.

*3.15. Synthesis of (2R,7S)-nonane-2,7-diol ((2R,7S)-**9**)*

Using a similar procedure for chiral diol (2*S*,7*S*)-**9**, the hydrogenation of (2*R*,7*S*)-non-4-yne-2,7-diol ((2*R*,7*S*)-**8**) (154.0 mg, 0.99 mmol, 1.0 equiv.) catalyzed with Pd/C (15.0 mg, 10%) yielded (2*R*,7*S*)-nonane-2,7-diol ((2*R*,7*S*)-**9**) (145.0 mg, 92% yield) as a colorless oil. $[\alpha]_D^{25}$ = +0.56 (c 2.15, $CHCl_3$). ^1H NMR (500 MHz, $CDCl_3$) δ 3.83–3.77 (m, 1H), 3.55–3.51 (m, 1H), 1.71 (br s, 2H), 1.53–1.40 (m, 8H), 1.37–1.30 (m, 2H), 1.19 (d, *J* = 6.2 Hz, 3H), 0.94 (t, *J* = 7.5 Hz, 3H). ^{13}C NMR (125 MHz, $CDCl_3$) δ 73.3, 68.1, 39.4, 36.9, 30.3, 25.9, 25.8, 23.6, 10.0. HRMS (ESI, *m*/*z*): calculated for $[M + H]^+$ $C_9H_{21}O_2$ 161.1536, found: 161.1538.

*3.16. Synthesis of (2S,7R)-nonane-2,7-diol ((2S,7R)-**9**)*

Using a similar procedure for chiral diol (2*S*,7*S*)-**9**, the hydrogenation of (2*S*,7*R*)-non-4-yne-2,7-diol ((2*S*,7*R*)-**8**) (180.0 mg, 1.15 mmol, 1.0 equiv.) catalyzed with Pd/C (18.0 mg, 10%) yielded (2*S*,7*R*)-nonane-2,7-diol ((2*S*,7*R*)-**9**) (170.0 mg, 92% yield) as a colorless oil. $[\alpha]_D^{25}$ = −1.24 (c 1.93, $CHCl_3$). ^1H NMR (500 MHz, $CDCl_3$) δ 3.75–3.69 (m, 1H), 3.46–3.43 (m, 1H), 1.85 (br s, 2H), 1.46–1.33 (m, 8H), 1.31–1.24 (m, 2H), 1.11 (d, *J* = 6.2 Hz, 3H), 0.87 (t, *J* = 7.5 Hz, 3H). ^{13}C NMR (125 MHz, $CDCl_3$) δ 73.3, 68.1, 39.3, 36.9, 30.3, 25.9, 25.8, 23.6, 10.0. HRMS (ESI, *m*/*z*): calculated for $[M + H]^+$ $C_9H_{21}O_2$ 161.1536, found: 161.1539.

3.17. Synthesis of (2*R*,7*R*)-nonane-2,7-diol ((2*R*,7*R*)-**9**)

Using a similar procedure for chiral diol (2*S*,7*S*)-**9**, the hydrogenation of (2*R*,7*R*)-non-4-yne-2,7-diol (2*R*,7*R*)-**8** (293.0 mg, 1.87 mmol, 1.0 equiv.) catalyzed with Pd/C (30.0 mg, 10%) yielded (2*R*,7*R*)-nonane-2,7-diol ((2*R*,7*R*)-**9**) (270.0 mg, 90% yield) as a colorless oil. $[\alpha]_D^{25} = -17.76$ (c 1.25, $CHCl_3$). ^1H NMR (500 MHz, $CDCl_3$) δ 3.76–3.71 (m, 1H), 3.48–3.43 (m, 1H), 1.65 (br s, 2H), 1.47–1.27 (m, 10H), 1.12 (d, *J* = 6.2 Hz, 3H), 0.87 (t, *J* = 7.5 Hz, 3H). ^{13}C NMR (125 MHz, $CDCl_3$) δ 73.3, 68.1, 39.3, 36.9, 30.3, 25.8, 25.7, 23.6, 10.0. HRMS (ESI, *m*/*z*): calculated for $[M + H]^+$ $C_9H_{21}O_2$ 161.1536, found: 161.1538.

3.18. Synthesis of (2*S*,7*S*)-nonane-2,7-diyl dibutyrate ((2*S*,7*S*)-**1**)

Under an argon atmosphere, butyric anhydride (**10**) (95.0 mg, 0.60 mmol, 3.0 equiv.) was added slowly to a stirred solution of (2*S*,7*S*)-nonane-2,7-diol ((2*S*,7*S*)-**9**) (32.0 mg, 0.20 mmol, 1.0 equiv.) in anhydrous pyridine (5 mL) at 0 °C. The reaction solution was allowed to warm to room temperature and stirred continuously overnight. The mixture was concentrated under reduced pressure to yield the crude product. The crude product was purified by silica gel column chromatography using an eluent of ethyl acetate and petroleum ether (1:50) to produce (2*S*,7*S*)-nonane-2,7-diyl dibutyrate ((2*S*,7*S*)-**1**) (56.0 mg, 94% yield) as a colorless oil. $[\alpha]_D^{25} = -2.41$ (c 1.16, $CHCl_3$). Lit. [20] $[\alpha]_D^{20} = -6.7$ (c 0.6, $CHCl_3$). ^1H NMR (500 MHz, $CDCl_3$) δ 4.92–4.86 (m, 1H), 4.84–4.79 (m, 1H), 2.26 (dt, *J* = 11.3, 7.5 Hz, 4H), 1.69–1.61 (m, 4H), 1.57–1.44 (m, 6H), 1.33–1.26 (m, 4H), 1.19 (d, *J* = 6.3 Hz, 3H), 0.97–0.87 (m, 6H), 0.87 (t, *J* = 7.5 Hz, 3H). ^{13}C NMR (125 MHz, $CDCl_3$) δ 173.7, 173.5, 75.2, 70.7, 36.8, 36.7, 36.0, 33.7, 27.1, 25.5, 25.3, 20.1, 18.8, 18.7, 13.83, 13.79, 9.7. HRMS (ESI, *m*/*z*): calculated for $[M + H]^+$ $C_{17}H_{33}O_4$ 301.2373, found: 301.2374.

3.19. Synthesis of (2*R*,7*S*)-nonane-2,7-diyl dibutyrate ((2*R*,7*S*)-**1**)

Using a similar procedure for pheromone (2*S*,7*S*)-**1**, the acylation of (2*R*,7*S*)-nonane-2,7-diol ((2*R*,7*S*)-**9**) (135.0 mg, 0.84 mmol, 1.0 equiv.) with butyric anhydride (**10**) (400.0 mg, 2.53 mmol, 3.0 equiv.) yielded (2*R*,7*S*)-nonane-2,7-diyl dibutyrate ((2*R*,7*S*)-**1**) (237.0 mg, 94% yield) as a colorless oil. $[\alpha]_D^{25} = -3.97$ (c 1.00, $CHCl_3$). ^1H NMR (500 MHz, $CDCl_3$) δ 4.92–4.86 (m, 1H), 4.84–4.79 (m, 1H), 2.26 (dt, *J* = 11.3, 7.4 Hz, 4H), 1.69–1.61 (m, 4H), 1.58–1.46 (m, 6H), 1.35–1.26 (m, 4H), 1.19 (d, *J* = 6.2 Hz, 3H), 0.97–0.93 (m, 6H), 0.87 (t, *J* = 7.5 Hz, 3H). ^{13}C NMR (125 MHz, $CDCl_3$) δ 173.7, 173.5, 75.2, 70.7, 36.8, 36.7, 36.0, 33.7, 27.1, 25.5, 25.3, 20.1, 18.8, 18.7, 13.83, 13.78, 9.7. HRMS (ESI, *m*/*z*): calculated for $[M + H]^+$ $C_{17}H_{33}O_4$ 301.2373, found: 301.2375.

3.20. Synthesis of (2*S*,7*R*)-nonane-2,7-diyl dibutyrate ((2*S*,7*R*)-**1**)

Using a similar procedure for pheromone (2*S*,7*S*)-**1**, the acylation of (2*S*,7*R*)-nonane-2,7-diol ((2*S*,7*R*)-**9**) (127.0 mg, 0.79 mmol, 1.0 equiv.) with butyric anhydride (**10**) (376.0 mg, 2.38 mmol, 3.0 equiv.) yielded (2*S*,7*R*)-nonane-2,7-diyl dibutyrate ((2*S*,7*R*)-**1**) (221.0 mg, 93% yield) as a colorless oil. $[\alpha]_D^{25} = +3.46$ (c 1.97, $CHCl_3$). ^1H NMR (500 MHz, $CDCl_3$) δ 4.92–4.86 (m, 1H), 4.84–4.79 (m, 1H), 2.26 (dt, *J* = 11.3, 7.4 Hz, 4H), 1.69–1.61 (m, 4H), 1.59–1.46 (m, 6H), 1.36–1.24 (m, 4H), 1.19 (d, *J* = 6.3 Hz, 3H), 0.97–0.93 (m, 6H), 0.87 (t, *J* = 7.5 Hz, 3H). ^{13}C NMR (125 MHz, $CDCl_3$) δ 173.7, 173.5, 75.1, 70.7, 36.74, 36.72, 36.0, 33.7, 27.1, 25.5, 25.3, 20.1, 18.8, 18.7, 13.82, 13.77, 9.7. HRMS (ESI, *m*/*z*): calculated for $[M + H]^+$ $C_{17}H_{33}O_4$ 301.2373, found: 301.2374.

3.21. Synthesis of (2*R*,7*R*)-nonane-2,7-diyl dibutyrate ((2*R*,7*R*)-**1**)

Using a similar procedure for pheromone (2*S*,7*S*)-**1**, the acylation of (2*R*,7*R*)-nonane-2,7-diol ((2*R*,7*R*)-**9**) (258.0 mg, 1.61 mmol, 1.0 equiv.) with butyric anhydride (**10**) (764.0 mg, 4.83 mmol, 3.0 equiv.) yielded (2*R*,7*R*)-nonane-2,7-diyl dibutyrate ((2*R*,7*R*)-**1**) (458.0 mg,

95% yield) as a colorless oil. $[\alpha]_D^{25}$ = +2.37 (c 1.52, $CHCl_3$). ^1H NMR (500 MHz, $CDCl_3$) δ 4.92–4.86 (m, 1H), 4.84–4.79 (m, 1H), 2.26 (dt, *J* = 11.3, 7.4 Hz, 4H), 1.69–1.61 (m, 4H), 1.59–1.44 (m, 6H), 1.32–1.28 (m, 4H), 1.19 (d, *J* = 6.3 Hz, 3H), 0.97–0.93 (m, 6H), 0.87 (t, *J* = 7.4 Hz, 3H). ^{13}C NMR (125 MHz, $CDCl_3$) δ 173.7, 173.5, 75.2, 70.7, 36.74, 36.72, 36.0, 33.7, 27.1, 25.5, 25.3, 20.1, 18.8, 18.7, 13.83, 13.78, 9.7. HRMS (ESI, *m*/*z*): calculated for $[M + H]^+$ $C_{17}H_{33}O_4$ 301.2373, found: 301.2366.

4. Conclusions

In summary, we have conducted a new and efficient synthesis of the sex pheromone of *S. mosellana* and its stereoisomers with overall yields of 59–64%. The key reactions were the ring opening of chiral epoxide with an alkynyllithium and the hydrogenation of a triple bond. Compared with the existing synthetic methods, our approach has the advantages of cheap materials, high total yields and being easily scaled. Furthermore, our synthesis would be useful in integrated pest management programs for the orange wheat blossom midge.

Supplementary Materials: The following supporting information can be downloaded at: https://www.mdpi.com/article/10.3390/molecules30030671/s1, Figures S1–S40 and Table S1. The ^1H NMR and ^{13}C NMR spectra for all the synthetic compounds, and comparison of NMR Data between the precious [20] and current synthesized (2*S*,7*S*)-**1**.

Author Contributions: Conceptualization, J.Z.; Methodology, J.W., X.L. and Y.Z.; Writing—review & editing, Q.B. All authors have read and agreed to the published version of the manuscript.

Funding: This research was funded by the National Key Technology Research and Development Program of China (No. 2023YFD1800900).

Institutional Review Board Statement: Not applicable.

Informed Consent Statement: Not applicable.

Data Availability Statement: The data presented in this article are available in the Supplementary Materials.

Conflicts of Interest: The authors declare no conflicts of interest.

References

1. Zhang, G.; Meng, L.; Chen, R.; Wang, W.; Jing, X.; Zhu-Salzman, K.; Cheng, W. Characterization of three glutathione *S*-transferases potentially associated with adaptation of the wheat blossom midge *Sitodiplosis mosellana* to host plant defense. *Pest Manage. Sci.* **2024**, *80*, 885. [CrossRef] [PubMed]
2. Weeraddana, C.D.S.; Wijesundara, R.; Hillier, W.; Swanburg, T.; Hillier, N.K.; Wang, H.V.; Faraone, N.; Wolfe, S.; McCartney, C.; Wist, T.; et al. Volatile organic compounds mediate host selection of wheat midge, *Sitodiplosis mosellana* (Gehin) (Diptera: Cecidomyiidae) between preanthesis and postanthesis stages of wheat. *J. Chem. Ecol.* **2024**, *50*, 237. [CrossRef] [PubMed]
3. Mingeot, D.; Chavalle, S.; Buhl, P.N.; Sonet, G.; Dubois, B.; Hautier, L. Molecular methods for the detection and identification of parasitoids within larval wheat midges. *Sci. Rep.* **2024**, *14*, 27770. [CrossRef] [PubMed]
4. Huang, Q.; Han, X.; Zhang, G.; Zhu-Salzman, K.; Cheng, W. Plant volatiles mediate host selection of *Sitodiplosis mosellana* (Diptera: Cecidomyiidae) among wheat varieties. *J. Agric. Food Chem.* **2022**, *70*, 10466. [CrossRef]
5. Ding, H.; Lamb, R.J. Oviposition and larval establishment of *Sitodiplosis mosellana* (Diptera: Cecidomyiidae) on wheat (Gramineae) at different growth stages. *Can. Entomol.* **1999**, *131*, 475. [CrossRef]
6. Echegaray, E.R.; Barbour, C.R.; Talbert, L.; Stougaard, R.N. Evaluation of *Sitodiplosis mosellana* (Diptera: Cecidomyiidae) infestation and relationship with agronomic traits in selected spring wheat cultivars in northwestern Montana, United States of America. *Can. Entomol.* **2018**, *150*, 675. [CrossRef]
7. Dufton, S.V.; Jorgensen, A.M.; Olfert, O.O.; Otani, J.K. *Sitodiplosis mosellana* (Gehin), orange wheat blossom midge/Cecidomyie du blé (Diptera: Cecidomyiidae). In *Biological Control Programmes in Canada, 2013–2023*; Vankosky, M.A., Martel, V., Eds.; CABI: Wallingford, UK, 2024; Volume 6, p. 359.

8. Shrestha, G.; Reddy, G.V.P. Field efficacy of insect pathogen, botanical, and jasmonic acid for the management of wheat midge *Sitodiplosis mosellana* and the impact on adult parasitoid *Macroglenes penetrans* populations in spring wheat. *Insect Sci.* **2019**, *26*, 523. [CrossRef]
9. El-Wakeil, N.E.; Abdel-Moniem, A.S.H.; Gaafar, N.; Volkmar, C. Effectiveness of some insecticides on wheat blossom midges in winter wheat. *Gesunde Pflanz.* **2013**, *65*, 7. [CrossRef]
10. Dufton, S.V.; Olfert, O.O.; Laird, R.A.; Floate, K.D.; Ge, X.; Otani, J.K. A global review of orange wheat blossom midge, *Sitodiplosis mosellana* (Gehin) (Diptera: Cecidomyiidae), and integrated pest management strategies for its management. *Can. Entomo.* **2022**, *154*, e30. [CrossRef]
11. Chavalle, S.; Jacquemin, G.; De Proft, M. Assessing cultivar resistance to *Sitodiplosis mosellana* (Gehin) (Diptera: Cecidomyiidae) using a phenotyping method under semi-field conditions. *J. Appl. Entomol.* **2017**, *141*, 780. [CrossRef]
12. Kumar, S.; Burt, A.; Green, D.; Humphreys, G.; McCallum, B.; Fetch, T.; Menzies, J.; Aboukhaddour, R.; Henriquez, M.A. AAC darby Canada western red spring wheat. *Can. J. Plant Sci.* **2024**. [CrossRef]
13. Rizvi, S.A.H.; George, J.; Reddy, G.V.P.; Zeng, X.; Guerrero, A. Latest developments in insect sex pheromone research and its application in agricultural pest management. *Insects* **2021**, *12*, 484. [CrossRef] [PubMed]
14. Souza, J.P.A.; Bandeira, P.T.; Bergmann, J.; Zarbin, P.H.G. Recent advances in the synthesis of insect pheromones: An overview from 2013 to 2022. *Nat. Prod. Rep.* **2023**, *40*, 866. [CrossRef]
15. Gries, R.; Gries, G.; Khaskin, G.; King, S.; Olfert, O.; Kaminski, L.-A.; Lamb, R.; Bennett, R. Sex pheromone of orange wheat blossom midge, *Sitodiplosis mosellana*. *Naturwissenschaften* **2000**, *87*, 450. [CrossRef] [PubMed]
16. Chavalle, S.; Censier, F.; Gomez, G.S.M.Y.; De Proft, M. Effect of trap type and height in monitoring the orange wheat blossom midge, *Sitodiplosis mosellana* (Gain) (Diptera: Cecidomyiidae) and its parasitoid, *Macroglenes penetrans* (Kirby) (Hymenoptera: Pteromalidae). *Crop Prot.* **2019**, *116*, 101. [CrossRef]
17. Senevirathna, K.M.; Guelly, K.N.; Mori, B.A. Management of the orange blossom wheat midge, *Sitodiplosis mosellana*, in western Canada. *Plant Health Cases* **2023**, *2023*, phcs20230002. [CrossRef]
18. Li, C.; Wu, Y.; Yin, X.; Gong, Z.; Xing, H.; Miao, J.; Wang, S.; Liu, J.; Na, R.; Li, Q. Modular synthesis of the pheromone (2*S*, 7*S*)-2, 7-nonanediyl dibutyrate and its racemate and their field efficacy to control orange wheat blossom midge, *Sitodiplosis mosellana* (Géhin)(Diptera: Cecidomyiidae). *Pest Manag. Sci.* **2023**, *79*, 97. [CrossRef]
19. Na, R.; Liu, J.; Song, N.; Yin, X.; Lu, S.; Wu, Y.; Tang, Q.; Li, W.; Liu, B.; Wang, C. Synthesis of Red Wheat Blossom Midge Sex Pheromone Precursor Compound and Red Wheat Blossom Midge Sex Pheromone. CN Patent 105085168, 29 March 2017.
20. Hooper, A.M.; Dufour, S.; Willaert, S.; Pouvreau, S.; Pickett, J.A. Synthesis of (2*S*,7*S*)-dibutyroxynonane, the sex pheromone of the orange wheat blossom midge, *Sitodiplosis mosellana* (Gehin) (Diptera: Cecidomyiidae), by diastereoselective silicon-tethered ring-closing metathesis. *Tetrahedron Lett.* **2007**, *48*, 5991. [CrossRef]
21. Bruce, T.J.; Hooper, A.M.; Ireland, L.; Jones, O.T.; Martin, J.L.; Smart, L.E.; Oakley, J.; Wadhams, L.J. Development of a pheromone trap monitoring system for orange wheat blossom midge, *Sitodiplosis mosellana*, in the UK. *Pest Manag. Sci.* **2007**, *63*, 49. [CrossRef]
22. Hernandez-Torres, G.; Mateo, J.; Urbano, A.; Carreno, M.C. The shortest (four-step) total synthesis of the eight-membered cyclic ether racemic- and (-)-cis-lauthisan. *Eur. J. Org. Chem.* **2013**, *2013*, 6259. [CrossRef]
23. Thiraporn, A.; Rukachaisirikul, V.; Iawsipo, P.; Somwang, T.; Tadpetch, K. Total synthesis and cytotoxic activity of 5′-hydroxyzearalenone and 5′β-hydroxyzearalenone. *Eur. J. Org. Chem.* **2017**, *2017*, 7133. [CrossRef]
24. Abad, J.-L.; Camps, F.; Fabrias, G. Substrate-dependent stereochemical course of the (*Z*)-13-desaturation catalyzed by the processionary moth multifunctional desaturase. *J. Am. Chem. Soc.* **2007**, *129*, 15007. [CrossRef]
25. Yan, F.; Moon, S.-J.; Liu, P.; Zhao, Z.; Lipscomb, J.D.; Liu, A.; Liu, H.-w. Determination of the substrate binding mode to the active site iron of (*S*)-2-hydroxypropylphosphonic acid epoxidase using ^{17}O-enriched substrates and substrate analogues. *Biochemistry* **2007**, *46*, 12628. [CrossRef] [PubMed]
26. Bhuniya, R.; Mahapatra, T.; Nanda, S. Klebsiellapneumoniae (NBRC 3319) mediated asymmetric reduction of α-substituted β-oxo esters and its application to the enantioiselective synthesis of small-ring carbocycle derivatives. *Eur. J. Org. Chem.* **2012**, *2012*, 1597. [CrossRef]
27. Pettersson, M.; Johnson, D.S.; Humphrey, J.M.; am Ende, C.W.; Butler, T.W.; Dorff, P.H.; Efremov, I.V.; Evrard, E.; Green, M.E.; Helal, C.J.; et al. Discovery of clinical candidate PF-06648671: A potent γ-secretase modulator for the treatment of Alzheimer's disease. *J. Med. Chem.* **2024**, *67*, 10248. [CrossRef] [PubMed]
28. Bucknam, A.R.; Micalizio, G.C. Asymmetric de novo synthesis of a cucurbitane triterpenoid: Total synthesis of octanorcucurbitacin B. *J. Am. Chem. Soc.* **2022**, *144*, 8493. [CrossRef] [PubMed]
29. Geerdink, D.; Horst, B.T.; Lepore, M.; Mori, L.; Puzo, G.; Hirsch, A.K.H.; Gilleron, M.; de Libero, G.; Minnaard, A.J. Total synthesis, stereochemical elucidation and biological evaluation of Ac2SGL; a 1,3-methyl branched sulfoglycolipid from *Mycobacterium tuberculosis*. *Chem. Sci.* **2013**, *4*, 709. [CrossRef]
30. Madasu, M.; Mohapatra, D.K. Total synthesis of okaspirodiol. *ChemistrySelect* **2023**, *8*, e202300352. [CrossRef]

31. Takahashi, N.; Hayashi, H.; Poznaks, V.; Kakeya, H. Total synthesis of verucopeptin, an inhibitor of hypoxia-inducible factor 1 (HIF-1). *Chem. Commun.* **2019**, *55*, 11956. [CrossRef]
32. Chen, D.; Evans, P.A. A concise, efficient and scalable total synthesis of thapsigargin and nortrilobolide from (*R*)-(-)-carvone. *J. Am. Chem. Soc.* **2017**, *139*, 6046. [CrossRef] [PubMed]
33. Zheng, J.F.; Lan, H.Q.; Yang, R.F.; Peng, Q.L.; Xiao, Z.H.; Tuo, S.C.; Hu, K.Z.; Xiang, Y.G.; Wei, Z.; Zhang, Z.; et al. Asymmetric syntheses of the sex pheromones of pine sawflies, their homologs and stereoisomers. *Helv. Chim. Acta* **2012**, *95*, 1799. [CrossRef]

molecules

MDPI

Article

Squaramide-Catalyzed Asymmetric Michael Addition/Cyclization Reaction for the Synthesis of Chiral Bisspiro Barbituric Acid–Oxindole Derivatives

De-Jun Qiao and Da-Ming Du *

Key Laboratory of Medicinal Molecule Science and Pharmaceutical Engineering, School of Chemistry and Chemical Engineering, Beijing Institute of Technology, Beijing 100081, China; 3220221413@bit.edu.cn
* Correspondence: dudm@bit.edu.cn

Abstract: An efficient stereoselective strategy for the synthesis of chiral bisspiro barbituric acid–oxindole derivatives was developed. The asymmetric Michael addition/cyclization tandem reaction between benzylidene barbituric acids and oxindolylmalonitriles was catalyzed by squaramide catalyst, and the corresponding spirocyclic products were obtained in good-to-high yields (up to 97%) with excellent stereoselectivities (up to >99% ee, >20:1 dr). At the same time, the practicality of the reaction was verified by the gram-scale preparation reaction.

Keywords: organocatalysis; asymmetric catalysis; barbituric acid; oxindole; Michael addition

1. Introduction

Barbituric acid was first discovered and named by German chemist Adolf von Baeyer in 1864 [1]. The methylene group at the C5 position of barbituric acid is highly reactive [$pka_{(DMSO)}$ 8.4] due to the influence of two adjacent electron-withdrawing carbonyl groups, so many chemical reactions take place at this position, such as the common Michael addition reaction, substitution reaction, chelation reaction, and Knoevenagel condensation reaction [2]. The biological activity of a series of barbiturates has attracted the attention of many scientists in the field of medicinal chemistry. Barbituric acid derivatives are widely used as anesthetics [3] and sedatives [4] and have anti-convulsant [5], anti-diabetic [6], anti-bacterial [7], anti-cancer [8], and other properties (Figure 1a). When the two H atoms of methylene at 5-position of barbituric acid are replaced by hydrocarbon groups or heterocycles, they can also be used as drug intermediates, such as barbiturates involved in the treatment of certain types of epilepsy [9]. Meanwhile, alkylidene barbituric acids are good Michael acceptors (Figure 1b), which can be applied for constructing many other barbituric acid derivatives [10]. In addition to being biologically active, the photophysical properties of barbiturate derivatives [11] have also been used for colorimetric or thermal detection [12] and have provided some promising dyes or fluorescent probes [13,14]. These applications indicate that barbiturate derivatives have very broad potential value.

As mentioned above, barbituric acid derivatives are easily deprotonated owing to their rather low pKa value. Catalytic asymmetric transformations of barbituric acid derivatives have received much attention in recent years for synthesis of chiral barbituric acid derivatives [10]. For example, Rawal et al. reported an enantioselective Michael addition of *N*,*N′*-disubstituted barbituric acid derivatives to β-nitro olefins using chiral thiosquaramide as a bifunctional organocatalyst (Scheme 1a) [15]. The addition products were obtained in high yields with excellent enantioselectivity at catalyst loading as low as 0.5 mol% in toluene at room temperature. Wang and co-workers [16] developed an enantioselective organocatalytic Michael addition of *N,N′*-dialkylbarbituric acid derivatives to enones using 10 mol% quinine-derived squaramide catalyst, a series of Michael adducts were obtained in 44–99% yields with 91–99% ee in *o*-xylene at room temperature (Scheme 1b). Chen and co-workers [17] developed a tertiary amine-thiourea-catalyzed domino Michael-oxa-Michael addition reaction of *N,N′*-dimethyl barbituric acid and Morita–Baylis-Hillman (MBH) acetate of nitroalkene; the corresponding tetrahydropyrano bicycles were obtained up to 95% yields with dr > 19:1 and up to 95% ee in CH_2Cl_2 at 25 °C (Scheme 1c).

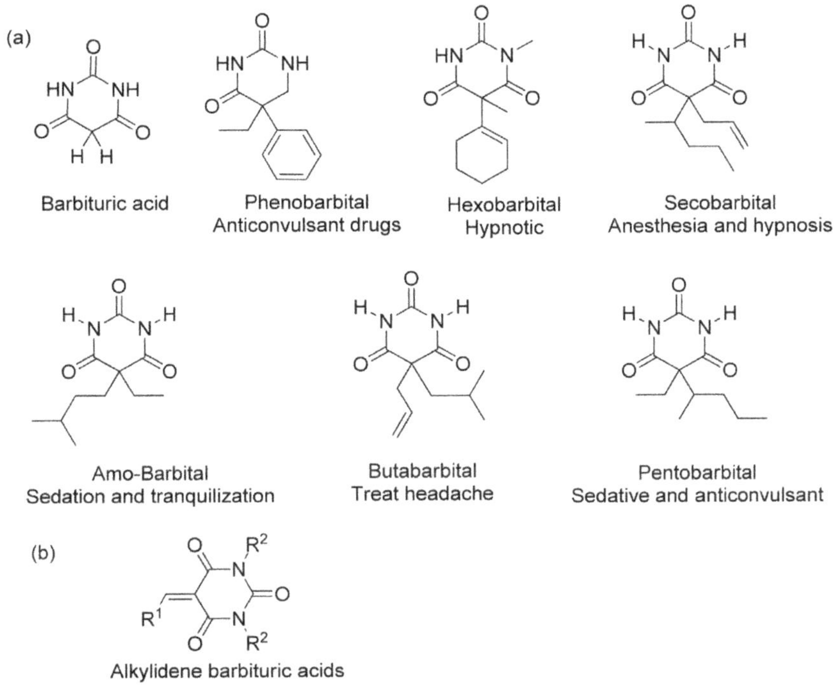

Figure 1. Examples of barbituric acid derivatives. (**a**) barbituric acid derivatives in pharmaceuticals, (**b**) alkylidene barbituric acids.

Alkylidene barbituric acids as reactive electron-poor alkene derivatives also attracted the attention of researchers in recent years; for example, Zhao's group reported on a racemic [3 + 2] cycloaddition between alkylidene barbiturates and 3-isothiocyanato oxindoles catalyzed by Et_3N [18]. In 2016, Zhao et al. developed an *epi*-quinine-thiourea-based thiourea-catalyzed enantioselective version in chloroform, and the corresponding spirobarbiturates were obtained in 80–99% yield with 9:1 to >20:1 dr and 18–99% ee [19]. Guo and co-worker an enantioselective phosphine-catalyzed [3 + 2] annulation of alkylidene barbiturates with MBH adducts, and spirobarbiturates were obtained in excellent diastereoselectivities (4;1–>20:1 dr) and high-to-excellent enantioselectivities (81–99% ee) (Scheme 1e) [20]. Our group developed a squaramide-catalyzed asymmetric Michael/Mannich [3 + 2] cycloaddition reaction of *N*-2,2,2-trifluoroethyl isatin ketimines and barbiturate-based olefins. The corresponding dispirobarbituric acid derivatives were obtained in excellent yields (up to 99% yield) and excellent stereoselectivities (up to 99:1 dr and >99% ee) (Scheme 1f) [21].

In the above-mentioned reports [19,21], the oxindole skeleton also played a crucial role, as these reagents are prone to undergo the tandem reaction with electron-deficient alkenes to construct spirooxindole derivatives. The catalytic asymmetric synthesis of chiral spirooxindoles has also received wide attention in recent years [22]. Continuing on our research project for squaramide-catalytic asymmetric reactions [23], herein, the asymmetric Michael addition/cyclization tandem reaction between barbituric acid derivatives and oxindole derivatives was developed using chiral squaramide as catalyst in order to obtain chiral bisspiro barbituric acid–oxindole derivatives, which may provide potential candidates for future drug design and biological activity research.

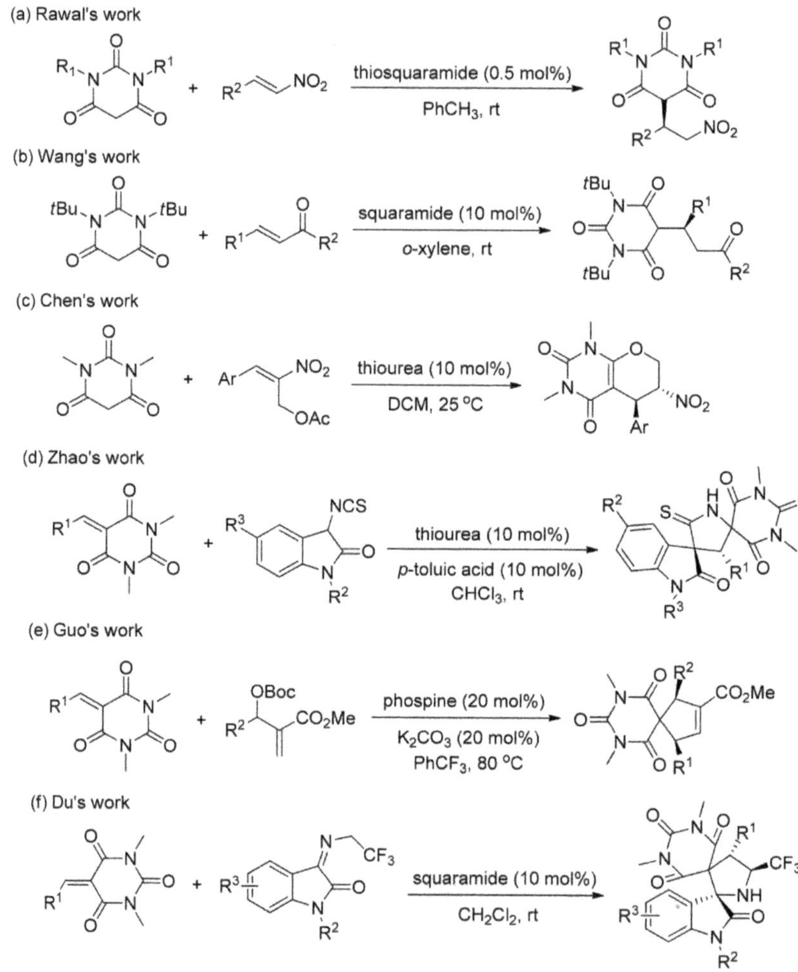

Scheme 1. Examples for asymmetric synthesis of barbituric acid derivatives. (**a**) Rawal's work [15], (**b**) Wang's work [16], (**c**) Chen's work [17], (**d**) Zhao's work [19], (**e**) Guo's work [20], and (**f**) Du's work [21].

2. Results and Discussion

2.1. Optimization of Reaction Conditions

To verify our hypothesis, the asymmetric Michael/cyclization reaction of substrates **1a** and **2a** in the presence of quinine-derived squaramide **C1** was selected as the model reaction. We were pleased to find that the asymmetric Michael/cyclization reaction could be completed within 8 h in the presence of 10 mol% **C1** at room temperature and obtained the desired product **3aa** in 88% yield with excellent stereoselectivity (>20:1 dr, 86% ee) (Table 1, entry 1). Encouraged by these excellent results, we screened several organocatalysts with different frameworks for this asymmetric Michael/cyclization reaction (Figure 2) (Table 1, entries 2–12). However, the catalytic yields using catalysts **C2**, **C3**, and **C9** were low (Figure 2) (Table 1, entries 2–12), and the remaining catalysts achieved high yields (>80%) and stereoselectivity (>20:1 dr, >75% ee) (Table 1, entry 2–12). Considering yield and stereoselectivity, **C5** has the best catalytic effect (Table 1, entry 5).

Table 1. Optimization of reaction conditions [a].

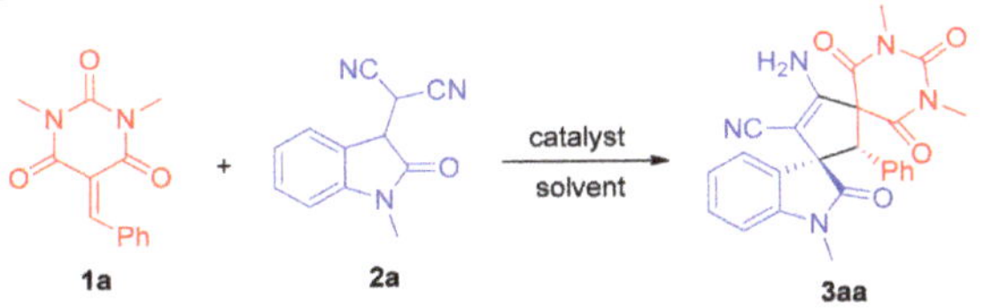

Entry [a]	Solvent	Catalyst	Yield [b] (%)	dr [c]	ee [d] (%)
1	CH_2Cl_2	**C1**	86	>20:1	88
2	CH_2Cl_2	**C2**	77	>20:1	65
3	CH_2Cl_2	**C3**	82	>20:1	50
4	CH_2Cl_2	**C4**	85	>20:1	83
5	CH_2Cl_2	**C5**	94	>20:1	>99
6	CH_2Cl_2	**C6**	83	>20:1	82
7	CH_2Cl_2	**C7**	75	>20:1	87
8	CH_2Cl_2	**C8**	88	>20:1	85
9	CH_2Cl_2	**C9**	92	>20:1	53
10	CH_2Cl_2	**C10**	85	>20:1	81
11	CH_2Cl_2	**C11**	80	>20:1	82
12	CH_2Cl_2	**C12**	77	>20:1	74
13	MeCN	**C5**	68	>20:1	25
14	PhMe	**C5**	72	>20:1	80
15	THF	**C5**	72	>20:1	47
16	$CHCl_3$	**C5**	92	>20:1	97
17	MTBE	**C5**	57	>20:1	5
18 [e]	CH_2Cl_2	**C5**	80	>20:1	88
19 [f]	CH_2Cl_2	**C5**	86	>20:1	94

[a] Unless otherwise specified, reactions were conducted with **1a** (0.24 mmol), **2a** (0.20 mmol), and catalyst (10 mol%) in solvent (1.5 mL) at room temperature under air for 8 h. [b] Isolated yield after column chromatography purification. [c] Determined by ^{1}H NMR analysis. [d] Enantiomeric excess (ee) was determined by HPLC analysis. [e] 5 mol% catalyst was used. [f] The reaction was performed at 0 °C for 24 h.

In order to further improve the reaction efficiency, squaramide **C5** was used as a catalyst to optimize other reaction conditions. The effects of solvent, catalyst loading, and reaction temperature on the reaction were evaluated in detail (Table 1, entries 13−19). Solvents play an integral role in the reaction, so we evaluated five other organic solvents, acetonitrile, toluene, THF, chloroform, and methyl tert-butyl ether (MTBE) (Table 1, entries 13−17). However, a series of results show that dichloromethane is still the best solvent. Then, we studied the effect of catalyst loading on the reaction. As the amount of catalyst was halved, the yield and stereoselectivity of the product unfortunately decreased (Table 1, entry 18). As the reaction temperature decreased, the yield of the product did not increase, and its enantioselectivity gradually decreased (Table 1, entry 19). By comparison, the optimal conditions were determined to be benzylidenebarbituric acid and oxindolylmalonitrile as raw materials in the molar ratio of 1.2:1, with 10 mol% **C5** as catalyst, in CH_2Cl_2 solvent, at room temperature for 8 h.

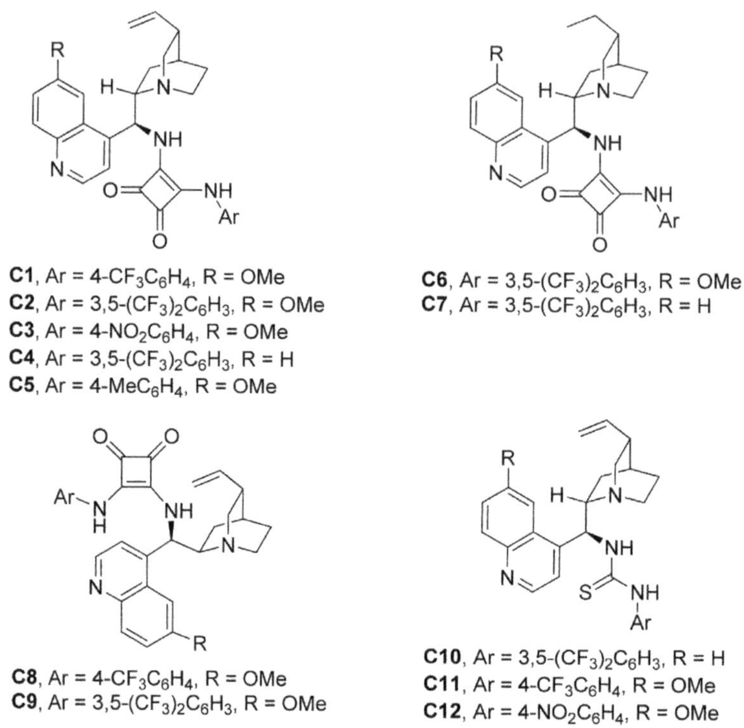

Figure 2. The screened squaramide and thiourea organocatalysts.

2.2. Substrate Scope

After the optimum reaction conditions were obtained, the applicability of different substrates to the asymmetric Michael/cyclization reaction was investigated. The result is shown in Scheme 2. Firstly, the effect of benzylidene barbituric acid substrate on the reaction was studied, and the effect of different substitution groups on the reaction was explored. The results show that steric hindrance had a great effect on the reaction. When benzene rings in benzylidenes were meta-substituted or para-substituted, the reaction occurred normally (**3aa–ja**), but when benzene rings in benzylidenes were ortho-substituted, the corresponding substrates did not react with **2a** (**3la–oa**). The electron effect of substituents also affected the course of the reaction. For electron-withdrawing groups, the substituted substrates bearing bromo- or chloro-substituents showed good yield and stereoselectivity. The substrate with a para-substituted fluoro group was an exception, and the stereoselectivity of the corresponding product was relatively poor, which was considered to be due to the strong electron-withdrawing withdrawing of fluorine. For electron-donating groups, methyl- or methoxy-substituted substrates behaved generally similarly as compared to the bromo-substituted substrate. In addition, the comparison of the yield and enantioselectivity of thienyl-, furanyl-, and pyridinyl-substituted substrates is very interesting. The reaction yield and enantioselectivity of thienyl-substituted substrate were very good (**3da**), the yield and enantioselectivity of furanyl-substituted substrate were moderate (**3fa**), and the reaction of pyridinyl-substituted substrate did not occur (**3na**). This may be ascribed to the electronic effect of these three different heterocycles.

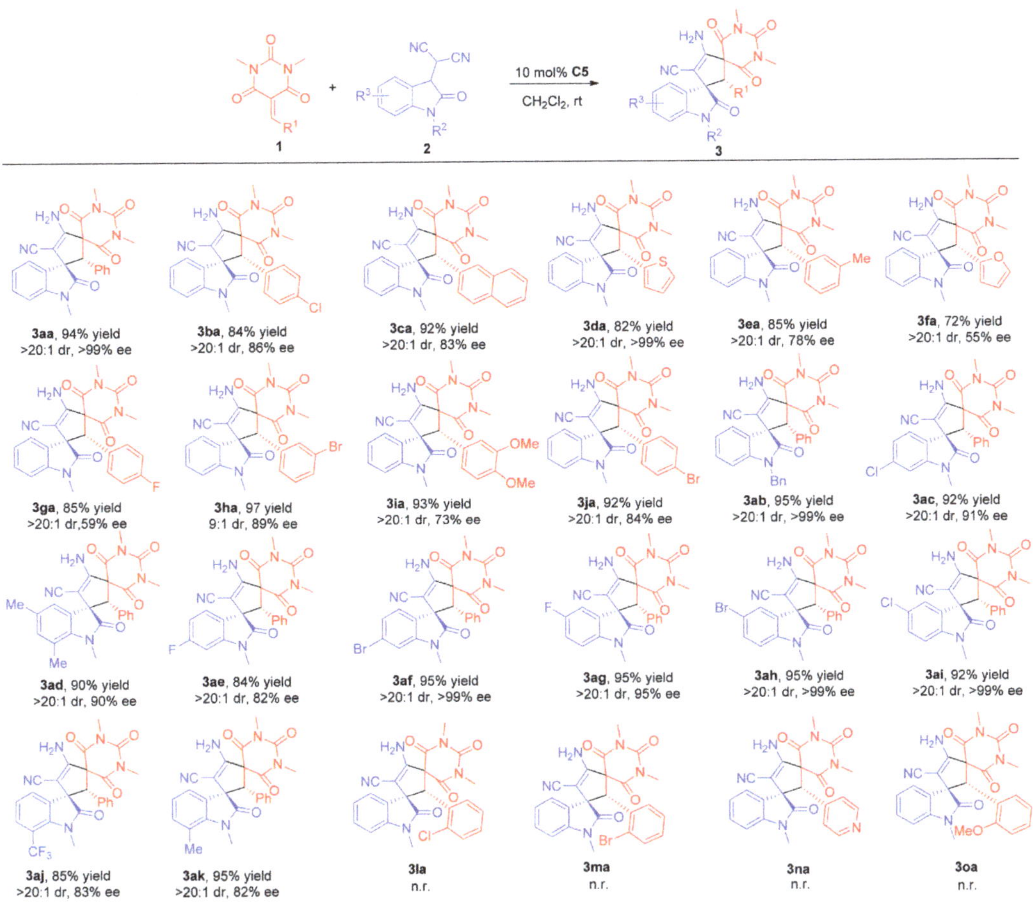

Scheme 2. The substrate scope for benzylidene barbituric acids and oxindoles. Unless otherwise noted, the reaction was performed in CH_2Cl_2 (1.5 mL) with **1** (0.24 mmol), **2** (0.20 mmol), and catalyst **C5** (0.02 mmol) at room temperature under air for 8 h. The dr values were determined by ^{1}H NMR, and the ee values were determined by HPLC.

After studying the effect of benzylidene substituents of benzylidene barbituric acids on the reaction, the effect of the *N*-protecting group of oxindolylmalonitriles on the reaction was studied. When R^1 was a benzyl group or a methyl group, the reaction maintained high selectivity and high efficiency (**3aa** and **3ab**). Subsequently, the different substituents on the phenyl rings of oxindolylmalonitriles were studied. We found that the stereoselectivity of substrates bearing electron-withdrawing substituted indole-phenyl rings was generally better than that of substrates bearing electron-donating substituents. Among the electron-withdrawing groups as substituents (**3ac**, **3ae–aj**), all of them maintained excellent yields and stereoselectivities except for the 2-fluoro-substituent. As for the electron-donating groups, the dimethyl-substituted substrate performed better than the monomethyl-substituted substrate (**3ad**, **3ak**).

2.3. Scaled-Up Synthesis

In order to further demonstrate the application value of this synthetic method, the gram-scale experiment was conducted under optimized conditions. As shown in Scheme 3, the gram-scale asymmetric Michael/cyclization reaction of **1a** and **2a** proceeded smoothly, and the product **3aa** was obtained in 88% yield with excellent stereoselectivity (>20:1 dr, 95% ee).

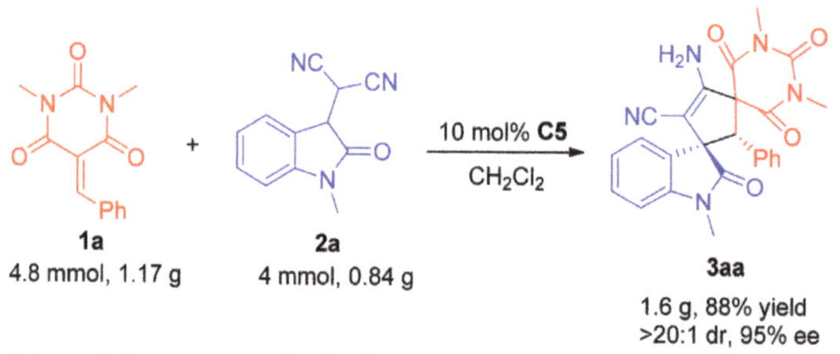

Scheme 3. Gram-scale synthesis of 3aa.

2.4. Absolute Configuration

The absolute configuration of the chiral product **3ha** was determined by X-ray diffraction analysis and was found to be (2′*R*,3*S*) (CCDC 2431390) (Figure 3). The absolute configurations of other chiral products were assigned by analogy.

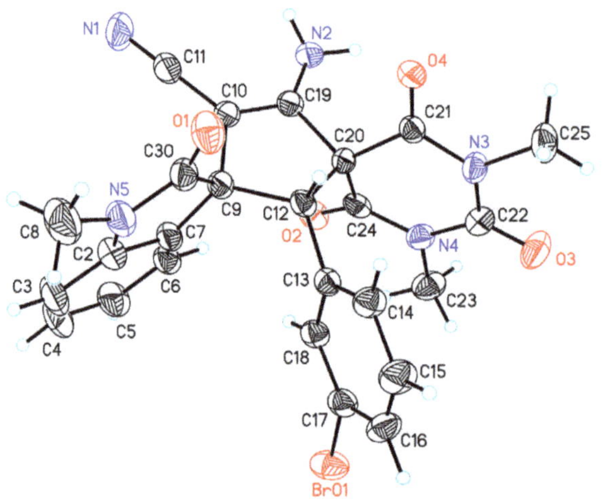

Figure 3. X-ray crystal structure of **3ha** (Displacement ellipsoids are drawn at the 50% probability level; the included solvent molecules were omitted for clarity).

2.5. Reaction Mechanism

According to the absolute configuration of the tandem product **3ha** and the catalytic mode of the chiral bifunctional squaramide for a similar reaction [24], a possible transition state model of the catalytic reaction was proposed (Scheme 4). On the one hand, oxindolylmalonitrile **2a** is partially deprotonated by the tertiary amine of catalyst **C5**. On the other hand, benzylidene barbituric acid 1a is activated by forming two hydrogen bonds in the N-H of the squaramide part. Subsequently, the deprotonated-activated oxindolylmalonitrile attacks the *Si*-face of the electron-deficient unsaturated barbituric acid **1a** through the transition state **A**, and the intermolecular Michael addition reaction occurs. At the same time, the resulting Michael addition intermediates undergo further intramolecular cyclization reaction to obtain **B**. Finally, the molecular isomerization reaction occurs to obtain the desired bisspirocyclic product **3aa**, and the bifunctional squaramide catalyst **C5** is regenerated to enter the next catalytic cycle of reaction.

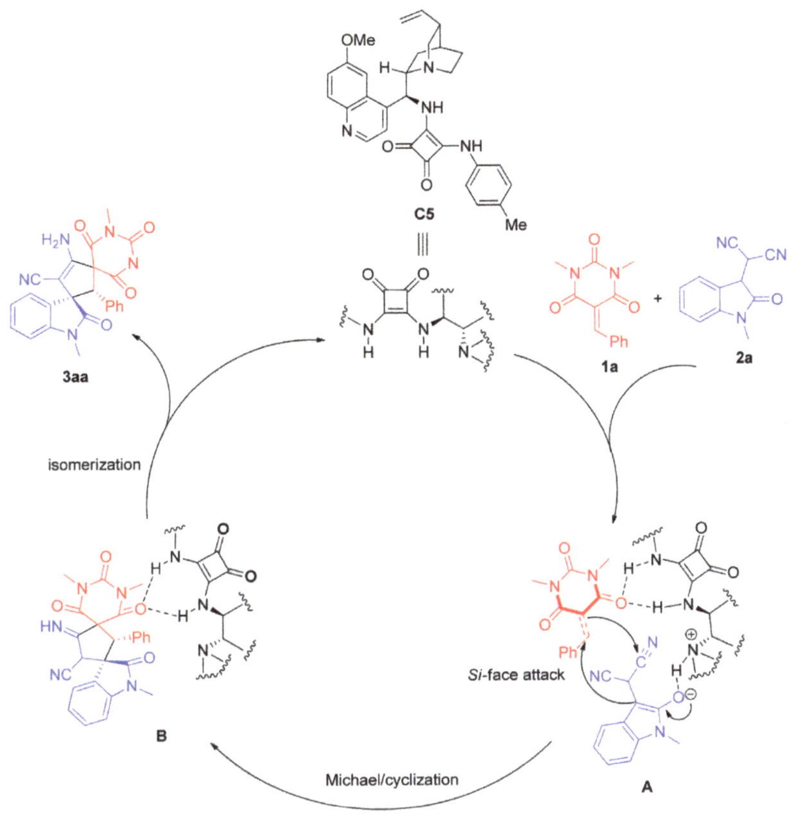

Scheme 4. Proposed reaction mechanism.

3. Materials and Methods

3.1. General Information

Commercially available compounds were used without further purification. Solvents were dried according to standard procedures. Column chromatography was performed with silica gel (200−300 mesh). Melting points were determined with a WRX-4 melting-point apparatus and were uncorrected. ^{1}H NMR spectra were measured with Bruker Ascend 400 MHz spectrometer and Bruker Ascend 700 MHz spectrometer (Bruker, Karlsurhe, Germany); chemical shifts were reported in δ (ppm) units relative to tetramethylsilane (TMS) as internal standard. ^{13}C NMR spectra were measured at 101 MHz with 400 MHz spectrometer and measured at 176 MHz with 700 MHz spectrometer; chemical shifts are reported in ppm relative to tetramethylsilane and referenced to solvent peak ($CDCl_3$, δC = 77.00; DMSO, δC = 39.43). High-resolution mass spectra (Electron spray ionization) were measured with an Agilent 6520 Accurate-Mass Q-TOF MS system (Agilent, Santa Clara, CA, USA) equipped with an electrospray ionization (ESI) source. Optical rotations were measured with a Krüss P8000 polarimeter (Krüss, Hamburg, Germany). Optical rotations at the indicated concentration with the units of g/100 mL. Enantiomeric excesses were determined by chiral HPLC analysis using an Agilent 1200 LC instrument (Agilent, Santa Clara, CA, USA) with a Daicel Chiralpak ADH, IA, or IC column (Daicel, Beijing, China).

3.2. Experimental Materials for Tandem Reactions

First, **1a**–**1j** were prepared according to literature reported by Neumann and co-workers [25]. Then, **2a**–**2k** were prepared according to literature reported by Lin and co-workers [24]. The chiral organocatalysts were prepared by following the reported procedures [26–29].

3.3. Procedure for the Synthesis of Racemates of 3

To a dried small vial, benzylidene barbituric acid **1** (0.24 mmol), oxindolylmalonitrile **2** (0.2 mmol), Et_3N (1.0 mg, 0.01 mmol, 0.05 equiv.), and CH_2Cl_2 (1 mL) were added. After stirring at room temperature under air without gas protection for 8 h, the reaction mixture was concentrated and directly purified by silica gel column chromatography to afford the racemates of **3**.

3.4. Procedure for the Asymmetric Michael/Cyclization Reaction

To a dried small vial, barbituric acid **1** (0.24 mmol), oxindolylmalonitrile **2** (0.2 mmol), chiral organocatalyst **C5** (5.08 mg, 0.01 mmol, 0.05 equiv), and CH_2Cl_2 (1.0 mL) were added. After stirring at room temperature under air without gas protection for 8 h, the reaction mixture was concentrated and directly purified by silica gel column chromatography (200–300 mesh) using ethyl acetate/petroleum ether (1:2) as eluent to afford the desired products **3**.

(2′*R*,3*S*)-4′-Amino-1,1″,3″-trimethyl-2,2″,4″,6″-tetraoxo-2′-phenyl-1″,3″,4″,6″-tetrahydro-2″*H*-dispiro[indoline-3,1′-cyclopentane-3′,5″-pyrimidin]-4′-ene-5′-carbonitrile (**3aa**). According to the general procedure from **1a** (58.6 mg, 0.24 mmol) and **2a** (42.2 mg, 0.2 mmol) to obtain 85.6 mg (94% yield) compound **3aa** as a yellow solid, m.p. 192−195 °C. HPLC (Daicel Chiralpak ADH, *n*-hexane/2-propanol = 70:30, flow rate 1.0 mL/min, detection at 254 nm): t_R = 9.2 min (major), >99% ee. $[\alpha]_D^{25}$ = +9.9 (c = 0.5, CH_2Cl_2). ^{1}H NMR (700 MHz, DMSO-d_6): δ 8.05 (dd, J_1 = 7.7 Hz, J_2 = 0.7 Hz, 1H), 7.28 (td, J_1 = 7.7 Hz, J_2 = 1.4 Hz 1H), 7.15–7.18 (m, 4H), 7.06 (t, J = 8.0 Hz, 2H), 6.86 (d, J = 7.7 Hz, 1H), 6.76 (d, J = 7.0 Hz, 2H), 4.24 (s, 1H), 3.07 (s, 3H), 2.99 (s, 3H), 2.87 (s, 3H) ppm. ^{13}C NMR (176 MHz, DMSO-d_6): δ 176.5, 168.6, 167.8, 158.4, 149.8, 143.0, 131.1, 129.4, 129.3, 129.0, 128.3, 128.0, 127.3, 122.3, 116.3, 108.7, 78.5, 68.1, 64.8, 63.4, 28.5, 28.3, 26.4 ppm. (see Supplementary Materials) HRMS (ESI): m/z calcd. for $C_{25}H_{22}N_5O_4$ $[M + H]^+$ 456.1666, found 456.1686.

(2′*R*,3*S*)-4′-Amino-2′-(4-chlorophenyl)-1,1″,3″-trimethyl-2,2″,4″,6″-tetraoxo-1″,3″,4″,6″-tetrahydro-2″*H*-dispiro[indoline-3,1′-cyclopentane-3′,5″-pyrimidin]-4′-ene-5′-carbonitrile (**3ba**). According to the general procedure from **1b** (66.7 mg, 0.24 mmol) and **2a** (42.2 mg, 0.2 mmol) to obtain 82.2 mg (84% yield) compound **3ba** as a yellow solid, m.p. 183–185 °C. HPLC (Daicel Chiralpak IC, *n*-hexane/2-propanol = 70:30, flow rate 1.0 mL/min, detection at 254 nm): t_R = 6.8 min (minor), t_R = 9.4 min (major), 86% ee. $[\alpha]_D^{25}$ = +12.2 (c = 0.34, CH_2Cl_2). ^{1}H NMR (400 MHz, $CDCl_3$): δ 8.12 (d, J = 7.2 Hz, 1H), 7.29–7.25 (m, 1H), 7.17 (td, J_1 = 7.6 Hz, J_2 = 0.8 Hz, 1H), 7.01 (d, J = 8.4 Hz, 2H), 6.78 (d, J = 8.8 Hz, 2H), 6.67 (d, J = 7.6 Hz, 1H), 5.65 (s, 2H), 4.37 (s, 1H), 3.20 (s, 3H), 3.03 (s, 3H), 2.98 (s, 3H) ppm. ^{13}C NMR (101 MHz, $CDCl_3$): δ 176.8, 168.6, 167.1, 158.0, 149.8, 143.9, 135.6, 131.1, 129.8, 129.3, 128.5, 128.2, 127.3, 123.2, 115.3, 108.7, 83.2, 68.6, 64.6, 63.9, 29.3, 28.9, 26.8 ppm. HRMS (ESI): m/z calcd. for $C_{25}H_{21}ClN_5O_4$ $[M + H]^+$ 490.1277, found 490.1301.

(2′*R*,3*S*)-4′-Amino-1,1″,3″-trimethyl-2′-(naphthalen-2-yl)-2,2″,4″,6″-tetraoxo-1″,3″,4″,6″-tetrahydro-2″*H*-dispiro[indoline-3,1′-cyclopentane-3′,5″-pyrimidin]-4′-ene-5′-carbonitrile (**3ca**). According to the general procedure from **1c** (70.6 mg, 0.24 mmol) and **2a** (42.2 mg, 0.2 mmol) to obtain 93.0 mg (92% yield) compound **3ca** as a yellow solid, m.p. 201–203 °C. HPLC (Daicel Chiralpak IC, *n*-hexane/2-propanol = 70:30, flow rate 1.0 mL/min, detection at 254 nm): t_R = 13.0 min (minor), t_R = 16.7 min (major), 83% ee. $[\alpha]_D^{25}$ = +16.5 (c = 0.5,

CH_2Cl_2). ^{1}H NMR (400 MHz, DMSO-d$_6$): δ 8.18 (d, J = 7.6 Hz, 1H), 7.85 (d, J = 8.0 Hz, 1H), 7.76 (t, J = 9.4 Hz, 2H), 7.61 (td, J_1 = 7.6 Hz, J_2 = 1.0 Hz, 1H), 7.50 (t, J = 7.4 Hz, 1H), 7.32–7.22 (m, 4H), 7.12 (t, J = 7.8 Hz, 1H), 7.01 (d, J = 7.2 Hz, 1H), 6.83 (d, J = 7.6 Hz, 1H), 5.49 (s, 1H), 2.88 (s, 3H), 2.81 (s, 3H), 2.60 (s, 3H) ppm. ^{13}C NMR (101 MHz, DMSO-d$_6$): δ 176.5, 168.7, 168.0, 158.5, 149.4, 143.1, 133.0, 131.9, 129.4, 129.3, 129.1, 128.5, 127.5, 127.2, 126.7, 125.9, 123.8, 122.5, 120.7, 116.3, 108.9, 79.1, 68.3, 63.6, 55.9, 28.3, 26.4 ppm. HRMS (ESI): m/z calcd. for $C_{29}H_{24}N_5O_4$ [M + H]$^+$ 506.1823, found 506.1818.

(2′*R*,3*S*)-4′-Amino-1,1″,3″-trimethyl-2,2″,4″,6″-tetraoxo-2′-(thiophen-2-yl)-1″,3″,4″,6″-tetrahydro-2″*H*-dispiro[indoline-3,1′-cyclopentane-3′,5″-pyrimidin]-4′-ene-5′-carbonitrile (**3da**). According to the general procedure from **1d** (60.0 mg, 0.24 mmol) and **2a** (42.2 mg, 0.2 mmol) to obtain 75.6 mg (82% yield) compound **3da** as a brown solid, m.p. 214–215 °C. HPLC (Daicel Chiralpak IA, *n*-hexane/2-propanol/ethyl acetate = 80:10:10, flow rate 1.0 mL/min, detection at 254 nm): t_R = 19.3 min (minor), t_R = 26.4 min (major); >99% ee. $[\alpha]_D^{25}$ = +18.5 (c = 0.4, CH_2Cl_2). ^{1}H NMR (700 MHz, DMSO-d$_6$): δ 8.06 (d, J = 7.7 Hz, 1H), 7.36–7.32 (m, 2H), 7.20–7.16 (m, 3H), 6.94 (d, J = 7.7 Hz, 1H), 6.78 (dd, J_1 = 4.9 Hz, J_2 = 4.2 Hz, 1H), 6.68 (d, J = 3.5 Hz, 1H), 4.62 (s, 1H), 3.18 (s, 3H), 3.05 (s, 3H), 2.85 (s, 3H) ppm. ^{13}C NMR (176 MHz, DMSO-d$_6$): δ 176.2, 168.3, 167.3, 158.0, 150.0, 143.4, 132.5, 129.8, 129.4, 128.2, 128.0, 127.8, 126.3, 122.4, 116.2, 108.7, 78.5, 68.4, 63.0, 59.7, 28.7, 28.6, 26.5 ppm. HRMS (ESI): m/z calcd. for $C_{23}H_{20}N_5O_4S$ [M + H]$^+$ 462.1231, found 462.1253.

(2′*R*,3*S*)-4′-Amino-1,1″,3″-trimethyl-2,2″,4″,6″-tetraoxo-2′-(m-tolyl)-1″,3″,4″,6″-tetrahydro-2″*H*-dispiro[indoline-3,1′-cyclopentane-3′,5″-pyrimidin]-4′-ene-5′-carbonitrile (**3ea**). According to the general procedure from **1e** (61.92 mg, 0.24 mmol) and **2a** (42.2 mg, 0.2 mmol) to obtain 79.7 mg (85% yield) compound **3ea** as a white solid, m.p. 194–196 °C. HPLC (Daicel Chiralpak ADH, *n*-hexane/2-propanol = 70:30, flow rate 1.0 mL/min, detection at 254 nm): t_R = 6.5 min (minor), t_R = 11.5 min (major), 78% ee. $[\alpha]_D^{25}$ = +14.8 (c = 0.5, CH_2Cl_2). ^{1}H NMR (400 MHz, DMSO-d$_6$): δ 8.06 (dd, J_1 = 7.6 Hz, J_2 = 0.8 Hz, 1H), 7.28 (td, J = 7.6 Hz, J_2 = 1.2 Hz, 1H), 7.17 (t, J = 7.2 Hz, 3H), 6.98–6.91 (m, 2H), 6.85 (d, J = 7.6 Hz, 1H), 6.58 (s, 1H), 6.54 (d, J = 7.2 Hz, 1H), 4.20 (s, 1H), 3.08 (s, 3H), 2.98 (s, 3H), 2.86 (s, 3H), 2.03 (s, 3H) ppm. ^{13}C NMR (101 MHz, DMSO-d$_6$): δ 176.5, 168.6, 167.8, 158.4, 149.8, 143.0, 137.2, 131.1, 130.1, 129.5, 129.2, 128.4, 127.8, 127.3, 126.4, 122.1, 116.3, 108.7, 78.5, 68.1, 64.7, 63.4, 28.5, 28.3, 26.4, 20.5. ppm. HRMS (ESI): m/z calcd. for $C_{26}H_{24}N_5O_4$ [M + H]$^+$ 470.1823, found 470.1812.

(2′*R*,3*S*)-4′-Amino-2′-(furan-2-yl)-1,1″,3″-trimethyl-2,2″,4″,6″-tetraoxo-1″,3″,4″,6″-tetrahydro-2″*H*-dispiro[indoline-3,1′-cyclopentane-3′,5″-pyrimidin]-4′-ene-5′-carbonitrile (**3fa**). According to the general procedure from **1f** (56.2 mg, 0.24 mmol) and **2a** (42.2 mg, 0.2 mmol) to obtain 64.1 mg (72% yield) compound **3fa** as a brown solid, m.p. 174–176 °C. HPLC (Daicel Chiralpak IA, *n*-hexane/2-propanol = 70:30, flow rate 1.0 mL/min, detection at 254 nm): t_R = 6.6 min (minor), t_R = 8.2 min (major), 55% ee. $[\alpha]_D^{25}$ = −2.0 (c = 0.33, CH_2Cl_2). ^{1}H NMR (700 MHz, DMSO-d$_6$): δ 7.93 (dd, J = 7.7 Hz, 1H), 7.38 (d, J = 1.4 Hz, 1H), 7.32 (td, J_1 = 7.7 Hz, J_2 = 0.7 Hz, 1H), 7.17–7.12 (m, 3H), 6.97 (d, J = 7.7 Hz, 1H), 6.15 (dd, J_1 = 3.2 Hz, J_2 = 1.8 Hz, 1H), 5.74 (d, J = 2.8 Hz, 1H), 4.41 (s, 1H), 3.19 (s, 3H), 3.10 (s, 3H), 2.94 (s, 3H) ppm. ^{13}C NMR (176 MHz, DMSO-d$_6$): δ 176.0, 168.2, 167.2, 157.7, 150.2, 145.9, 144.0, 142.9, 129.3, 128.2, 127.4, 122.3, 116.0, 110.7, 110.4, 108.6, 78.8, 66.9, 61.7, 56.5, 28.7, 28.6, 26.5 ppm. HRMS (ESI): m/z calcd. for $C_{23}H_{20}N_5O_5$ [M + H]$^+$ 446.1459, found 446.1472.

(2′*R*,3*S*)-4′-Amino-2′-(4-fluorophenyl)-1,1″,3″-trimethyl-2,2″,4″,6″-tetraoxo-1″,3″,4″,6″-tetrahydro-2″*H*-dispiro[indoline-3,1′-cyclopentane-3′,5″-pyrimidin]-4′-ene-5′-carbonitrile (**3ga**). According to the general procedure from **1g** (62.9 mg, 0.24 mmol) and **2a** (42.2 mg, 0.2 mmol) to obtain 80.4 mg (85% yield) compound **3ga** as a pink solid, m.p. 178–179 °C. HPLC (Daicel Chiralpak IA, *n*-hexane/2-propanol/ethyl acetate = 80:15:5, flow rate 1.0 mL/min, detection at 254 nm): t_R = 13.4 min (minor), t_R = 21.9 min (major); 59% ee.

$[\alpha]_D^{25}$ = +9.0 (c = 0.8, CH_2Cl_2). ^1H NMR (700 MHz, DMSO-d_6): δ 8.03 (d, J = 7.0 Hz, 1H), 7.30 (t, J = 7.7 Hz, 1H), 7.20–7.16 (m, 3H), 6.94–6.88 (m, 3H), 6.83–6.80 (m, 2H), 4.25 (s, 1H), 3.09 (s, 3H), 3.01 (s, 3H), 2.91 (s, 3H) ppm. ^{13}C NMR (176 MHz, DMSO-d_6): δ 176.4, 168.5, 167.7, 162.1 ($^1J_{C-F}$ = 246.9 Hz), 158.3, 149.9, 143.0, 131.5 ($^3J_{C-F}$ = 8.3 Hz), 129.4, 128.1, 127.3 ($^4J_{C-F}$ = 2.6 Hz), 127.2, 122.4, 116.2, 115.0 ($^2J_{C-F}$ = 21.5 Hz), 108.8, 78.3, 67.9, 63.7, 63.4, 28.6, 28.4, 26.4 ppm. ^{19}F NMR (659 MHz, DMSO-d_6): δ −112.1. HRMS (ESI): m/z calcd. for $C_{25}H_{21}FN_5O_4$ [M + H]$^+$ 474.1572, found 474.1587.

(2′*R*,3*S*)-4′-Amino-2′-(3-bromophenyl)-1,1″,3″-trimethyl-2,2″,4″,6″-tetraoxo-1″,3″,4″,6″-tetrahydro-2″*H*-dispiro[indoline-3,1′-cyclopentane-3′,5″-pyrimidin]-4′-ene-5′-carbonitrile (**3ha**). According to the general procedure from **1h** (77.3 mg, 0.24 mmol) and **2a** (42.2 mg, 0.2 mmol) to obtain 103.4 mg (97% yield) compound **3ha** as a pink solid, m.p. 194–195 °C. HPLC (Daicel Chiralpak ADH, n-hexane/2-propanol = 75:25, flow rate 1.0 mL/min, detection at 254 nm): t_R = 11.8 min (minor), t_R = 14.5 min (major); 89% ee. $[\alpha]_D^{25}$ = +14.9 (c = 0.5, CH_2Cl_2). ^1H NMR (700 MHz, DMSO-d_6): δ 8.03 (d, J = 7.7 Hz, 1H), 7.39 (d, J = 8.4 Hz, 1H), 7.32 (t, J = 7.7 Hz, 1H), 7.23–7.18 (m, 3H), 7.07 (t, J = 8.0 Hz, 1H), 6.93 (s, 1H), 6.90 (d, J = 7.7 Hz, 1H), 6.85 (d, J = 7.7 Hz, 1H), 4.20 (s, 1H), 3.10 (s, 3H), 3.01 (s, 3H), 2.92 (s, 3H) ppm. ^{13}C NMR (176 MHz, DMSO-d_6): δ 176.2, 168.2, 167.7, 158.2, 149.8, 143.0, 133.5, 131.9, 131.6, 130.3, 129.5, 129.1, 127.9, 127.1, 122.3, 121.0, 116.2, 109.0, 78.2, 67.9, 63.6, 63.3, 28.6, 28.3, 26.4 ppm. HRMS (ESI): m/z calcd. for $C_{25}H_{21}{}^{79}BrN_5O_4$ [M + H]$^+$ 534.0771, found 534.0786; calcd. for $C_{25}H_{21}{}^{81}BrN_5O_4$ [M + H]$^+$ 536.0751, found 536.0770.

(2′*R*,3*S*)-4′-Amino-2′-(3,4-dimethoxyphenyl)-1,1″,3″-trimethyl-2,2″,4″,6″-tetraoxo-1″,3″,4″,6″-tetrahydro-2″*H*-dispiro[indoline-3,1′-cyclopentane-3′,5″-pyrimidin]-4′-ene-5′-carbonitrile (**3ia**). According to the general procedure from **1i** (73.0 mg, 0.24 mmol) and **2a** (42.2 mg, 0.2 mmol) to obtain 90.2 mg (93% yield) compound **3ia** as a yellow solid, m.p. 203–205 °C. HPLC (Daicel Chiralpak ADH, n-hexane/2-propanol = 80:20, flow rate 1.0 mL/min, detection at 254 nm): t_R = 19.0 min (minor), t_R = 21.3 min (major), 73% ee. $[\alpha]_D^{25}$ = +49.1 (c = 0.7, CH_2Cl_2). ^1H NMR (700 MHz, DMSO-d_6): δ 8.10 (d, J = 7.7 Hz, 1H), 7.31 (t, J = 7.7 Hz, 1H), 7.18 (t, J = 7.7 Hz, 3H), 6.90 (d, J = 7.7 Hz, 1H), 6.65 (d, J = 8.4 Hz, 1H), 6.37 (dd, J_1 = 8.0 Hz, J_2 = 1.4 Hz, 1H), 6.21 (s, 1H), 4.17 (s, 1H), 3.61 (s, 3H), 3.33 (s, 3H), 3.11 (s, 3H), 3.00 (s, 3H), 2.87 (s, 3H) ppm. ^{13}C NMR (176 MHz, DMSO-d_6): δ 176.6, 168.7, 167.9, 158.5, 150.0, 148.9, 147.5, 143.2, 129.2, 128.8, 127.2, 123.0, 122.8, 122.1, 116.4, 111.5, 110.7, 108.9, 78.3, 68.2, 64.5, 63.5, 55.1, 55.0, 28.6, 28.4, 26.4 ppm. HRMS (ESI): m/z calcd. for $C_{27}H_{26}N_5O_6$ [M + H]$^+$ 516.1878, found 516.1895.

(2′*R*,3*S*)-4′-Amino-2′-(4-bromophenyl)-1,1″,3″-trimethyl-2,2″,4″,6″-tetraoxo-1″,3″,4″,6″-tetrahydro-2″*H*-dispiro[indoline-3,1′-cyclopentane-3′,5″-pyrimidin]-4′-ene-5′-carbonitrile (**3ja**). According to the general procedure from **1j** (77.3 mg, 0.24 mmol) and **2a** (42.2 mg, 0.2 mmol) to obtain 98.1 mg (92% yield) compound **3ja** as a pink solid, m.p. 217–219 °C. HPLC (Daicel Chiralpak IA, n-hexane/2-propanol = 70:30, flow rate 1.0 mL/min, detection at 254 nm): t_R = 7.5 min (minor), t_R = 12.4 min (major); 84% ee. $[\alpha]_D^{25}$ = +27.4 (c = 0.9, CH_2Cl_2). ^1H NMR (400 MHz, DMSO-d_6): δ 8.01 (dd, J_1 = 7.6 Hz, J_2 = 0.8 Hz, 1H), 7.32–7.27 (m, 3H), 7.20–7.15 (m, 3H), 6.88 (d, J = 8.0 Hz, 1H), 6.72 (d, J = 8.8 Hz, 2H), 4.22 (s, 1H), 3.09 (s, 3H), 3.01 (s, 3H), 2.93 (s, 3H) ppm. ^{13}C NMR (176 MHz, DMSO-d_6): δ 176.4, 168.5, 167.8, 158.4, 149.9, 143.0, 131.5, 131.1, 130.5, 129.5, 128.0, 127.3, 122.6, 122.5, 116.3, 108.9, 78.4, 67.8, 63.9, 63.4, 28.7, 28.5, 26.5 ppm. HRMS (ESI): m/z calcd. for $C_{25}H_{21}{}^{79}BrN_5O_4$ [M + H]$^+$ 534.0771, found 534.0793; calcd. for $C_{25}H_{21}{}^{81}BrN_5O_4$ [M + H]$^+$ 536.0751, found 536.0777.

(2′*R*,3*S*)-4′-Amino-1-benzyl-1″,3″-dimethyl-2,2″,4″,6″-tetraoxo-2′-phenyl-1″,3″,4″,6″-tetrahydro-2″*H*-dispiro[indoline-3,1′-cyclopentane-3′,5″-pyrimidin]-4′-ene-5′-carbonitrile (**3ab**). According to the general procedure from **1a** (58.6 mg, 0.24 mmol) and **2b** (57.4 mg, 0.2 mmol) to obtain 100.9 mg (95% yield) compound **3ab** as a yellow solid, m.p. 165–167 °C. HPLC (Daicel Chiralpak IC, n-hexane/2-propanol = 70:30, flow rate 1.0 mL/min, detection at 254 nm): t_R = 40.3 min (minor), t_R = 24.9 min (major), >99% ee. $[\alpha]_D^{25}$ = −11.4 (*c* = 0.4, CH_2Cl_2). ^{1}H NMR (400 MHz, $CDCl_3$): δ 8.17 (dd, J_1 = 6.4 Hz, J_2 = 1.4 Hz, 1H), 7.24–7.20 (m, 1H), 7.17–7.00 (m, 7H), 6.82 (d, *J* = 7.6 Hz, 2H), 6.61 (d, *J* = 7.6 Hz, 2H), 6.41 (d, *J* = 7.2 Hz, 1H), 5.60 (s, 2H), 5.02 (d, *J* = 16.4 Hz, 1H), 4.49 (d, *J* = 16.4 Hz, 1H), 4.44 (s, 1H), 3.13 (s, 3H), 2.93 (s, 3H) ppm. ^{13}C NMR (101 MHz, $CDCl_3$): δ 177.2, 168.6, 167.2, 158.2, 149.8, 142.3, 134.4, 130.5, 123.0, 129.55, 129.52, 128.5, 128.30, 128.27, 127.7, 127.2, 126.1, 123.1, 115.5, 109.5, 83.3, 68.8, 66.1, 64.1, 43.8, 29.1, 28.8 ppm. HRMS (ESI): *m*/*z* calcd. for $C_{31}H_{26}N_5O_4$ [M + H]$^+$ 532.1979, found 532.2000.

(2′*R*,3*S*)-4′-Amino-6-chloro-1,1″,3″-trimethyl-2,2″,4″,6″-tetraoxo-2′-phenyl-1″,3″,4″,6″-tetrahydro-2″*H*-dispiro[indoline-3,1′-cyclopentane-3′,5″-pyrimidin]-4′-ene-5′-carbonitrile (**3ac**). According to the general procedure from **1a** (58.6 mg, 0.24 mmol) and **2c** (49.0 mg, 0.2 mmol) to obtain 90.0 mg (92% yield) compound **3ac** as a yellow solid, m.p. 172–174 °C. HPLC (Daicel Chiralpak ADH, *n*-hexane/2-propanol = 70:30, flow rate 1.0 mL/min, detection at 254 nm): t_R = 8.0 min (minor), t_R = 15.6 min (major), 91% ee. $[\alpha]_D^{25}$ = −30.9 (*c* = 0.4, CH_2Cl_2). ^{1}H NMR (700 MHz, DMSO-d_6): δ 8.04 (d, *J* = 8.4 Hz, 1H), 7.25 (dd, J_1 = 8.0 Hz, J_2 = 2.2 Hz, 3H), 7.19 (t, *J* = 7.4 Hz, 1H), 7.11 (t, *J* = 7.7 Hz, 2H), 7.04 (d, *J* = 1.4 Hz, 1H), 6.75 (d, *J* = 7.7 Hz, 2H), 4.23 (s, 1H), 3.07 (s, 3H), 3.01 (s, 3H), 2.89 (s, 3H) ppm. ^{13}C NMR (176 MHz, DMSO-d_6): δ 176.6, 168.5, 167.9, 158.7, 149.8, 144.5, 133.8, 130.8, 129.3, 129.2, 128.6, 128.2, 127.2, 122.1, 116.1, 109.3, 77.9, 67.9, 64.7, 63.2, 28.6, 28.4, 26.6 ppm. HRMS (ESI): *m*/*z* calcd. for $C_{25}H_{21}ClN_5O_4$ [M + H]$^+$ 490.1277, found 490.1300.

(2′*R*,3*S*)-4′-amino-1,1″,3″,5,7-pentamethyl-2,2″,4″,6″-tetraoxo-2′-phenyl-1″,3″,4″,6″-tetrahydro-2″*H*-dispiro[indoline-3,1′-cyclopentane-3′,5″-pyrimidin]-4′-ene-5′-carbonitrile (**3ad**). According to the general procedure from **1a** (58.6 mg, 0.24 mmol) and **2d** (47.8 mg, 0.2 mmol) to obtain 86.9 mg (90% yield) compound **3ad** as a white solid, m.p. 181–182 °C. HPLC (Daicel Chiralpak IC, *n*-hexane/2-propanol = 70:30, flow rate 1.0 mL/min, detection at 254 nm): t_R = 56.3 min (minor), t_R = 40.3 min (major), 90% ee. $[\alpha]_D^{25}$ = −40.6 (*c* = 0.4, CH_2Cl_2). ^{1}H NMR (700 MHz, DMSO-d_6): δ 7.76 (s, 1H), 7.16 (t, *J* = 7.4 Hz, 1H), 7.13 (s, 2H), 7.07 (t, *J* = 7.7 Hz, 2H), 6.83 (s, 1H), 6.76 (d, *J* = 7.7 Hz, 2H), 4.22 (s, 1H), 3.23 (s, 3H), 3.07 (s, 3H), 2.84 (s, 3H), 2.33 (s, 3H), 2.31 (s, 3H) ppm. ^{13}C NMR (176 MHz, DMSO-d_6): δ 177.1, 168.7, 167.8, 158.2, 149.8, 138.5, 133.3, 131.3, 130.8, 129.5, 129.1, 128.9, 128.0, 126.0, 119.3, 116.5, 79.2, 68.2, 65.0, 63.0, 29.6, 28.5, 28.4, 20.7, 18.3 ppm. HRMS (ESI): *m*/*z* calcd. for $C_{27}H_{26}N_5O_4$ [M + H]$^+$ 484.1979, found 484.1992.

(2′*R*,3*S*)-4′-Amino-6-fluoro-1,1″,3″-trimethyl-2,2″,4″,6″-tetraoxo-2′-phenyl-1″,3″,4″,6″-tetrahydro-2″*H*-dispiro[indoline-3,1′-cyclopentane-3′,5″-pyrimidin]-4′-ene-5′-carbonitrile (**3ae**). According to the general procedure from **1a** (58.6 mg, 0.24 mmol) and **2e** (45.8 mg, 0.2 mmol) to obtain 79.5 mg (84% yield) compound **3ae** as a white solid, m.p. 173–175 °C. HPLC (Daicel Chiralpak IC, *n*-hexane/2-propanol = 70:30, flow rate 1.0 mL/min, detection at 254 nm): t_R = 9.4 min (minor), t_R = 11.8 min (major), 82% ee. $[\alpha]_D^{25}$ = −62.8 (*c* = 1, CH_2Cl_2). ^{1}H NMR (700 MHz, DMSO-d_6): δ 8.04 (dd, J_1 = 7.7 Hz, J_2 = 5.6 Hz, 1H), 7.23 (s, 2H), 7.18 (t, *J* = 7.4 Hz, 1H), 7.10 (t, *J* = 7.4 Hz, 2H), 7.01–6.98 (m, 1H), 6.86 (d, *J* = 9.1 Hz, 1H), 6.75 (d, *J* = 8.4 Hz, 2H), 4.22 (s, 1H), 3.07 (s, 3H), 3.00 (s, 3H), 2.88 (s, 3H) ppm. ^{13}C NMR (176 MHz, DMSO-d_6): δ 176.9, 168.6, 167.9, 162.8 (d, $^1J_{C\text{–}F}$ = 243.8 Hz), 158.5, 149.8, 144.8 (d, $^2J_{C\text{–}F}$ = 12.1 Hz), 130.9, 129.4, 129.1, 128.8 (d, $^3J_{C\text{–}F}$ = 9.7 Hz), 128.1, 124.0 (d, $^4J_{C\text{–}F}$ = 2.5 Hz), 116.2, 108.4 (d, $^2J_{C\text{–}F}$ = 22.2 Hz), 97.5 (d, $^2J_{C\text{–}F}$ = 27.6 Hz), 78.1, 68.0, 64.7, 63.1, 28.6, 28.4, 26.7 ppm. ^{19}F NMR (659 MHz, DMSO-d_6): δ −110.9. HRMS (ESI): *m*/*z* calcd. for $C_{25}H_{21}FN_5O_4$ [M + H]$^+$ 474.1572, found 474.1602.

(2′*R*,3*S*)-4′-Amino-6-bromo-1,1″,3″-trimethyl-2,2″,4″,6″-tetraoxo-2′-phenyl-1″,3″,4″,6″-tetrahydro-2″*H*-dispiro[indoline-3,1′-cyclopentane-3′,5″-pyrimidin]-4′-ene-5′-carbonitrile (**3af**). According to the general procedure from **1a** (58.6 mg, 0.24 mmol) and **2f** (57.8 mg, 0.2 mmol) to obtain 101.3 mg (95% yield) compound **3af** as a white solid, m.p. 161–163 °C.

HPLC (Daicel Chiralpak IC, *n*-hexane/2-propanol = 70:30, flow rate 1.0 mL/min, detection at 254 nm): t_R = 17.9 min (major); >99% ee. $[\alpha]_D^{25}$ = −42.4 (*c* = 0.5, CH_2Cl_2). ^{1}H NMR (700 MHz, DMSO-d$_6$): *δ* 7.97 (d, *J* = 7.7 Hz, 1H), 7.39 (dd, J_1 = 8.0 Hz, J_2 = 1.8 Hz, 1H), 7.25 (s, 2H), 7.19 (t, *J* = 7.4 Hz, 1H), 7.16 (d, *J* = 1.4 Hz, 1H), 7.11 (t, *J* = 7.7 Hz, 2H), 6.75 (d, *J* = 7.7 Hz, 2H), 4.23 (s, 1H), 3.07 (s, 3H), 3.01 (s, 3H), 2.89 (s, 3H) ppm. ^{13}C NMR (176 MHz, DMSO-d$_6$): *δ* 176.4, 168.5, 167.8, 158.7, 149.8, 144.6, 130.8, 129.3, 129.2, 129.0, 128.2, 127.6, 125.0, 122.2, 116.1, 112.1, 77.9, 67.9, 64.3, 63.2, 28.6, 28.4, 26.6 ppm. HRMS (ESI): *m*/*z* calcd. for $C_{25}H_{21}{}^{79}BrN_5O_4$ [M + H]$^+$ 534.0771, found 534.0785; calcd. for $C_{25}H_{21}{}^{81}BrN_5O_4$ [M + H]$^+$ 536.0751, found 536.0765.

(2′*R*,3*S*)-4′-Amino-5-fluoro-1,1″,3″-trimethyl-2,2″,4″,6″-tetraoxo-2′-phenyl-1″,3″,4″,6″-tetrahydro-2″*H*-dispiro[indoline-3,1′-cyclopentane-3′,5″-pyrimidin]-4′-ene-5′-carbonitrile (**3ag**). According to the general procedure from **1a** (58.6 mg, 0.24 mmol) and **2g** (45.8 mg, 0.2 mmol) to obtain 89.9 mg (95% yield) compound **3ag** as a white solid, m.p. 201–203 °C. HPLC (Daicel Chiralpak ADH, *n*-hexane/2-propanol = 70:30, flow rate 1.0 mL/min, detection at 254 nm): t_R = 18.7 min (minor), t_R = 20.2 min (major); 95% ee. $[\alpha]_D^{25}$ = −58.6 (*c* = 0.5, CH_2Cl_2). ^{1}H NMR (700 MHz, DMSO-d$_6$): *δ* 7.91 (dd, J_1 = 8.8 Hz, J_2 = 2.4 Hz, 1H), 7.28 (s, 2H), 7.20–7.15 (m, 2H), 7.12 (t, *J* = 7.7 Hz, 2H), 6.90 (dd, J_1 = 8.4 Hz, J_2 = 4.2 Hz, 1H), 6.76 (d, *J* = 7.7 Hz, 2H), 4.26 (s, 1H), 3.07 (s, 3H), 3.00 (s, 3H), 2.91 (s, 3H) ppm. ^{13}C NMR (176 MHz, DMSO-d$_6$): *δ* 176.2, 168.5, 168.0, 158.8, 158.3 ($^1J_{C-F}$ = 237.4 Hz), 149.8, 139.4, 130.8, 130.2 ($^3J_{C-F}$ = 8.3 Hz), 129.23, 129.19, 128.3, 116.1, 115.7 ($^2J_{C-F}$ = 23.4 Hz), 114.8 ($^2J_{C-F}$ = 25.7 Hz), 109.8 ($^3J_{C-F}$ = 8.3 Hz), 78.1, 67.9, 64.6, 63.8, 29.9, 28.6, 28.4, 26.6 ppm. ^{19}F NMR (659 MHz, DMSO-d$_6$) δ −120.1. HRMS (ESI): *m*/*z* calcd. for $C_{25}H_{21}FN_5O_4$ [M + H]$^+$ 474.1572, found 474.1595.

(2′*R*,3*S*)-4′-Amino-5-bromo-1,1″,3″-trimethyl-2,2″,4″,6″-tetraoxo-2′-phenyl-1″,3″,4″,6″-tetrahydro-2″*H*-dispiro[indoline-3,1′-cyclopentane-3′,5″-pyrimidin]-4′-ene-5′-carbonitrile (**3ah**). According to the general procedure from **1a** (58.6 mg, 0.24 mmol) and **2h** (57.8 mg, 0.2 mmol) to obtain 101.3 mg (95% yield) compound **3ah** as a white solid, m.p. 192–194 °C. HPLC (Daicel Chiralpak IA, *n*-hexane/2-propanol = 80:20, flow rate 1.0 mL/min, detection at 254 nm): t_R = 14.1 min (minor), t_R = 11.3 min (major), >99% ee. $[\alpha]_D^{25}$ = −211.6 (*c* = 0.8, CH_2Cl_2). ^{1}H NMR (700 MHz, DMSO-d$_6$): *δ* 8.23 (d, *J* = 2.1 Hz, 1H), 7.50 (dd, J_1 = 8.4 Hz, J_2 = 2.1 Hz, 1H), 7.30 (s, 2H), 7.19 (t, *J* = 7.4 Hz, 1H), 7.12 (t, *J* = 7.7 Hz, 2H), 6.87 (d, *J* = 8.4 Hz, 1H), 6.75 (d, *J* = 7.0 Hz, 2H), 4.24 (s, 1H), 3.07 (s, 3H), 2.99 (s, 3H), 2.90 (s, 3H) ppm. ^{13}C NMR (176 MHz, DMSO-d$_6$): *δ* 176.0, 168.4, 167.9, 158.8, 149.7, 142.3, 132.0, 130.8, 130.7, 130.0, 129.23, 129.18, 128.3, 116.1, 114.2, 110.8, 77.9, 67.8, 64.6, 63.5, 28.6, 28.4, 26.6 ppm. HRMS (ESI): *m*/*z* calcd. for $C_{25}H_{21}{}^{79}BrN_5O_4$ [M + H]$^+$ 534.0771, found 534.0786; calcd. for $C_{25}H_{21}{}^{81}BrN_5O_4$ [M + H]$^+$ 536.0751, found 536.0769.

(2′*R*,3*S*)-4′-Amino-5-chloro-1,1″,3″-trimethyl-2,2″,4″,6″-tetraoxo-2′-phenyl-1″,3″,4″,6″-tetrahydro-2″*H*-dispiro[indoline-3,1′-cyclopentane-3′,5″-pyrimidin]-4′-ene-5′-carbonitrile (**3ai**). According to the general procedure from **1a** (58.6 mg, 0.24 mmol) and **2i** (49.0 mg, 0.2 mmol) to obtain 90.0 mg (92% yield) compound **3ai** as a white solid, m.p. 201–204 °C. HPLC (Daicel Chiralpak IA, *n*-hexane/2-propanol = 90:10, flow rate 1.0 mL/min, detection at 254 nm): t_R = 45.7 min (minor), t_R = 32.5 min (major); >99% ee. $[\alpha]_D^{25}$ = −100.6 (*c* = 0.5, CH_2Cl_2). ^{1}H NMR (700 MHz, DMSO-d$_6$): *δ* 8.11 (d, *J* = 2.8 Hz, 1H), 7.37 (dd, J_1 = 8.4 Hz, J_2 = 2.8 Hz, 1H), 7.30 (s, 2H), 7.19 (t, *J* = 7.7 Hz, 1H), 7.12 (t, *J* = 7.7 Hz, 2H), 6.92 (d, *J* = 7.7 Hz, 1H), 6.75 (dd, J_1 = 8.4 Hz, J_2 = 1.4 Hz, 2H), 4.25 (s, 1H), 3.08 (s, 3H), 3.00 (s, 3H), 2.91 (s, 3H) ppm. ^{13}C NMR (176 MHz, DMSO-d$_6$): *δ* 176.1, 168.4, 168.0, 158.8, 149.74, 142.0, 130.7, 130.4, 129.23, 129.19, 128.3, 127.2, 126.5, 116.1, 110.4, 77.9, 67.8, 64.6, 63.6, 28.6, 28.4, 26.6 ppm. HRMS (ESI): *m*/*z* calcd. for $C_{25}H_{21}ClN_5O_4$ [M + H]$^+$ 490.1277, found 490.1304.

(2′*R*,3*S*)-4′-Amino-1,1″,3″-trimethyl-2,2″,4″,6″-tetraoxo-2′-phenyl-7-(trifluoromethyl)-1″,3″,4″,6″-tetrahydro-2″*H*-dispiro[indoline-3,1′-cyclopentane-3′,5″-pyrimidin]-4′-ene-5′-carbonitrile (**3aj**). According to the general procedure from **1a** (58.6 mg, 0.24 mmol) and **2j** (55.8 mg, 0.2 mmol) to obtain 88.9 mg (85% yield) compound **3aj** as a white solid, m.p. 177–179 °C. HPLC (Daicel Chiralpak IA, *n*-hexane/2-propanol = 70:30, flow rate 1.0 mL/min, detection at 254 nm): t_R = 5.6 min (minor), t_R = 7.6 min (major); 83% ee. $[\alpha]_D^{25}$ = −60.8 (c = 0.5, CH_2Cl_2). ^1H NMR (700 MHz, DMSO-d_6): δ 8.37 (d, J = 7.0 Hz, 1H), 7.64 (dd, J_1 = 8.0 Hz, J_2 = 1.0 Hz, 1H), 7.38 (t, J = 8.0 Hz, 3H), 7.17 (t, J = 7.4 Hz, 1H), 7.07 (t, J = 7.7 Hz, 2H), 6.67 (d, J = 7.0 Hz, 2H), 4.22 (s, 1H), 3.14 (s, 3H), 3.08 (s, 3H), 2.89 (s, 3H) ppm. ^{13}C NMR (176 MHz, DMSO-d_6): δ 178.0, 168.5, 167.8, 159.3, 149.8, 140.6, 131.4, 131.3, 130.4, 129.2, 128.1, 127.3, 127.2 (q, $^3J_{C\text{-}F}$ = 5.5 Hz), 123.2 (q, $^1J_{C\text{-}F}$ = 271.2 Hz), 122.2, 116.0, 110.9 (q, $^2J_{C\text{-}F}$ = 32.7 Hz), 77.2, 67.8, 65.3, 62.2, 29.0 (q, $J_{C\text{-}F}$ = 5.8 Hz), 28.6, 28.4 ppm. ^{19}F NMR (659 MHz, DMSO-d_6): δ −52.1. HRMS (ESI): m/z calcd. for $C_{26}H_{21}F_3N_5O_4$ $[M + H]^+$ 524.1540, found 524.1559.

(2′*R*,3*S*)-4′-Amino-1,1″,3″,7-tetramethyl-2,2″,4″,6″-tetraoxo-2′-phenyl-1″,3″,4″,6″-tetrahydro-2″*H*-dispiro[indoline-3,1′-cyclopentane-3′,5″-pyrimidin]-4′-ene-5′-carbonitrile (**3ak**). According to the general procedure from **1a** (58.6 mg, 0.24 mmol) and **2k** (45.0 mg, 0.2 mmol) to obtain 89.1 mg (95% yield) compound **3ak** as a white solid, m.p. 182–185 °C. HPLC (Daicel Chiralpak IA, *n*-hexane/2-propanol = 70:30, flow rate 1.0 mL/min, detection at 254 nm): t_R = 7.5 min (minor), t_R = 9.3 min (major); 82% ee. $[\alpha]_D^{25}$ = 23.4 (c = 0.4, CH_2Cl_2). ^1H NMR (700 MHz, DMSO-d_6): δ 7.92 (d, J = 6.3 Hz, 1H), 7.18–7.14 (m, 3H), 7.08–7.01 (m, 3H), 6.75 (d, J = 7.7 Hz, 2H), 4.23 (s, 1H), 3.27 (s, 3H), 3.07 (s, 3H), 2.84 (s, 3H), 2.38 (s, 3H) ppm. ^{13}C NMR (176 MHz, DMSO-d_6): δ 177.2, 168.7, 167.8, 158.3, 149.8, 140.8, 132.9, 131.2, 129.5, 129.0, 128.0, 125.4, 122.1, 119.7, 116.4, 79.0, 68.2, 65.0, 62.9, 29.6, 28.5, 28.3, 18.4 ppm. HRMS (ESI): m/z calcd. for $C_{26}H_{24}N_5O_4$ $[M + H]^+$ 470.1823, found 470.1834.

3.5. *Gram-Scale Synthesis of 3aa*

To a dried 50 mL round-bottom flask, benzylidene barbituric acid **1a** (1.17 g, 4.8 mmol), oxindolylmalonitrile **2a** (0.84 g, 4.0 mmol), chiral organocatalyst **C5** (101.6 mg, 0.2 mmol, 0.05 equiv), and CH_2Cl_2 (20 mL) were added. After stirring at room temperature for 8 h, the reaction mixture was concentrated and directly purified by silica gel column chromatography (200–300 mesh) using ethyl acetate/petroleum ether (1:2) as eluent to afford the desired product **3aa** (1.6 g, 88% yield).

4. Conclusions

In summary, we developed an efficient and practical asymmetric Michael/cyclization reaction of benzylidene barbituric acids with oxindolylmalonitriles at room temperature. Using a squaramide catalyst, a series of chiral bisspiro barbituric acid derivatives were obtained in high yields (72–97%) with high-to-excellent stereoselectivities (up to >99% ee and >20:1 dr). In addition, the practicability of the reaction was verified by the preparation of the product at the gram-scale.

Supplementary Materials: The following supporting information can be downloaded at https://www.mdpi.com/article/10.3390/molecules30092000/s1, Copies of ^1H and ^{13}C NMR spectra, HPLC chromatograms for all new compounds.

Author Contributions: D.-J.Q. wrote the preliminary manuscript and performed the experiments and acquired and analyzed the original data; D.-M.D. designed the research plan, supervised the experiments, modified all figures and schemes, analyzed and checked all the data, and revised this manuscript. All authors have read and agreed to the published version of the manuscript.

Funding: This research received no external funding.

Institutional Review Board Statement: Not applicable.

Informed Consent Statement: Not applicable.

Data Availability Statement: Data are contained within this article and Supplementary Materials.

Acknowledgments: We thank the Analysis and Testing Center of the Beijing Institute of Technology for the measurement of NMR and mass spectrometry.

Conflicts of Interest: Authors declare no conflict of interest.

References

1. Carter, M.K. The story of Barbituric acid. *J. Chem. Edu.* **1951**, *28*, 524–526. [CrossRef]
2. Hu, X.; Long, Q. The application of barbituric acid in drug synthesis. *Chin. J. Synth. Chem.* **1994**, *2*, 5–12.
3. Sandberg, F. Anaesthetic properties of some new N-substituted and N,N′-disubstituted derivatives of 5,5-diallyl-barbituric acid. *Acta Phys. Scand.* **1951**, *24*, 7–26. [CrossRef] [PubMed]
4. Kliethermes, C.L.; Metten, P.; Belknap, J.K.; Buck, K.J.; Crabbe, J.C. Selection for pentobarbital withdrawal severity: Correlated differences in withdrawal from other sedative drugs. *Brain Res.* **2004**, *1009*, 17–25. [CrossRef]
5. Archana; Srivastava, V.K.; Kumar, A. Synthesis of some newer derivatives of substituted quinazolinonyl-2-oxo/thiobarbituric acid as potent anticonvulsant agents. *Bioorg. Med. Chem.* **2004**, *12*, 1257–1264. [CrossRef]
6. Hese, S.V.; Meshram, R.J.; Kamble, R.D.; Mogle, P.P.; Patil, K.K.; Kamble, S.S.; Gacche, R.N.; Dawane, B.S. Antidiabetic and allied biochemical roles of new chromeno-pyrano pyrimidine compounds: Synthesis, in vitro and in silico analysis. *Med. Chem. Res.* **2017**, *26*, 805–818. [CrossRef]
7. Dhorajiya, B.D.; Dholakiya, B.Z.; Mohareb, R.M. Hybrid probes of aromatic amine and barbituric acid: Highly promising leads for anti-bacterial, anti-fungal and anti-cancer activities. *Med. Chem. Res.* **2014**, *23*, 3941–3952. [CrossRef]
8. Hu, Y.Q.; Gao, C.; Zhang, S.; Xu, L.; Xu, Z.; Feng, L.S.; Wu, X.; Zhao, F. Quinoline hybrids and their antiplasmodial and antimalarial activities. *Eur. J. Med. Chem.* **2017**, *139*, 22–47. [CrossRef]
9. Bialer, M. How did phenobarbital's chemical structure affect the development of subsequent antiepileptic drugs (AEDs)? *Epilepsia* **2012**, *53*, 3–11. [CrossRef]
10. Segovia, C.; Lebrêne, A.; Levacher, V.; Oudeyer, S.; Brière, J.F. Enantioselective catalytic transformations of barbituric acid derivatives. *Catalysts* **2019**, *9*, 131. [CrossRef]
11. Gomes, R.F.A.; Coelho, J.A.S.; Afonso, C.A.M. Synthesis and applications of stenhouse salts and derivatives. *Chem.-Eur. J.* **2018**, *24*, 9170–9186. [CrossRef] [PubMed]
12. Ziarani, G.M.; Aleali, F.; Lashgari, N. Recent applications of barbituric acid in multicomponent reactions. *RSC Adv.* **2016**, *6*, 50895–50922. [CrossRef]
13. Schade, A.; Schreiter, K.; Rüffer, T.; Lang, H.; Spange, S. Interactions of enolizable barbiturate dyes. *Chem.-Eur. J.* **2016**, *22*, 5734–5748. [CrossRef] [PubMed]
14. Freeman, F. Properties and reactions of ylidenemalononitriles. *Chem. Rev.* **1980**, *80*, 329–350. [CrossRef]
15. Rombola, M.; Sumaria, C.S.; Montgomery, T.D.; Rawal, V.H. Development of chiral, bifunctional thiosquaramides: Enantioselective Michael additions of barbituric acids to nitroalkenes. *J. Am. Chem. Soc.* **2017**, *139*, 5297–5300. [CrossRef]
16. Liu, Y.; Zhang, Y.; Duan, H.X.; Wanyan, D.Y.; Wang, Y.Q. Enantioselective organocatalytic Michael additions of N,N′-dialkylbarbituric acids to enones. *Org. Biomol. Chem.* **2017**, *15*, 8669–8679. [CrossRef]
17. Zhang, J.; Yin, G.; Du, Y.; Yang, Z.; Li, Y.; Chen, L. Michael–Michael addition reactions promoted by secondary amine-thiourea: Stereocontrolled construction of barbiturate-fused tetrahydropyrano scaffolds and pyranocoumarins. *J. Org. Chem.* **2017**, *82*, 13594–13601. [CrossRef]
18. Zhao, H.W.; Tian, T.; Li, B.; Yang, Z.; Pang, H.L.; Meng, W.; Song, X.Q.; Chen, X.J. Diastereoselective synthesis of dispirobarbiturates through et3n-catalyzed [3 + 2] cycloaddition of barbiturate-based olefins with 3-isothiocyanato oxindoles. *J. Org. Chem.* **2015**, *80*, 10380–10385. [CrossRef]
19. Zhao, H.W.; Tian, T.; Pang, H.L.; Li, B.; Chen, X.Q.; Yang, Z.; Meng, W.; Song, X.-Q.; Zhao, Y.D.; Liu, Y.Y. Organocatalytic [3 + 2] cycloadditions of barbiturate-based olefins with 3-isothiocyanato oxindoles: Highly diastereoselective and enantioselective synthesis of dispirobarbiturates. *Adv. Synth. Catal.* **2016**, *358*, 2619–2630. [CrossRef]
20. Liu, Y.; Yang, W.; Wu, Y.; Mao, B.; Gao, X.; Liu, H.; Sun, Z.; Xiao, Y.; Guo, H. Asymmetric construction of highly functionalized spirobarbiturate-cyclopentenes through chiral phosphine-catalyzed [3 + 2] annulation of Morita–Baylis–Hillman carbonates with barbiturate-derived alkenes. *Adv. Synth. Catal.* **2016**, *358*, 2867–2872. [CrossRef]
21. An, T.L.; Du, D.M. Chiral squaramide catalyzed asymmetric [3 + 2] cycloaddition reaction for synthesis of trifluoromethylated barbituric acid derivatives. *Chem. Sel.* **2019**, *4*, 11302–11306. [CrossRef]

22. Alexander, J.; Boddy, A.J.; Bull, J.A. Stereoselective synthesis and applications of spirocyclic oxindoles. *Org. Chem. Front.* **2021**, *8*, 1026–1084.
23. Zhao, B.L.; Li, J.H.; Du, D.M. Squaramide-catalyzed asymmetric reactions. *Chem. Record* **2017**, *17*, 994–1018. [CrossRef] [PubMed]
24. Lin, Y.; Zhao, B.L.; Du, D.M. Bifunctional squaramide-catalyzed asymmetric [3 + 2] cyclization of 2-(1-methyl-2-oxoindolin-3-yl)malononitriles with unsaturated pyrazolones to construct spirooxindole-fused spiropyrazolones. *J. Org. Chem.* **2019**, *84*, 10209–10220. [CrossRef]
25. Neumann, D.M.; Cammarata, A.; Backes, G.; Palmer, G.E.; Jusic, B.S. Synthesis and antifungal activity of substituted 2,4,6-pyrimidinetrione carbaldehyde hydrazones. *Bioorg. Med. Chem.* **2014**, *22*, 813–826. [CrossRef]
26. Zhu, Y.; Malerich, J.P.; Rawal, V.H. Squaramide-catalyzed enantioselective Michael addition of diphenyl phosphite to nitroalkenes. *Angew. Chem. Int. Ed.* **2010**, *49*, 153–156. [CrossRef]
27. Yang, W.; Du, D.M. Highly enantioselective Michael addition of nitroalkanes to chalcones using chiral squaramides as hydrogen bonding organocatalysts. *Org. Lett.* **2010**, *12*, 5450–5453. [CrossRef]
28. Yang, W.; Du, D.M. Chiral squaramide-catalyzed highly enantioselective Michael addition of 2-hydroxy-1,4-naphthoquinones to nitroalkenes. *Adv. Synth. Catal.* **2011**, *353*, 1241–1246. [CrossRef]
29. Vakulya, B.; Varga, S.; Csámpai, A.; Soós, T. Highly enantioselective conjugate addition of nitromethane to chalcones using bifunctional cinchona organocatalysts. *Org. Lett.* **2005**, *7*, 1967–1969. [CrossRef]

molecules

MDPI

Article

Natural Low-Eutectic Solvent Co-Culture-Assisted Whole-Cell Catalyzed Synthesis of Ethyl (*R*)-4-Chloro-3-Hydroxybutyrate

Yanni Wang [1], Bo Liu [2], Yanmei Dai [1], Zijuan Tao [1], Lan Tang [1] and Zhimin Ou [1,*]

[1] College of Pharmaceutical Science, Zhejiang Key Laboratory of Green Manufacturing Technology for Chemical Drugs, Zhejiang University of Technology, Hangzhou 310014, China; 13329103632@163.com (Y.W.); 211122070126@zjut.ecu.cn (Y.D.); 211122070059@zjut.edu.cn (Z.T.); tanglan@zjut.edu.cn (L.T.)

[2] College of Biological & Environmental Sciences, Zhejiang Wanli University, Ningbo 315199, China; liubottbf666@163.com

* Correspondence: oozzmm@zjut.edu.cn; Tel./Fax: +86-571-88320320

Abstract

In this study, CGMCC NO:28566, a strain that can efficiently convert Ethyl 4-chloroacetoacetate (COBE) to (*R*)-4-chloro-3-hydroxybutyrate((*R*)-CHBE), was screened by soil-sieving bacteria. In order to improve the transformation effect of the strain, the natural low-eutectic solvent (NADES), which can alter the cell permeability, was utilized for assisted catalysis, and a better catalytic effect was achieved. This study was carried out using a co-culture of strains with NADES and secondary addition of NADES on the basis of co-culture, and 10 NADESs were screened at the same time. The co-catalytic effect of 0.5% (w/v) choline chloride: urea (1:2) (ChCl:U (1:2)) was found to be the most significant, with a yield of (*R*)-CHBE reaching 89.1%, which was 58.2% higher than that of the control group, with a 99% ee value. Furthermore, the catalytic results demonstrated that the co-culture of the strain with NADES during fermentation yielded superior outcomes to the secondary addition of NADES during the reaction buffer. Furthermore, the catalytic effect of ChCl:U (1:2) was demonstrated to be superior to that of its individual components or single-component blends, due to its distinctive valence bonding advantage. The results indicate that the addition of 0.5% (w/v) ChCl:U (1:2) during the co-culture process has the effect of improving cell permeability to a certain extent, thereby increasing the contact between the substrate and the enzyme during the whole-cell catalytic reactions.

Keywords: NADES; co-culture; cell permeability; whole-cell catalysis

1. Introduction

Natural low-eutectic solvents (NADES) are composed of natural hydrogen-bond acceptors (HBAs) and natural hydrogen-bond donors (HBDs) [1–3]. The discovery of NADES offers a promising avenue for addressing the limitations of ionic liquids [4]. NADES can be used for extraction, solubility enhancement, and stabilization of bioactives [5]. In terms of composition, the natural hydrogen bond acceptors in NADES are predominantly quaternary ammonium salts, including betaine, dihydrocholine citrate, and choline chloride. In contrast, the hydrogen bond donors are primarily alcohols, sugars, organic acids, and urea [6].

The majority of pro-chiral ketone substrates are organic compounds that are insoluble in water. In order to facilitate the dissolution of the substrate, organic solvents such as isopropanol, glycerol, and ethanol are often employed to enhance the dissolution process [7]. Nevertheless, the addition of a considerable number of organic solvents can potentially cause damage to the cells. Consequently, researchers have focused their attention on the NADES, which are less toxic and have superior solubility properties [4]. Furthermore, NADES pre-treatment can enhance catalytic efficiency to a certain extent by modifying cell permeability, thereby increasing enzyme-substrate contact [8,9]. The advantages of NADES-assisted catalysis include NADES media ensuring higher activity for a wide range of enzymes, increased substrate solubility, and altered cell membrane permeability [10].

Isolation and purification of enzymes is a time-consuming process that will also result in the loss of enzyme content and the reduction in enzyme activity during purification, thus leading to a higher production cost [11]. In contrast, whole-cell catalysis enables the catalysis of the substrate without the destruction of the cells, which reduces the production cost and simplifies the production process [12]. Nevertheless, the principal obstacle to whole-cell catalysis is the challenge of mass transfer between intracellular enzymes and substrates in solvent systems [13]. The cell membrane of microorganisms exhibits a greater degree of selectivity in regulating the permeability of substances entering and exiting the cell [14]. In contrast, NADES can achieve changes in catalytic efficiency by altering the permeability of cell membranes [15]. The novel approach of co-culture with NADES represents a new type of assisted catalysis that pre-treats the permeability of bacterial cells during the enrichment stage of the bacterium. The method described by Qian et al. resulted in a notable enhancement in the production of chiral alcohols [9].

Ethyl 4-chloroacetoacetate (COBE) can be reduced asymmetrically to the enantiomers (*R*)-CHBE and (*S*)-CHBE [16]. (*R*)-CHBE can be employed as a pivotal intermediate in the synthesis of numerous pharmaceuticals. For instance, (*R*)-CHBE is utilized in the synthesis of antibiotics, macrolides, γ-amino-hydroxybutyric acid, cyclohexanone, and other compounds. Furthermore, it is employed in the synthesis of a plethora of pharmaceuticals, including antibacterial and anti-inflammatory agents, treatments for Alzheimer's disease, cardiovascular disorders, and agents that address the dietary fatigue link and other related conditions [17,18].

In this study, the CGMCC NO:28566 strain, which is capable of efficiently converting COBE to (*R*)-CHBE, was screened by soil-sieving bacteria. Ten NADESs were then used to assist the catalysis of the CGMCC NO:28566 strain. The yields and ee% of (*R*)-CHBE obtained by catalysis of COBE by whole-cell strains were employed as reference standards, from which NADES that could effectively enhance the catalytic effect were identified. In parallel, the quantity of the preferred NADES added, the method of addition, the individual components and blends, as well as the pivotal reaction conditions, and the cell membrane permeability were subjected to further investigation.

2. Results and Discussion

2.1. Identification of the Best Strains

The preferred strain, yielding (*R*)-CHBE at 95.7% and ee at 99%, was finally screened and named *Burkholderia cepacia* WZ-5. *Burkholderia cepacia* WZ-5 was preserved in the China General Microbiological Culture Collection Center and designated CGMCC No:28566. Microscopic observation, agarose nucleic acid electrophoresis of 16S rDNA sequences, and the evolutionary tree of the species are shown in Figure 1.

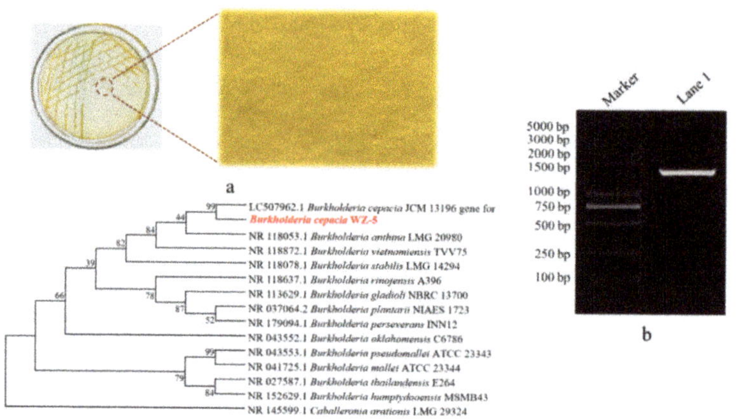

Figure 1. Identification of the best strains. (**a**) Microscope picture of strain No:28566; (**b**) Agarose nucleic acid electrophoresis of 16S rDNA sequences; (**c**) Evolutionary tree of the species.

2.2. Substrate Conversion Analysis

The substrate profile used by the CGMCC NO:28566 strain to reduce prochiral ketones was analyzed. As can be seen from Figure 2, the CGMCC NO:28566 strain has the ability to transform a variety of substrates, including a, b, c, and j, which are fluorinated aromatic ketones, and b has a 95.2% yield and 90.1% ee value. In addition, the strain also converted ester substrates such as e and i, and had a 95.7% yield and 99% ee value for COBE. COBE will be chosen as a substrate for further studies.

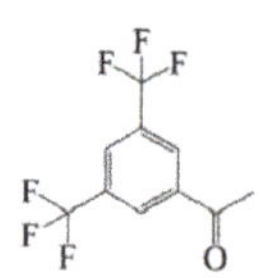

a
89.4% Yield
56.6% ee(*S*)

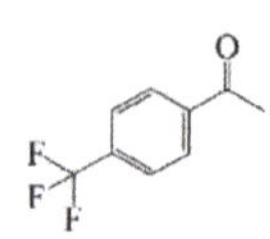

b
95.2% Yield
90.1% ee(*S*)

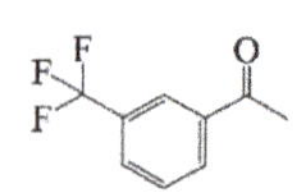

c
80.1% Yield
95.6% ee(*S*)

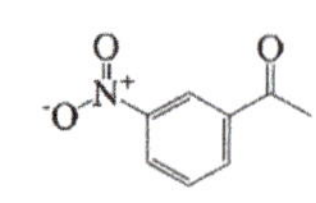

d
30.8% Yield
91.2% ee(*R*)

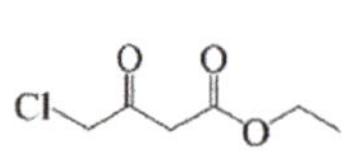

e
95.7% Yield
99% ee(*R*)

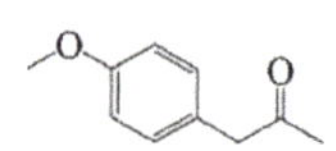

f
39.7% Yield
97.2% ee(*R*)

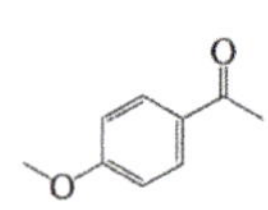

g
10.4% Yield
47.2% ee(*R*)

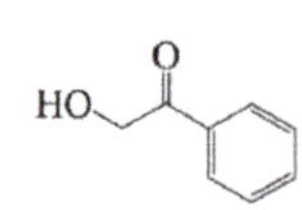

h
8.7% Yield
52.1% ee(*R*)

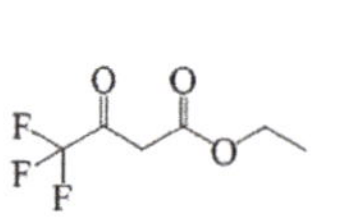

i
95.2% Yield
54.9% ee(*R*)

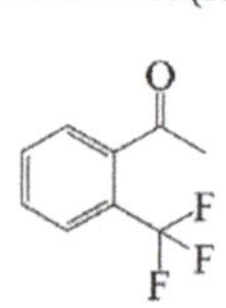

j
5.3% Yield
80.1% ee(*S*)

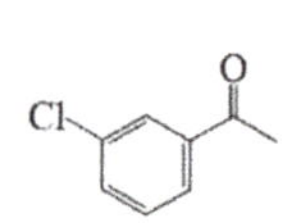

k
68.4% Yield
56.2% ee(*R*)

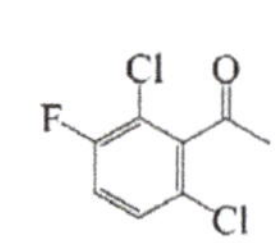

l
2.9% Yield
67.3% ee(*R*)

Figure 2. CGMCC NO:28566 strain catalytic substrate analysis. Reaction conditions: 10 mL PB buffer (pH 7.0), 120 g/L wet cells, 100 g/L glucose as co-substrate, 15 mM COBE, 200 rpm, and 25 °C for 30 h.

2.3. Effect of Medium Type on the Catalytic Activity of CGMCC NO:28566 Strain

As shown in Figure 3, LB medium, modified LB medium, complete medium, and TSB medium were selected as the basis for screening. Under the same reaction conditions, the yields of (*R*)-CHBE were TSB medium > complete medium > modified LB medium or LB medium, respectively, and the TSB medium had an 85.7% yield and a 99% ee value. It was postulated that the TSB medium was rich in carbon and nitrogen sources, as well as inorganic salts, which allowed the CGMCC NO:28566 strain to grow rapidly and demonstrate enhanced transformation effects. Consequently, TSB liquid medium was selected as the seed medium and fermentation medium for the CGMCC NO:28566 strain.

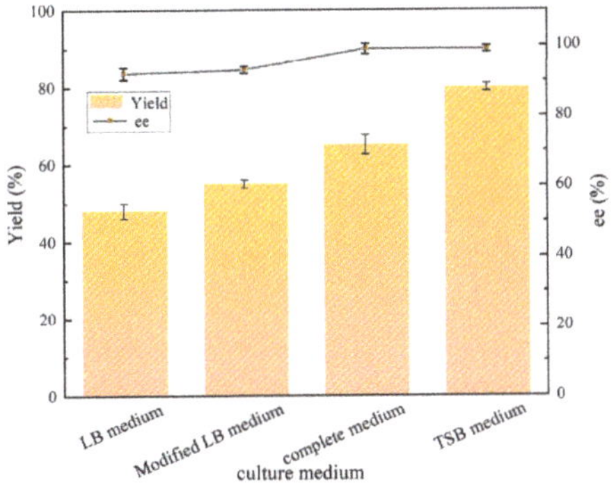

Figure 3. Effect of medium type on the catalytic activity of CGMCC NO:28566. Reaction conditions: 5 mL PB buffer (pH 7.0), 100 g/L wet cells, 100 g/L glucose as co-substrate, 15 mM COBE, 200 rpm, and 25 °C for 30 h.

2.4. Effect of NADESs and the Method of Addition of NADES on the Reduction Activity of CGMCC NO:28566

As illustrated in Figure 4, all 10 NADESs (Table 1) exhibited enhanced activity compared to the control. For instance, betaine:urea (1:2), proline:urea (1:2), choline chloride:glycerol (1:2), and choline chloride:urea (1:2) were observed to facilitate catalysis, resulting in 42.8%, 37.5%, and 33.9% higher yields of (*R*)-CHBE compared to the control, respectively. The assisted catalysis of ChCl:U (1:2) had the most significant effect, with the yield of (*R*)-CHBE reaching 89.1%, which was 58.2% higher than that of the control group. Conversely, the co-culture-based secondary addition of NADES resulted in lower yields than the control group in all experimental groups, with the lowest (*R*)-CHBE yield reaching 29.6%. It was observed that, following the addition of two NADES solutions, the yield of the experimental group in which ChCl:U (1:2) was employed was found to be significantly lower. In this study, two methods of adding NADES were employed for reduction. The first method was co-culture treatment with 0.5% (w/v) ChCl:U (1:2). The second method involved the re-addition of 0.5% (w/v) ChCl:U (1:2) to the reaction buffer under co-culture conditions. The results indicated no significant differences in ee values between the two NADES addition methods. However, co-culture supplementation significantly enhanced (*R*)-CHBE yield compared to buffer re-addition.

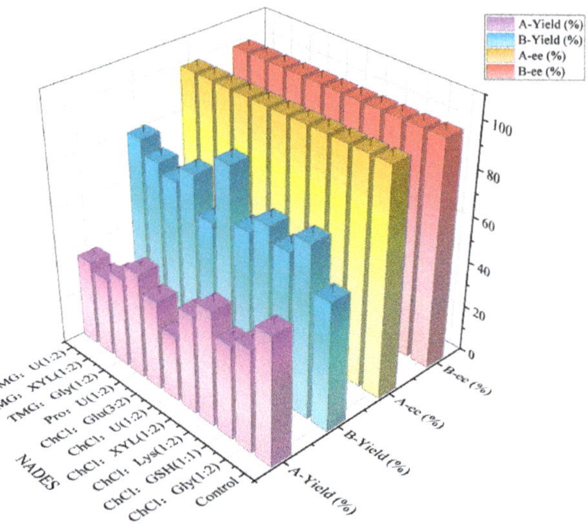

Figure 4. Effect of two NADES addition methods on reduction reaction. A: re-addition of 0.5% (*w*/*v*) ChCl:U (1:2) to the reaction buffer under co-culture conditions. B: co-culture treatment with 0.5% (*w*/*v*) ChCl:U (1:2).

Table 1. NADES for experiments.

Component I	Component II	Proportion (mol)
betaine	glycerol	1:2
betaine	xylitol	1:2
betaine	urea	1:2
choline chloride	urea	1:2
choline chloride	xylitol	1:2
choline chloride	glucose	3:2
choline chloride	glycerol	1:2
choline chloride	lysine	1:2
choline chloride	glutathione	1:1
proline	urea	1:2

When choline chloride is combined with urea in a 1:2 molar ratio to create a low-eutectic solvent, the resulting fluidic liquid contains anionic, cationic, and neutral molecules [15]. Upon the introduction of ChCl:U (1:2) into an aqueous solution, the presence of water molecules results in a disruption of the solvent's structural integrity. This may result in the partial breaking of hydrogen bonds and the partial dissociation of the two components, choline chloride and urea [19]. In contrast, CGMCC NO:28566 was a Gram-negative bacterium. The cytoplasmic membrane of this bacterium contained a substantial quantity of peptidoglycan outside the membrane, which was employed to maintain the structural stability of the cell wall. The partially dissociated choline chloride cation is capable of interacting with the polysaccharide backbone and peptide chains in peptidoglycan, a process that is facilitated by hydrogen bonding or electrostatic forces [20]. It is also possible that Cl^- may interact with the cell membrane in a manner that could result in alterations to the cell wall or cell membrane structure [20]. The alteration may be slight or drastic, but the addition of urea as a hydrogen bond donor at this time alters or reduces the damaging effect. This can be explained by the fact that the addition of urea forms a hydrogen bond with the salt anion [21]. This mechanism reduces the toxicity of choline chloride to the cell while concomitantly modifying the structural and functional permeability of the cell wall and cell membrane [20]. The series of interactions described above results in a more stable binding of the substrate COBE to the enzyme, thus facilitating the completion of the catalytic process.

The results demonstrated that the solvent ChCl:U (1:2) exerted a more pronounced effect on the bacterial structure, yet caused greater damage when reintroduced during the substrate transformation phase [22]. It was therefore concluded that it would be more beneficial to assist the process of catalysis by affecting the bacterial structure during the growth period of the strain.

2.5. Effect of ChCl:U (1:2) Content on the Reduction Reaction

The results of the addition analysis showed that the concentration of ChCl:U (1:2) solvent required further optimization. Figure 5 illustrates that the addition of ChCl:U (1:2) had no significant effect on the ee value of (*R*)-CHBE. The results indicated that the addition of 0.1% (*w*/*v*) did not have a significant effect on the co-catalysis, with the yield of (*R*)-CHBE increasing from 50.9% to 53.6%. However, the (*R*)-CHBE yield increased from 53.6% to 85.4% at 0.5% (*w*/*v*) ChCl:U (1:2), indicating that this additive concentration significantly affected cellular structure. The yield of (*R*)-CHBE exhibited a notable decline at additive dosages between 1% (*w*/*v*) and 2% (*w*/*v*). This further demonstrates the low toxicity of the low-eutectic solvent as well as the investigability of the addition method. Consequently, adding 0.5% (*w*/*v*) ChCl:U (1:2) was more beneficial for enhancing (*R*)-CHBE yield in the strain's fermentation culture.

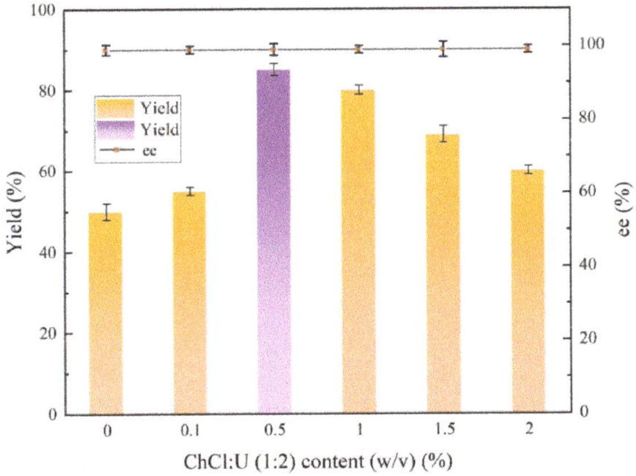

Figure 5. Effect of ChCl:U (1:2) content on the reduction reaction.

2.6. Comparison of Optimization of Key Reaction Conditions Before and After Co-Culture Treatment

2.6.1. Effect of Temperature on Reduction Reaction Before and After Co-Culture Treatment

As illustrated in Figure 6, the yield of (*R*)-CHBE exhibited a notable reduction from 85.9% to 76.4% when the reaction temperature was elevated from 25 °C to 30 °C following co-culture treatment. Further increases in temperature, up to 35 °C and 45 °C, resulted in a yield of only 65.7%. This indicates that as the temperature rises, the activity of the cells, which are already partially altered in their cellular structure, is destroyed by the higher temperature. Nevertheless, the observed decline in ee value at temperatures between 40 and 45 °C may be attributed to a reduction in enzyme activity and stereoselectivity. Prior to and following the co-culture treatment, there was no significant decrease in yield, indicating that the CGMCC NO:28566 strain exhibits enhanced high-temperature resistance.

The optimal reaction temperature for CGMCC NO:28566 in the absence of the co-culture treatment was 35 °C. In contrast, the co-culture treatment resulted in a reduction in the optimum selection temperature to 25 °C.

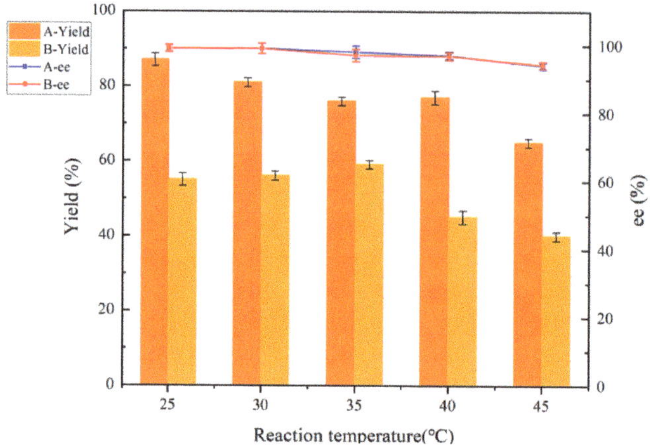

Figure 6. Effect of temperature on reduction reaction before and after co-culture treatment. A: co-culture treatment with 0.5% (*w*/*v*) ChCl:U (1:2). B: no co-culture treatment.

2.6.2. Effect of Reaction Time on Reduction Reaction Before and After Co-Culture Treatment

The results of the analysis in Figure 7 demonstrate that the co-culture treatment had no significant effect on the ee value. The yield of (*R*)-CHBE exhibited a rapid increase, reaching 85.7% within the range of 6 h–24 h. Upon extending the reaction time to 30 h, the yield remained largely unaltered under co-culture conditions. At this time, the reaction time was extended to 48 h, and the yields of (*R*)-CHBE exhibited a decline to varying degrees. It is postulated that the observed decline in yield can be attributed to the accumulation of conversion products as the reaction time increases, a reduction in the conversion rate, and potentially other by-products. The optimal reaction time of CGMCC NO:28566 without co-culture treatment was 30 h, while co-culture treatment reduced the optimal reaction time to 24 h. Co-culture treatments increase yields while saving time.

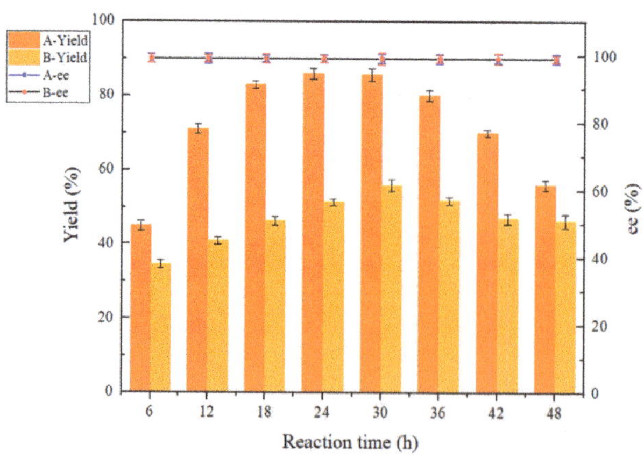

Figure 7. Effect of reaction time on reduction reaction before and after co-culture treatment. A: co-culture treatment with 0.5% (*w*/*v*) ChCl:U (1:2). B: no co-culture treatment.

2.6.3. Effect of pH on Reduction Reaction Before and After Co-Culture Treatment

The alteration in the cell structure under co-culture treatment conditions also affects the pH tolerance of the cells. As illustrated in Figure 8, the yields were found to be lower under more acidic conditions (pH 5.0–6.0) and more alkaline conditions (pH 9.0–10.0). The enzymes that play a significant role in this process were also subjected to more acidic or alkaline conditions, where enzyme activity was affected, and whole-cell catalytic efficiency was affected. The reaction was more favorable under neutral conditions, resulting in a yield of 86.7%. The optimal pH conditions were 7.0 before and after co-culture treatments.

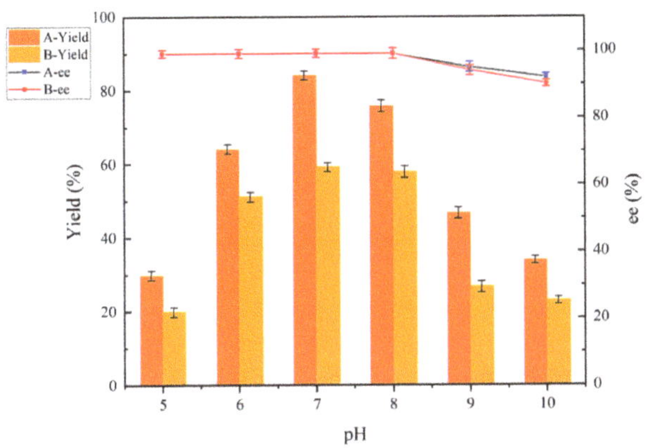

Figure 8. Effect of pH on reduction reaction before and after co-culture treatment. A: co-culture treatment with 0.5% (*w*/*v*) ChCl:U (1:2). B: no co-culture treatment.

2.7. Effect of ChCl:U (1:2) and Its Individual Components on the Reduction Reaction

The results of the analyses presented in Figure 9 indicate that neither the additional components nor the method of addition had a significant effect on the ee value of (*R*)-CHBE. The co-culture treatments of urea, choline chloride, or a simple co-mixture of choline chloride and urea resulted in increased yields of (*R*)-CHBE, with values of 66.3%, 71.2% and 71.2%, respectively. It can be analyzed that the co-culture treatments of urea, choline chloride, and a simple co-mixture of both did somehow increase the conversion capacity of the cells, but this effect is less obvious. However, this advantage is only apparent in the formation of ChCl:U (1:2), where the yield of (*R*)-CHBE increases from 58.1% to 83.2%.

However, the results from the repeated addition of choline chloride, urea, the simple two-component mixture, and ChCl:U (1:2) to the reaction buffer showed different decreases in (*R*)-CHBE yields. In addition, the effect of the repeated addition of ChCl:U (1:2) was also more pronounced in the above components, reducing the yield of (*R*)-CHBE to 34.8% compared to the control yield of 54.2%. It has been shown that low-eutectic solvents with choline chloride have some degree of biotoxicity [19], but that this toxicity may vary depending on the cellular system. The study also suggested that low-eutectic solvents based on choline chloride and urea are more biotoxic than either component alone or simple mixtures of the two, due to the high likelihood of ammonia production during the use phase of ChCl:U (1:2) [22].

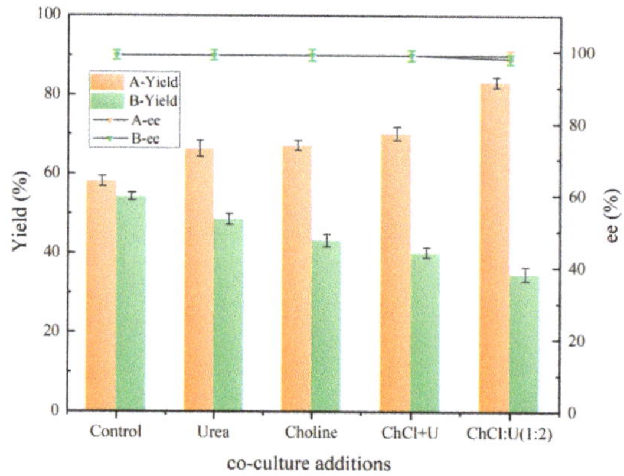

Figure 9. Effect of ChCl:U (1:2) and its individual components on the reduction reaction A: co-culture treatment with 0.5% (w/v) ChCl:U (1:2). B: re-addition of 0.5% (w/v) ChCl:U (1:2) to the reaction buffer under co-culture conditions.

2.8. Effect of ChCl:U (1:2) and Its Single Component on Cell Membrane Permeability

The process of improving catalytic efficiency by the degree of solvent action on the cell membrane, the components of the ChCl:U (1:2) solvent may also have a similar effect on the cell membrane. Typically, OD260 indicates the cellular nucleic acid leakage value, and OD280 indicates the cellular protein leakage value [23], so this subsection indirectly examines the change in cellular membrane permeability by testing the OD260 and OD280 indicators.

As illustrated in Table 2, the OD260 or OD280 values observed following 24 h of co-culture were found to be lower than those observed following 24 h of re-addition of the components presented in the table. In the case of secondary addition, the cell membrane structure may be more affected. The effect was significantly higher for ChCl:U (1:2) than for the single component and simple mixtures. Furthermore, the results of the assay after secondary addition of ChCl:U (1:2) to the reaction buffer reinforced this conclusion. Moreover, the measured OD260 or OD280 values corroborate the catalytic effect of Section 2.7.

Table 2. Quantitative analysis of the OD260 and OD280 values following the application of various treatments to ChCl:U (1:2) and its components.

Groups	Net OD260 nm	Net OD280 nm
Control	0.122 ± 0.015	0.158 ± 0.010
Urea * 24 h	0.135 ± 0.002	0.166 ± 0.012
Choline chloride * 24 h	0.145 ± 0.019	0.171 ± 0.011
Urea + Choline chloride *24 h	0.173 ± 0.023	0.191 ± 0.009
ChCl:U (1:2) * 24 h	0.351 ± 0.022	0.385 ± 0.024
Urea # 24 h	0.175 ± 0.016	0.193 ± 0.007
Choline chloride # 24 h	0.183 ± 0.013	0.199 ± 0.020
Urea + Choline chloride # 24 h	0.211 ± 0.021	0.341 ± 0.005
ChCl:U (1:2) # 24 h	0.583 ± 0.007	0.671 ± 0.014

Note: * represents co-culture; # represents re-addition in reaction buffer.

2.9. FTIR Analysis of ChCl:U (1:2) and Its Single Components

From the infrared stacking diagrams in Figure 10, it can be observed that the infrared spectra of the ChCl:U (1:2) at the wavelength range of 4000–500 cm^{-1} combine the infrared spectral features of the choline chloride and urea. On the surface, the map appears to be a superposition of the maps of both choline chloride and urea. In comparison to the telescopic vibration absorption peaks of NH2 in urea at the wavelengths of 3346 cm^{-1} and 3264 cm^{-1}, the absorption bands of ChCl:U (1:2) at this wavelength exhibit a broadening. Concurrently, the bending vibration of NH2 results in a shift of the absorption bands at the wavelengths of 1679 cm^{-1} and 1628 cm^{-1} towards the wavelengths of 1665 cm^{-1} and 1620 cm^{-1}, respectively. The observed formation of hydrogen bonding between choline chloride and urea is indicative of the absorption peak at 1474 cm^{-1}, which is attributed to the wobbling vibration of the CH3 group in the choline chloride molecule. In contrast, the absorption peak formed by the stretching vibration of the CCO group at 955 cm^{-1} in the spectrum of choline chloride remained after the formation of ChCl:U (1:2) low-eutectic solvent, indicating that the structure of Ch+ was not destroyed in the ChCl:U (1:2) solvent system.

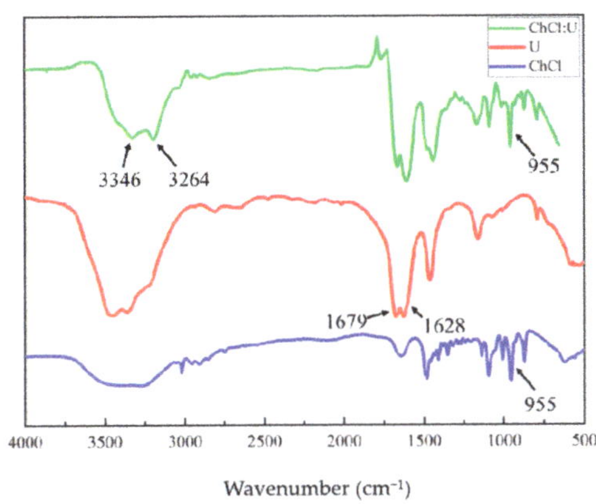

Figure 10. Infrared spectra of ChCl:U (1:2) and its single components.

Furthermore, the substantial number of hydrogen bonds that are formed during the preparation of this system indicates excellent flexibility properties [24]. The assertion that the interaction of choline chloride and urea forms hydrogen bonds during the formation of ChCl:U (1:2) is still the subject of some controversy. Nevertheless, the final conclusions indicate that this low-eutectic solvent possesses distinctive advantages over choline chloride or urea.

2.10. FCM Analysis Under Different Treatment Conditions of ChCl:U (1:2)

Propidium iodide (PI) [25,26] staining is a technique for staining cellular membranes that cannot penetrate the cell membrane of living cells. However, it can penetrate the cell membrane of broken cells, enabling nuclear staining [27]. The device exhibited a detection flux of 10,000 cell counts. The results of the flow cytometry (FCM) analysis, as presented in Figure 11, demonstrate that the cells treated with the three different approaches exhibited distinct outcomes. In particular, the 24 h co-culture treatment (PE-A+ 0.14) exhibited a notable difference in comparison to the control (PE-A+ 0.37). It can be surmised that during the period of co-cultivation with ChCl:U (1:2), the degree of alteration to the cell membrane is small, which confirms that the higher assisted catalysis results under this condition are based on lower damage to the cells. Furthermore, the result (PE-A+ 4.89) with ChCl:U (1:2) re-addition for 24 h on the basis of the co-culture (Figure 11c) appeared to be more pronounced than the control, indicating that this treatment condition caused greater damage to the cells. Consequently, the results demonstrate that the assisted catalysis effect was significantly diminished under the specified treatment conditions. The findings were in alignment with the outcomes of the OD260 and OD280 tests presented in Section 2.8.

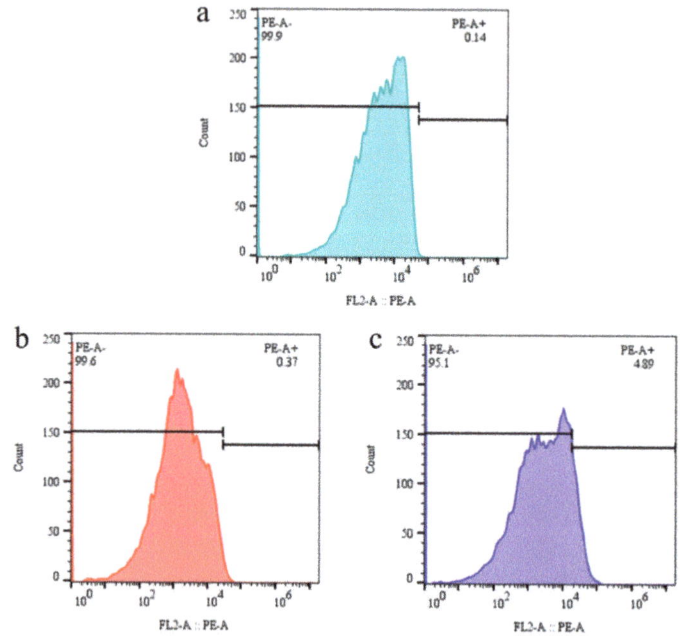

Figure 11. FCM analysis under different treatment conditions with ChCl:U (1:2). (**a**) Cells without co-culture treatment; (**b**) co-culture treatment with 0.5% (*w*/*v*) ChCl:U (1:2) for 24 h; (**c**) re-addition of 0.5% (*w*/*v*) ChCl:U (1:2) for 24 h in reaction buffer.

2.11. SEM Analysis of CGMCC NO:28566 Strain Cells Under Different Treatment Conditions of ChCl:U (1:2)

Scanning electron microscopy was conducted on CGMCC NO:28566 cells following a series of treatment conditions involving ChCl:U (1:2) (voltage settings were 15 kV, magnification of Figure 12a was 10,000, and the remainder of the magnification was 80,000). As illustrated in Figure 12a, the electron microscope images of the CGMCC NO:28566 strain revealed an overall rod-like (or stick-shaped) morphology, which was consistent with the observations made using light microscopy (Figure 1). The length of the cells was approximately 1.39 μm. In comparison to the results obtained in the absence of ChCl:U (1:2) (Figure 12a), the surface of the cell exhibited varying degrees of alteration. Cellular depressions were observed in cells co-cultured with ChCl:U (1:2) for 24 h (Figure 12c). In contrast, cells that had been re-added ChCl:U (1:2) in reaction buffer for 24 h demonstrated an increased area of cell breakage and deformation (Figure 12d). Consequently, this also elucidates the catalytic outcomes delineated above, as well as the varying degrees of elevation of OD260 and OD280.

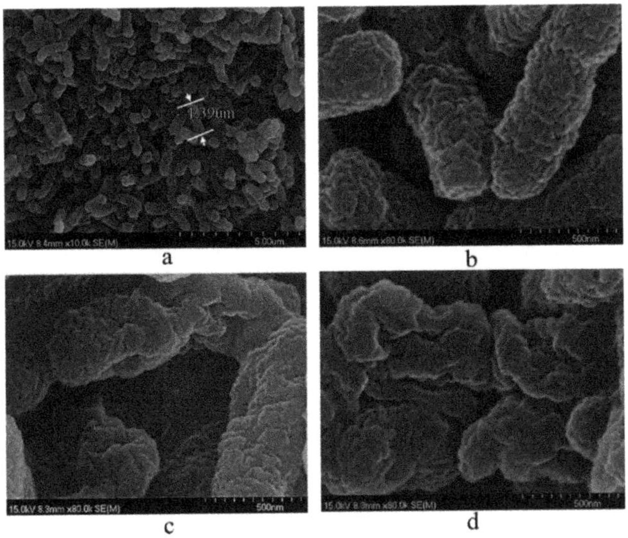

Figure 12. SEM analysis of CGMCC NO:28566 cells under different processing conditions with ChCl:U (1:2). (**a**) Appearance of CGMCC NO:28566 cells. (**b**) Cells without co-culture treatment. (**c**) Co-culture treatment with 0.5% (*w*/*v*) ChCl:U (1:2) for 24 h. (**d**) Re-addition of 0.5% (*w*/*v*) ChCl:U (1:2) for 24 h in reaction buffer.

3. Materials and Methods

3.1. Materials and Reagents

The strains employed in the experiment were derived from a soil screening process. All NADESs (Table 1) were supplied by Shanghai Chengjie Chemical Co., Ltd. (Shanghai, China). Media-related reagents were obtained from Sinopharm Chemical Reagent Co., Ltd. (Ningbo, China). COBE (>98% purity, HPLC) and (*R*)-CHBE (>99% purity, HPLC) were procured from Rinn Technology Development Co., Ltd. (Shanghai, China).

3.2. Screening, Identification, and Culture of Microorganisms

The following culture media are relevant: enrichment liquid medium, glucose 25 g/L, yeast extract 3 g/L, ammonium sulphate 5 g/L, magnesium sulphate 0.25 g/L, dipotassium hydrogen phosphate trihydrate 1.5 g/L, and potassium dihydrogen phosphate 1.5 g/L. The pH was adjusted to approximately 7.0 with a 1 M sodium hydroxide solution. The restriction liquid medium comprised ammonium sulphate 5 g/L, magnesium sulphate 0.25 g/L, dipotassium hydrogen phosphate trihydrate 1.5 g/L, potassium dihydrogen phosphate 5 g/L, 10 mM COBE, deionized water as a solvent, and a final pH adjustment to 7.0 with 1 M sodium hydroxide solution. The solid medium was restricted, containing 5 g/L ammonium sulphate and 0.25 g/L magnesium sulphate. The medium consisted of 25 g/L dipotassium hydrogen phosphate trihydrate, 1.5 g/L potassium dihydrogen phosphate, 1.5 g/L potassium hydrogen phosphate, 10 mM COBE, 16 g/L agar, and deionized water as a solvent. The pH was finally adjusted to approximately 7.0 with 1 M sodium hydroxide solution.

The soil samples were collected from the provinces of Zhejiang, Jiangsu, and Shandong in China. The soil sieving procedure was as follows: approximately 0.5 g of each soil sample was dissolved in 5 mL of 0.9% saline solution, vortexed and shaken thoroughly for approximately 2–3 min, and then left to stand for 30 min. Two milliliters of the supernatant was then transferred to the enrichment medium. Subsequently, 1 mL of the enriched bacterial solution and 10 mM COBE were added to the restriction-screening liquid medium (30 mL). The culture was incubated at 25 °C and 180 rpm for 2–5 days until it became turbid. Subsequently, turbid restrictive liquid cultures were inoculated onto solid medium for single colony culture. The resulting single colonies were subjected to incubation and biotransformation experiments. The strains that exhibited the highest efficiency in transforming COBE into (*R*)-CHBE were finally screened, and the strains were conserved.

Total DNA was extracted from the best strains, and then the sequence was amplified as bacterial 16S rDNA using universal primers (upstream primer 5′ AGTTTGATCMTGGCTCAG 3′, downstream primer 5′ GGTTACCTTGTTACGACTT 3′), and the results were used for strain identification. The PCR products were subjected to 0.9% agarose gel electrophoresis, and then sequenced by Beijing Prime Biotechnology Co., Ltd. (Beijing, China)

3.3. Asymmetric Bioreduction Process

The best strains of wet cells were obtained and resuspended in a test tube containing 50 mL of pH 7.0 phosphate buffer and 0.1 g/mL glucose as a co-substrate. COBE was pre-dissolved in 5% (*v*/*v*) isopropanol, and the buffer was subsequently added to 5 mL. The reaction was incubated at 25 °C and 180 rpm for 30 h, centrifuged at 9000 rpm for 10 min, and the supernatant was extracted with an equal volume of ethyl acetate three times, and dried over anhydrous sodium sulfate. The obtained reduction product was dried until free of liquid, followed by successive addition of 10 μL acetic anhydride and 10 μL pyridine. The mixture was then boiled for 30 min. Finally, the resolved chiral product was dissolved in an appropriate amount of ethyl acetate. The sample was redissolved in 300 μL of ethyl acetate and analyzed by GC.

3.4. Analytical Methods

A gas chromatography Agilent CP7502 J&W CP ChirasilDex CB fitted with a chiral column (Machery Nagel; 25 m × 0.25 mm × 0.25 mm) was employed for the detection process. The inlet temperature was 250 °C, the column temperature was 110 °C, and the detector temperature was 250 °C. A hydrogen ion flame detector was employed with a split ratio of 1:15 and a flow rate of 1 mL/min. The retention time for COBE was 9 min, while those for (*R*)-CHBE and (*S*)-CHBE were 19 min and 19.8 min, respectively. Equations (1) and (2) evaluated the yield (X) and enantiomeric excess of *R*-CHBE (eep).

$$X(\%) = \frac{\mathrm{p} \times \mathrm{Ms}}{\mathrm{q} \times \mathrm{Mp}} \times 100\% \qquad (1)$$

Ms and Mp are the molecular weights of the substrate and the product, respectively. p and q represent the mass of the product at the end of the reaction and the initial mass of the substrate, respectively.

$$eep = \frac{\mathrm{C_R} - \mathrm{C_s}}{\mathrm{C_R} + \mathrm{C_s}} \times 100\% \qquad (2)$$

C_R and C_S represent the concentrations of *R*-CHBE and *S*-CHBE, respectively.

3.5. Selection of the Most Suitable Culture Medium

Screening for the optimal medium was conducted by studying the effect of the type of medium on the catalytic activity of the bacterium. LB medium (tryptone: 10 g/L; yeast extract: 5 g/L; sodium chloride: 10 g/L), modified LB medium (tryptone: 10 g/L; yeast extract: 10 g/L; sodium chloride: 10 g/L), TSB medium (tryptone: 17 g/L, soya peptone: 3.0 g/L, sodium chloride: 5 g/L, glucose: 2.5 g/L, chlorodimethyl phosphate: 2.5 g/L), and complete medium (glucose: 25 g/L, yeast extract: 3 g/L, ammonium sulphate: 5 g/L, magnesium sulphate: 0.25 g/L, dipotassium hydrogen phosphate trihydrate: 1.5 g/L, potassium dihydrogen phosphate: 1.5 g/L) were used for the fermentation of the engineered bacteria. The rest of the procedure is described in Section 3.3.

3.6. Effect of Different NADESs on Biotransformation of COBE

On the basis of the base medium screened in Section 3.5, the seed solution was inoculated into 150 mL of liquid fermentation medium at an inoculation rate of 3%. At the same time, 10 NADESs were added at 0.5% (*w*/*v*) of the fermentation broth (150 mL), respectively. The best strain wet cells (240 g/L) obtained from the co-culture were resuspended in test tubes with 0.1 M phosphate buffer, pH 7.0, and a final concentration of 40 mM COBE was added. The remainder of the procedure was carried out as in Section 3.3.

3.7. Effect of Secondary Addition of NADESs on the Biotransformation of COBE

The wet cells of the best strain (240 g/L) obtained by co-cultivation were resuspended in test tubes with 0.1 M phosphate buffer, pH 7.0, and 10 NADES were added again. The amount of NADES added was 0.5% (*w*/*v*) of the reaction system (5 mL), and the final concentration of COBE added was 40 mM. The rest of the procedure was the same as in Section 3.6.

3.8. Optimization of Biotransformation Conditions

Depending on the preferred NADES, adjust the addition gradient from 0.1% to 2% (*w*/*v*) according to the preferred addition method and proceed as in Section 3.6.

The best strain of wet cells enriched under NADES co-culture conditions was subjected to the reduction reaction at a temperature of 25–45 °C, a reaction time of 12–60 h, and buffer pH 5–10. At the same time, the best strain of wet cells without co-culture treatment was subjected to reduction reactions under the different parameters mentioned above. The rest of the procedure was the same as in Section 3.6.

3.9. Comparison of Assisted Catalysis of Related Components in Preferred NADES by Two Addition Methods

Addition method I: During the fermentation period, the best strain cells were co-cultured with the preferred NADES, single-component (choline chloride/urea) and two-component simple co-mixtures (choline chloride and urea) of the preferred NADES, respectively. The above components were added at 0.5% (*w*/*v*) of the fermentation broth (150 mL), respectively. The rest of the procedure was the same as in Section 3.6.

Addition method II: The best strain wet cells under co-culture conditions were collected, and then the preferred NADES, single-component (choline chloride/urea) and two-component simple co-mixtures (choline chloride and urea) of the preferred NADES, were added again in reaction buffer. The above components were added again at 0.5% (*w*/*v*) of the reaction system (5 mL). The rest of the procedure was performed as in Section 3.7.

3.10. Characterization of Cell Permeability Under Different Addition Conditions of NADES

The seed liquid of the best strain was inoculated into the fermentation broth (150 mL) at an inoculum rate of 3%. Meanwhile, three portions of the inoculated fermentation broths were added with 0.5% (*w*/*v*) of the preferred NADES: single component I (choline chloride), single component II (urea), and two-component simple co-mixtures (choline chloride and urea). The last portion of the fermentation broth was used as a blank control (only the seed liquid was added). After 24 h of incubation at 25 °C, 180 rpm, the appropriate amount of supernatant was obtained by centrifugation at 9000 rpm for 10 min, and then the OD260 and OD280 indexes were measured under UV. The bacterial precipitates were further re-collected in 5 mL of reaction buffer, and the above components were added separately. The supernatants were then collected under the same incubation conditions for testing.

3.11. FTIR Characterization of ChCl:U (1:2) and Its Single Components

The FTIR (Vertex 70, Bruker Corporation, Berlin, Germany) spectra of individual components of the preferred NADES were analyzed for spectral variations in the 4000–400 cm^{-1} wavelength region. The preferred NADES and its single component were structurally analyzed.

3.12. FCM Characterization of Cells Obtained by Different Treatment Conditions

An appropriate amount of fermentation broth without NADES treatment, with NADES co-culture treatment for 24 h, and with NADES treatment in buffer for an additional 24 h was taken. After centrifugation at 9000 rpm for 10 min, the supernatant was discarded and washed twice with 0.1 M pH 7.0 PB buffer, then resuspended in 1 mL of 0.1 M pH 7.0 phosphate buffer. Under light avoidance conditions, 200 μL of diluted propidium iodide (PI) staining solution was gently aspirated or beaten into the above resuspended bacterial solution, and then flow cytometry single stain detection was performed.

3.13. SEM Observation of Cells Obtained Under Different Treatment Conditions

Strain CGMCC No:28566 was subjected to the following treatments: (1) control (untreated cells without co-cultivation); (2) co-cultivation with 0.5% (*w*/*v*) ChCl:U (1:2) for 24 h; and (3) additional supplementation with 0.5% (*w*/*v*) ChCl:U (1:2) in the reaction buffer, followed by continued cultivation for 24 h. Adequate amounts of wet cells from the best strain under each treatment condition were collected and slowly added to pre-cooled 2.5% glutaraldehyde solution, fixed overnight, and centrifuged to remove the fixative. The samples were then dehydrated with 70%, 80%, 90%, and 100% ethanol, respectively, lyophilized, and subjected to gold spraying treatment. The structure was observed and analyzed under the scanning electron microscope (SEM) (JSM-6360LV, JEOL Ltd., Tokyo, Japan).

4. Conclusions

This study investigated the catalytic reactions under different conditions based on coculture-assisted catalysis of NADSE. The ten NADESs were co-cultured with the CGMCC NO:28566 strain in fermentation broth for 24 h. The wet cell obtained under these conditions was subjected to a reduction reaction. Concurrently, the NADES, which possesses the qualities of enhancing substrate solubility and stabilizing enzyme activity, was incorporated into the reduction reaction for a period of 24 h, during which time the process was facilitated by the NADES. The outcomes of the two methodologies for catalytic additions demonstrated that the enantiomeric excess values were largely unperturbed. The former approach, in which ChCl:U (1:2) [28–30] assisted catalysis was found to be the most effective, with a 58.2% increase in (*R*)-CHBE yield. However, in the latter case, the lowest yield of (*R*)-CHBE was found to have decreased from 54.2% to 29.6%. It is likely that the observed effects were due to the action of NADES on the bacterial cells during the co-culture stage, resulting in changes to their membrane permeability. The addition of NADES to this environment may have further increased the extent of these changes, potentially impairing cellular activity. Consequently, we have elected to employ the use of NADES during the enrichment phase of fermentation, with the intention of facilitating catalysis.

Furthermore, the objective of this investigation is to determine whether the observed effect of ChCl:U (1:2) can be attributed to the individual components present in the solvent or whether a simple blend of single components can achieve the same result. It was thus demonstrated that a single component or simple co-mixture exerted a certain degree of assisted catalysis effect. Furthermore, the auxiliary effect of simple co-mixture was found to be slightly higher than that of a single component. However, the effect was considerably less pronounced than that observed with ChCl:U (1:2) solvent assistance. The OD260 and OD280 values were determined, and the results demonstrated that the OD260 and OD280 values were indeed elevated in comparison to the control due to the single component or simple co-mixture. However, the elevation of the two values due to ChCl:U (1:2) was more significant. The trends observed in the OD260 and OD280 values were also corroborated by the results of the catalysis of the components.

The selected wild-type enzyme possesses an intrinsic catalytic advantage due to its native multi-enzyme synergy system, enabling superior substrate specificity compared to artificially engineered strains. Moreover, the wild-type strain demonstrates enhanced growth kinetics and operational stability, eliminating the need for genetic-instability-prone regulatory interventions required in engineered counterparts. The screened wild strain exhibits remarkable environmental tolerance and a broad substrate spectrum. These inherent enzymatic virtues establish a robust biological foundation for the NADES system investigated herein, while providing a diverse enzymatic repertoire for discovering pivotal biocatalysts.

Author Contributions: Writing—methodology, original draft, and data curation, Y.W.; methodology and writing—review and editing, B.L.; methodology, Y.D.; software, Z.T.; supervision, L.T.; funding acquisition and review and editing, Z.O. All authors have read and agreed to the published version of the manuscript.

Funding: This work was supported by the Zhejiang Provincial Science and Technology Plan Project (2024C03014), the National Natural Science Foundation of China (Project No. 21978267).

Institutional Review Board Statement: Not applicable.

Informed Consent Statement: Not applicable.

Data Availability Statement: All data generated or analyzed during this study are included in this article.

Conflicts of Interest: The authors declare no conflicts of interest.

References

1. Ren, S.H.; Mu, T.C.; Wu, W.Z. Advances in Deep Eutectic Solvents: New Green Solvents. *Processes* **2023**, *11*, 1920. [CrossRef]
2. Plotka-Wasylka, J.; Guardia, M.D.L.; Andruch, V.; Vilková, M. Deep eutectic solvents vs ionic liquids: Similarities and differences. *Microchem. J.* **2020**, *159*, 105539. [CrossRef]
3. Svigelj, R.; Zanette, F.; Toniolo, R. Electrochemical Evaluation of Tyrosinase Enzymatic Activity in Deep Eutectic Solvent and Aqueous Deep Eutectic Solvent. *Sensors* **2023**, *23*, 3915. [CrossRef]
4. Vanda, H.; Dai, Y.; Wilson, E.G.; Verpoorte, R.; Choi, Y.H. Green solvents from ionic liquids and deep eutectic solvents to natural deep eutectic solvents. *Comptes Rendus Chim.* **2018**, *21*, 628–638. [CrossRef]
5. Liu, Y.; Friesen, J.B.; Mcalpine, J.B.; Lankin, D.C.; Chen, S.N.; Pauli, G.F. Natural Deep Eutectic Solvents: Properties, Applications, and Perspectives. *J. Nat. Prod.* **2018**, *81*, 679–690. [CrossRef]
6. Zhang, Y.L.; Ren, H.W.; Maarof, H.; Udin, S.M.; Liu, Y.Z.; Li, M.Y.; Alias, H.; Duan, E. The effect of water content on lignin solubilization in deep eutectic solvents. *J. Mol. Liq.* **2023**, *374*, 121271. [CrossRef]
7. Kumar, A.; Dhar, K.; Kanwar, S.S.; Arora, P.K. Lipase catalysis in organic solvents: Advantages and applications. *Biol. Proced. Online* **2016**, *18*, 2. [CrossRef]
8. Yu, T.; Manman, Z.; Hu, T.T.; Liu, C.G. Natural deep eutectic solvent—A novel green solvent for protein stabilization. *Int. J. Biol. Macromol.* **2023**, *247*, 125477. [CrossRef]
9. Qian, F.; Liu, H.Y.; Yu, S.R.; Zhang, Y.; Wang, P. New Strategy for Effective Biosynthesis of Chiral Aryl Alcohols: Co-Cultivation Microbe with Natural Deep-Eutectic Solvent. *ACS Sustain. Chem. Eng.* **2023**, *11*, 4441–4449. [CrossRef]
10. Długosz, O. Natural Deep Eutectic Solvents in the Synthesis of Inorganic Nanoparticles. *Materials* **2023**, *16*, 627. [CrossRef]
11. Ye, M.Q.; Ye, Y.Q.; Du, Z.J.; Chen, G.J. Cell-surface engineering of yeasts for whole-cell biocatalysts. *Bioprocess Biosyst. Eng.* **2021**, *44*, 1003–1019. [CrossRef] [PubMed]
12. Garzón-Posse, F.; Becerra-Figueroa, L.; Hernández-Arias, J.; Gamba-Sánchez, D. Whole Cells as Biocatalysts in Organic Transformations. *Molecules* **2018**, *23*, 1265. [CrossRef] [PubMed]
13. Wang, D.G.; Chen, M.Q.; Zeng, X.; Li, W.J.; Liang, S.L.; Lin, Y. Improving the catalytic performance of Pichia pastoris whole-cell biocatalysts by fermentation process. *RSC Adv.* **2023**, *11*, 36329–36339. [CrossRef]

14. Sadiq, I.Z.; Muhammad, A.; Mada, S.B.; Ibrahim, B.; Umar, U.A. Biotherapeutic effect of cell-penetrating peptides against microbial agents: A review. *Tissue Barriers* **2022**, *10*, 1995285. [CrossRef]
15. Yang, T.X.; Zhao, L.Q.; Wang, J.; Song, G.L.; Liu, H.M.; Cheng, H.; Yang, Z. Improving Whole-Cell Biocatalysis by Addition of Deep Eutectic Solvents and Natural Deep Eutectic Solvents. *ACS Sustain. Chem. Eng.* **2017**, *5*, 5713–5722. [CrossRef]
16. Zhu, Y.H.; Liu, C.Y.; Cai, S.; Guo, L.B.; Kim, I.W.; Kalia, V.C.; Lee, J.K.; Zhang, Y.W. Cloning, expression and characterization of a highly active alcohol dehydrogenase for production of ethyl (*S*)-4-chloro-3-hydroxybutyrate. *Indian J. Microbiol.* **2019**, *59*, 225–233. [CrossRef]
17. Zhu, J.F.; Bai, Y.J.; Fan, T.P.; Zheng, X.H.; Cai, Y.J. Characterization of acid-resistant aldo–keto reductases capable of asymmetric synthesis of (*R*)-CHBE from Lactobacillus plantarum DSM20174. *Syst. Microbiol. Biomanufacturing* **2023**, *3*, 634–646. [CrossRef]
18. Chen, X.; Liu, Z.Q.; Lin, C.P.; Zheng, Y.G. Efficient biosynthesis of ethyl (*R*)-4-chloro-3-hydroxybutyrate using a stereoselective carbonyl reductase from *Burkholderia gladioli*. *BMC Biotechnol.* **2016**, *16*, 70. [CrossRef]
19. Marcus, Y. Unconventional deep eutectic solvents: Aqueous salt hydrates. *ACS Sustain. Chem. Eng.* **2017**, *5*, 11780–11787. [CrossRef]
20. Wen, Q.; Chen, J.X.; Tang, Y.L.; Wang, J.; Yang, Z. Assessing the toxicity and biodegradability of deep eutectic solvents. *Chemosphere* **2015**, *132*, 63–69. [CrossRef]
21. Monhemi, H.; Housaindokht, M.R.; Moosavi-Movahedi, A.A.; Bozorgmehr, M.R. How a protein can remain stable in a solvent with high content of urea: Insights from molecular dynamics simulation of Candida antarctica lipase B in urea: Choline chloride deep eutectic solvent. *Phys. Chem. Chem. Phys.* **2014**, *16*, 14882–14893. [CrossRef] [PubMed]
22. Jung, D.; Jung, J.B.; Kang, S.; Li, K.; Hwang, I.; Jeong, J.H.; Kim, H.S.; Lee, J. Toxico-metabolomics study of a deep eutectic solvent comprising choline chloride and urea suggests in vivo toxicity involving oxidative stress and ammonia stress. *Green Chem.* **2021**, *23*, 1300–1311. [CrossRef]
23. Gao, X.; Liu, J.; Li, B.; Xie, J. Antibacterial Activity and Antibacterial Mechanism of Lemon Verbena Essential Oil. *Molecules* **2023**, *28*, 3102. [CrossRef]
24. Ashworth, C.R.; Matthews, R.P.; Welton, T.; Hunt, P.A. Doubly ionic hydrogen bond interactions within the choline chloride-urea deep eutectic solvent. *Phys. Chem. Chem. Phys.* **2016**, *18*, 18145–18160. [CrossRef]
25. Rosenberg, M.; Azevedo, N.F.; Ivask, A. Propidium iodide staining underestimates viability of adherent bacterial cells. *Sci. Rep.* **2019**, *9*, 6483. [CrossRef]
26. Sharma, T.; Kavita Mishra, B.B.; Variyar, P.S. Detection of Gamma Radiation Processed Onion During Storage Using Propidium Iodide Based Fluorescence Microscopy. *Food Chem.* **2022**, *398*, 133928. [CrossRef]
27. Zhang, N.; Fan, Y.; Li, C.; Wang, Q.M.; Leksawasdi, N.; Li, F.L.; Wang, S.A. Cell permeability and nuclear DNA staining by propidium iodide in basidiomycetous yeasts. *Appl. Microbiol. Biotechnol.* **2018**, *10*, 4183–4191. [CrossRef] [PubMed]
28. Soares, G.A.; Alnoch, R.C.; Santos, L.A.D.; Mafra, M.R.; Mitchell, D.A.; Krieger, N. High hydrolytic activity of the metagenomic lipase LipC12 in deep eutectic solvents. *J. Mol. Liq.* **2023**, *391*, 123383. [CrossRef]
29. Ghazali, Z.; Yarmo, M.A.; Hassan, N.H.; Teh, L.P.; Othaman, R. New Green Adsorbent for Capturing Carbon Dioxide by Choline Chloride: Urea-Confined Nanoporous Silica. *Arab. J. Sci. Eng.* **2020**, *45*, 4621–4634. [CrossRef]
30. Stefanovic, R.; Ludwig, M.; Webber, G.B.; Atkin, R.; Page, A.J. Nanostructure, hydrogen bonding and rheology in choline chloride deep eutectic solvents as a function of the hydrogen bond donor. *Phys. Chem. Chem. Phys.* **2017**, *19*, 3297. [CrossRef]

molecules

MDPI

Article

Competition of the Addition/Cycloaddition Schemes in the Reaction Between Fluorinated Nitrones and Arylacetylenes: Comprehensive Experimental and DFT Study

Szymon Jarzyński [1], Andrzej Krempiński [2], Anna Pietrzak [3], Radomir Jasiński [4,*] and Emilia Obijalska [2,*]

1 Faculty of Chemistry, Department of Organic Chemistry, University of Lodz, Tamka 12, 91-403 Lodz, Poland; szymon.jarzynski@chemia.uni.lodz.pl
2 Faculty of Chemistry, Department of Organic and Applied Chemistry, University of Lodz, 91-403 Lodz, Poland; andrzej.krempinski@edu.uni.lodz.pl
3 Institute of General and Ecological Chemistry, Lodz University of Technology, Żeromskiego 116, 90-924 Lodz, Poland; anna.pietrzak.1@p.lodz.pl
4 Cracow University of Technology, Department of Organic Chemistry and Technology, Warszawska 24, 31-155 Kraków, Poland
* Correspondence: radomir.jasinski@pk.edu.pl (R.J.); emilia.obijalska@chemia.uni.lodz.pl (E.O.)

Abstract

The course of the reactions of acetylenes with fluorinated nitrones in the presence of $Zn(OTf)_2$ and Et_2Zn was investigated. The formation of hydroxylamines and/or 1,2-oxazolines as products was observed. The desired hydroxylamines were formed as main products if reactions were carried out with the usage of Et_2Zn. In order to explain the obtained results, quantum mechanical calculations of the reaction paths leading to both products were carried out. Further research allowed us to develop the enantioselective variant of described reactions with the usage of enantiomerically pure AziPhenol ligand bearing chiral aziridine scaffold.

Keywords: fluorinated nitrones; acetylenes; enantioselective synthesis; cycloaddition; DFT calculation

1. Introduction

The synthesis of organofluorine compounds is an intensively developed field of organic chemistry. Compounds containing fluorine atoms (in particular small fluoroalkyl substituents) in their structure exhibit interesting physicochemical and biological properties [1–4]. On the other hand, nitrones are an extremely important class of building blocks that have been used for the synthesis of various organic derivatives [5,6]. For example, propargyl *N*-hydroxylamines obtained by the addition of acetylenes to nitrones constitute an interesting class of compounds that can be easily transformed into valuable compounds e.g., 4-isoxazolines, isoxazoles, pyrimidines, acylaziridines, α,β-unsaturated compounds [7–13]. Fluorinated nitrones, which are now easily synthesized from commercially available reagents [14–16], are powerful building blocks that can be used to incorporate CF_3 and CHF_2 substituents into organic molecules. To date, nitrones have not been widely explored in organic synthesis. Examples of their use are cycloadditions of nitrones with alkenes, alkynes leading to the formation of oxazolines [17], oxazolidines [18] and β-lactams [15]. Other reactions described in the literature are the addition of Grignard reagents to obtain the corresponding hydroxylamines [19]. Additionally, since data on enantioselective protocols of additions of nucleophiles to fluorinated nitrones have not been

 https://doi.org/10.3390/molecules30234578

reported. The only example of an enantioselective reaction using nitrones derived from fluorinated aldehydes is the Kinugasa reaction but reported enantioselectivities were very low [15]. Several examples described in the literature concern enantioselective additions of nucleophiles to fluorinated imines [20].

This article describes the results of research on the addition of various acetylenes to nitrones derived from trifluoro- and difluoroacetaldehydes. Contrary to such reactions with non-fluorinated substrates [21], additions to fluorinated nitrones have not been studied. In the extension of the research, it was also decided to develop an enantioselective variant of the reaction.

2. Results and Discussion

The key starting materials, i.e., the nitrones **3a,b**, were obtained according to a previously published procedure involving the reaction of the appropriate hydroxylamine with trifluoroacetaldehyde hydrate or difluoroacetaldehyde ethyl hemiacetal (Scheme 1) [14–16].

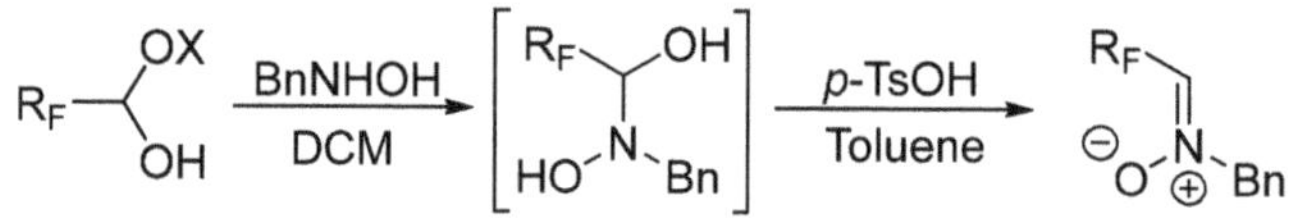

1a: $R_F = CF_3$; X = H
1b: $R_F = CHF_2$; X = Et
2a-b
3a: $R_F = CF_3$; 54%
3b: $R_F = CHF_2$; 82%

Scheme 1. Synthesis of fluorinated nitrones.

2.1. Additions of Acetylenes to Fluorinated Nitrones

The goal of the studies was to investigate the scope and application of additions of diverse acetylenes to nitrones derived from fluorinated aldehydes and the elaboration of enantioselective protocol of this reaction. In the course of experimental work two procedures previously described in the literature for non-fluorinated substrates were applied [7,22]. Method A was based on the reaction of an appropriate nitrone with acetylene in the presence of zinc triflate and triethylamine [22]. The optimization of the amount of reagents used was performed for model substrates **3a** and **4a**. As described in the literature, it was noted that to obtain satisfactory reaction yields, a 2–3-fold excess of acetylene **4a** should be used.

The optimal amount of $Zn(OTf)_2$ was 0.5 equivalents relative to the amount of used nitrone of type **3**. The application of larger amounts (eg. 1.2 equiv) of this catalyst did not significantly change the yield or the proportion of obtained products. Reactions of the *N*-alkyl-C-(fluoroalkyl) nitrones **3a,b** with the acetylenes **4a-g** were performed under optimized conditions (Scheme 2). The reaction time was monitored by TLC (the disappearance of the spot from the substrate was followed) and it varied (1.5–12 h) depending on the type of substituents in the used substrates. Longer reaction times were observed for substrates containing sterically hindered substituents or an electron-donating group on the aromatic ring. The change of the CF_3 substituent to CHF_2 in the molecule of the starting nitrone **3** did not change the reaction time, but the hydroxylamine **5ba** was isolated with a higher yield than product **5aa**. In the case of using phenylacetylene **4b** having an OMe substituent in the aryl ring, isoxazoline **6ab** and the expected hydroxylamine **5ab** were isolated as products in similar amounts. The incorporation of the electron-withdrawing Cl or CF_3 group to the phenyl ring resulted in obtaining isoxazolines **6ac** and **6ad** as a major or sole product. A similar result was observed in the case of using methyl propiolate (**4g**) as a substrate. In the case of acetylene bearing a *tert*-butyl substituent, adduct **5ae** was isolated in comparable yield to that obtained in the reaction with **4a**. Using acetylenes **4h-k** with more complex substituents ($CH(OMe)_2$, $P(O)(OEt)_2$, piryd-2-yl, CH_2OSiMe_3), the expected products were obtained in low yields, or no product could be identified or isolated in pure form.

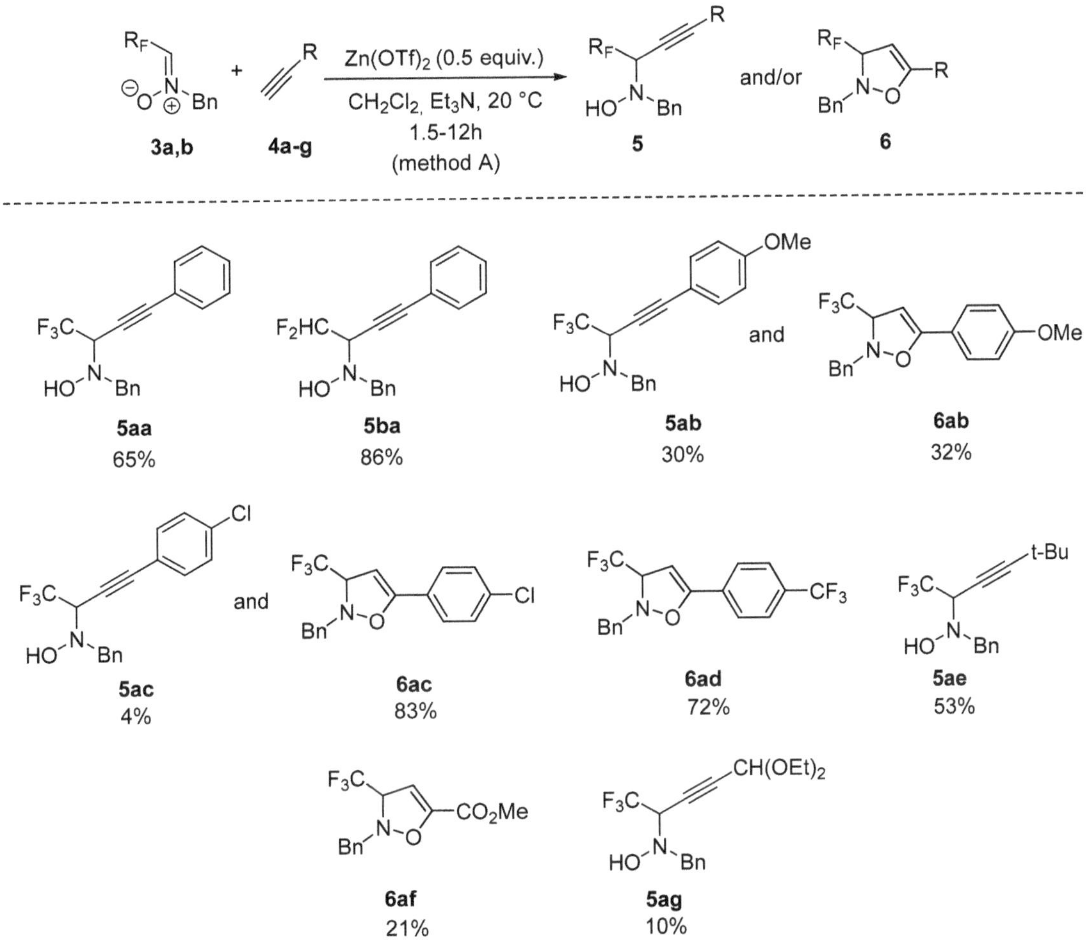

Scheme 2. Additions of acetylenes to fluorinated nitrones in the presence of $Zn(OTf)_2$.

In the summary of this part of the research, it can be stated that the additions of acetylenes **4** to fluorinated nitrones **3** performed in the presence of zinc triflate did not lead to hydroxylamines **5** satisfactory results (in terms of hydroxylamine formation). Only in the case of using acetylenes with an alkyl or an unsubstituted phenyl ring attached to the carbon atom of a triple bond, the yields were satisfying. Therefore, carrying out the reaction under these conditions cannot be considered as a good general method for obtaining the expected fluorinated hydroxylamines of type **5**.

Due to unsatisfactory results of the reaction between C-(fluoroalkyl) nitrones of type **3** and acetylenes **4** obtained in the presence of a catalytic amount of $Zn(OTf)_2$, it was decided to test other reaction conditions [7]. Additions were performed with a slight excess of substituted alkynylzinc generated using a stoichiometric amount of diethylzinc and the corresponding acetylenes **4a-k** (Scheme 3). It has been observed that the reactions conducted under these conditions give better results in terms of selectivity of obtained hydroxylamines **5** from aromatic acetylenes. It was noticed that in the case of the usage of acetylenes **4c,d** bearing electron-withdrawing substituents (Cl, CF_3) in the aryl ring as substrates, the hydroxylamines **5** were isolated with lower yields than in reactions with acetylenes **4a,b** bearing an unsubstituted phenyl ring or electron-donating group attached to aryl ring. Unfortunately, the reaction carried out with the usage of the acetylene having a sterically hindered *t*-Bu substituent led to the formation of desired product **5ae** in trace amount. In the case of using of acetylenes containing substituents such as $CH(OEt)_2$, CH_2OSiMe_3, CO_2Me, $P(O)(OEt)_2$, piryd-2-yl attached to the carbon atoms of a triple bond formation of unidentified decomposition products was observed.

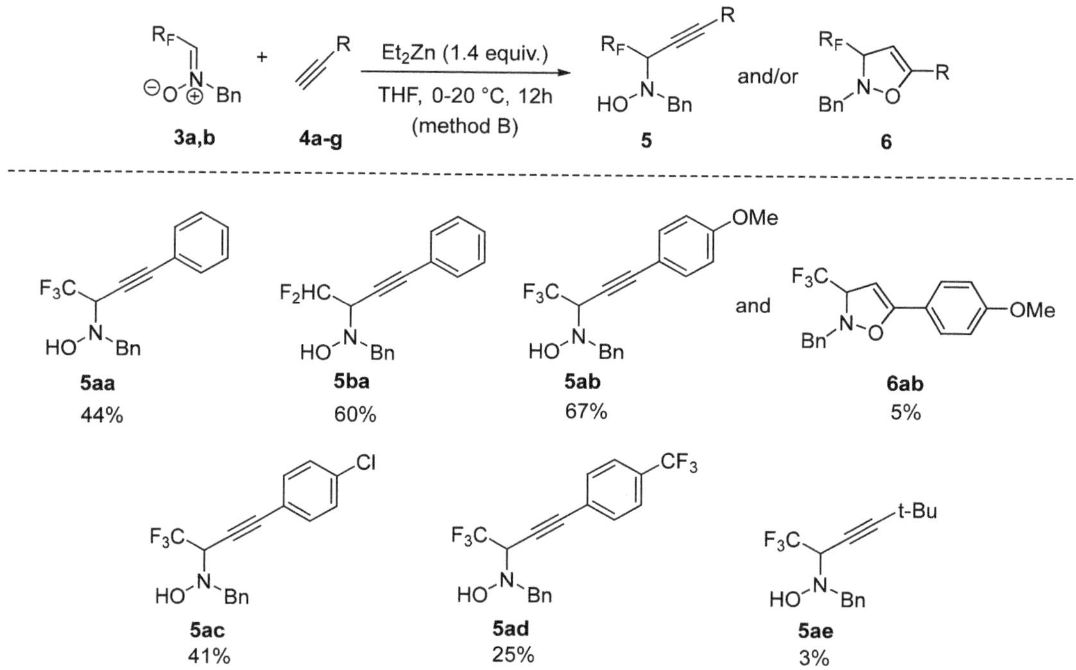

Scheme 3. Addition of acetylenes to fluorinated nitrones in the presence of Et_2Zn.

The observed composition of postreaction mixtures can be explained on the basis of the comprehensive quantum chemical analysis of the reaction mechanism (Figures 1 and 2; Scheme 4). For this purpose, results from the wb97xd/6-311+G(d) (PCM) computational study were used. Within these considerations, the addition process between nitrone **3a** and (2-phenyletynyl)zinc (Scheme 4) was used. This step is crucial because it is known from the literature that the energy of formation of (2-phenyletynyl)zinc from phenylacetylene **4a** and diethylzinc is negligible and does not limit the addition process [21,23].

Independently of the reaction path, the initial interactions between reagents lead to the formation of respective pre-reaction complex (**MCA** and **MCB** for paths **A** and **B**, respectively). This is connected with the decreasing of the enthalpy of reaction system about few kcal/mol. The entropic factor determines however, positive values of the Gibbs free energy for considered transformations. This excludes the existence of MCs as relatively stable intermediates. Within MCs, substructures of addents adopts the orientation determined the further course of transformation. It should be underlined, that any new bonds are not formed however at this stage. Next, the key interatomic distances exist beyond of the range typical for respective bonds within transition states [24]. Lastly, the electron density transfer (GEDT) between structures is not observed within both MCs. So, the localized structures can be considered as orientation, but not charge-transfer complexes [25]. The further transformation of MCs along the reaction coordinate leads to the area associated with the existence of the transition state (**TSA** and **TSB** for paths **A** and **B,** respectively). The clear increase in the energy of the reaction system is a consequence of this process. It is interesting that the favored from the kinetic point of view is the formation of the isoxazoline-type adduct via **TSA**. The structure of this TS is typical for the one-step [3 + 2] cycloaddition process with the participation of bent-type TACs [26–28] and exhibits a moderately polar nature (GEDT = 0.10e). For the contrast, the less kinetic favored **TSB** is evidently more polar (GEDT = 0.24e). Both TSs are connected directly with respective valleys of adducts. So, all new sigma-bonds must be formed at this stage. This was confirmed by the IRC analysis. Analysis of thermochemistry of products derived from IRC experiments show clearly that, from a thermodynamic point of view, the more favored is not [3 + 2] cycloaddition product (**PA**), but hydroxylamine derivative **PB**. So, in the light of our DFT computational study, the isoxazoline-type adduct should be treated as the primary reaction product, which is further converted to the more thermodynamically stable hydroxylamine-type product. It should be underlined-that the conversion of PA in PB is realized via dissociation to the individual nitrone–acetylene pair, and next, via the secondary addition according to the path **B**. All attempts for the localization of reaction channel leading directly from the P**A** in **PB** were not successful.

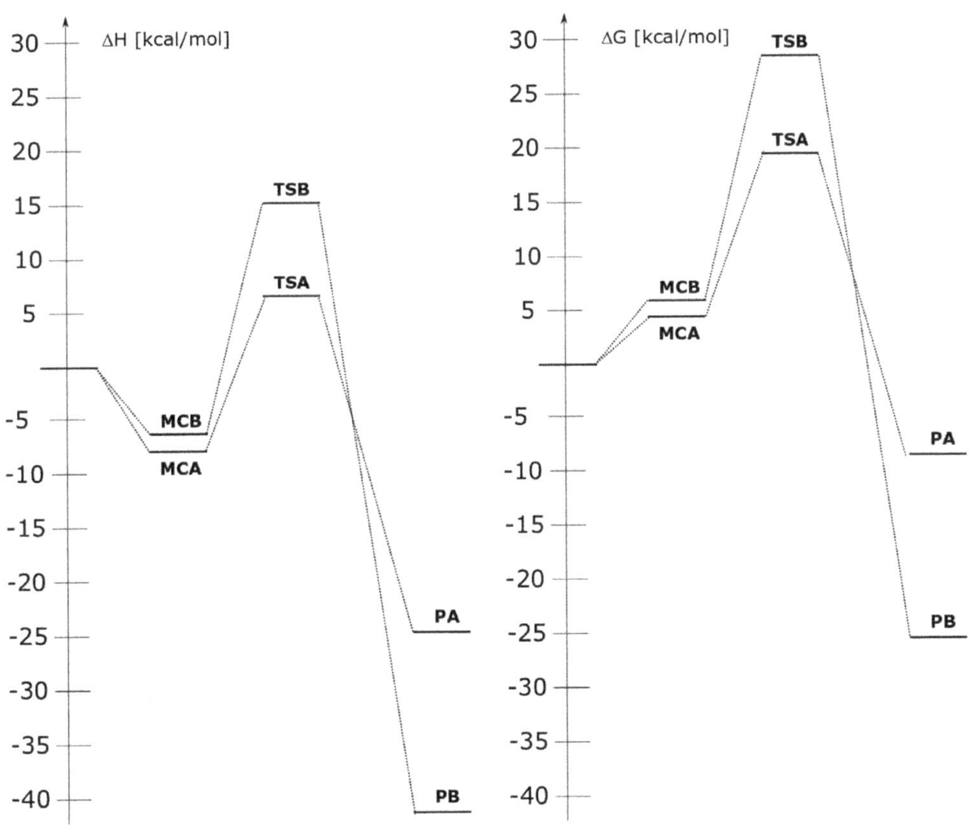

Figure 1. The enthalpy and Gibbs free energy profiles for the addition processes between nitrone **3a** and (2-phenyletynyl)zinc in the light of the wb97xd/6-311+G(d) (PCM) calculations.

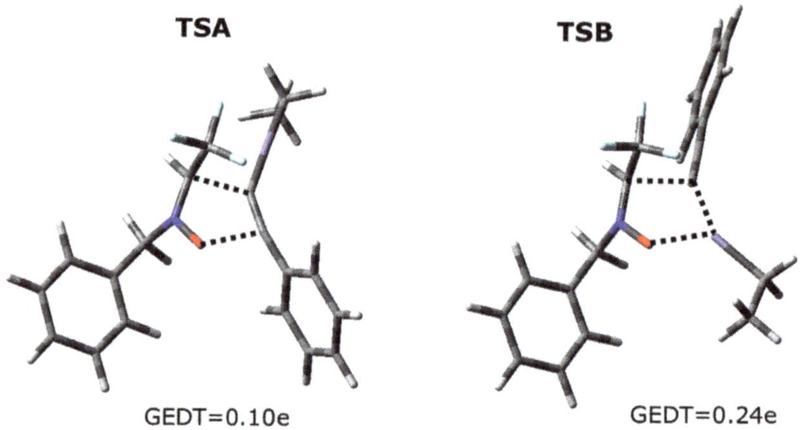

Figure 2. Views of TSs for the addition processes between nitrone **3a** and (2-phenyletynyl)zinc in the light of the wb97xd/6-311+G(d) (PCM) calculations.

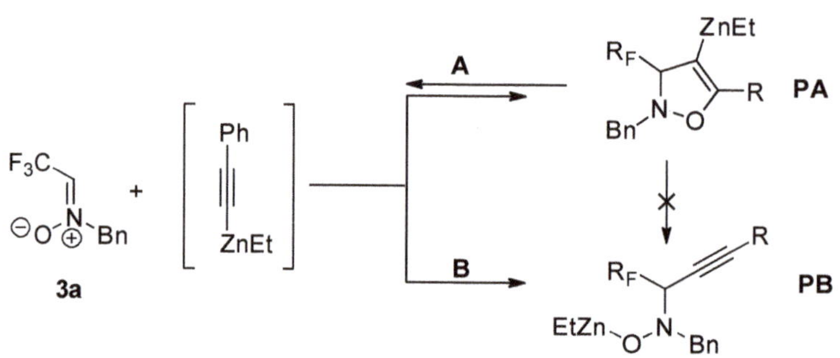

Scheme 4. Competitive paths of the reaction between nitrone **3a** and (2-phenyletynyl)zinc considered in the framework of the DFT study.

The structures of all products were confirmed by spectroscopic methods. For example, a shift in the characteristic quartet derived from the hydrogen atom of the CH group from 6.87 ppm (**3a**) to 4.33 ppm (**5aa**) was observed in the ^{1}H-NMR spectrum. Additionally, a singlet derived from the dynamic proton of the OH group (5.12 ppm) and two doublets (4.01 and 4.27 ppm) from the diastereotopic CH_2 protons of the benzyl group in ^{1}H-NMR spectrum of **5aa** appeared. Also, in the ^{13}C-NMR spectrum, a shift in the quartet derived from the carbon atom of the C=N bond of the nitrone **3a** (122.6 ppm) to the region characteristic for the signals of the C-sp^3 atom in the product molecule **5aa** (60.7 ppm) was observed. In the case of the ^{1}H-NMR spectrum of an exemplary oxazoline **6ab**, the lack of a singlet derived from a proton of OH group was noticed. Instead, there was a doublet derived from olefinic proton of oxazoline ring (5.01 ppm). In addition, the signal of the proton of the $CHCF_3$ group appeared in the form of a doublet of quartet at 4.43 ppm, which proves that there is a coupling with three fluorine atoms and with an olefinic proton. Also in the ^{13}C-NMR spectrum recorded for **6ab** was a shift in the signals from carbon atoms of C≡C bond of hydroxylamine **5ab** occurring in the region characteristic for the signals of the C-sp atoms (75.0 and 89.5 ppm) to the region characteristic for the signals of the C-sp^2 atoms (83.4 and 157.4 ppm) in the product molecule **6ab**. Eventually, the structure of the obtained hydroxylamines of type **5** was confirmed by an X-ray structure registered for one product **5aa** (Figure 3).

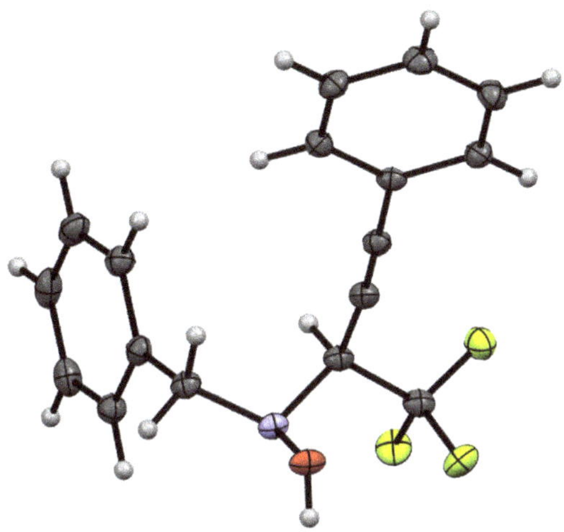

Figure 3. Single crystal X-ray analysis of **5aa**. Displacement ellipsoids are drawn at a 50% probability level.

2.2. Enantioselective Protocol

Due to the better results of reactions carried out with the use of Et_2Zn, the enantioselective protocol was also optimized with the use of this reagent. To this end, several ligands previously proven to be efficient in asymmetric catalysis were examined [29–32]. Our initial investigation began by testing several catalysts bearing the chiral aziridine ring reacting in situ with $ZnEt_2$ to evaluate their catalytic ability. The optimization of the enantioselective addition of phenylacetylene **4a** to nitrone **3a** started with the screening of several chiral aziridine alcohols **L1-L5** which were synthesized as previously described (Scheme 5) [32–34]. Table 1 summarizes the comparison of the results obtained for various chiral ligands. The simple β-amino alcohol catalysts **L1-L4** exhibited poor enantioselectivity. Fortunately, the ligand screening compromised that the AziPhenol ligand **L5** provided the best results in terms of both the yield and enantioselectivity, and the desired product was isolated in 67% yield and 44% *ee*. When the metal source was changed from $ZnEt_2$ to $Zn(OTf)_2$, a pronounced decrease in both the yield and enantiomeric excess was observed (Table 1, entry 6). Consequently, the most efficient ligand **L5** was selected for further optimization involving solvent, temperature, catalyst loading, and additive effects.

Table 1. Effects of ligand and metal reagent on the asymmetric additions of phenylacetylene **4a** to nitrone **3a** under the indicated conditions [a].

Entry	Ligand	M	Yield [%] [b]	e.r. [%] [c]
1	L1	Et_2Zn	51	54:46
2	L2	Et_2Zn	58	59:41
3	L3	Et_2Zn	38	51:49
4	L4	Et_2Zn	60	57:43
5	L5	Et_2Zn	67	72:28
6 [d]	L5	$Zn(OTf)_2$	21	52:48

[a] All reactions were carried out with **3a** (0.1 mmol), **4a** (0.15 mmol), ligand (0.01 mmol) and Et_2Zn (0.17 mmol) in THF (2.0 mL) under nitrogen at 20 °C for 3 h. [b] Isolated yield after silica gel chromatography. [c] Determined by chiral HPLC. [d] $Zn(OTf)_2$ (0.05 mmol).

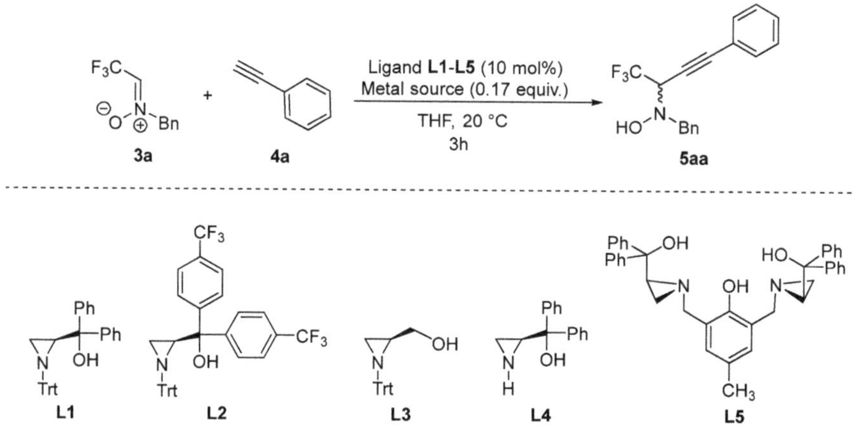

Scheme 5. Additions of acetylene **4a** to nitrone **3a** in the presence of diverse chiral aziridine ligands.

The initial studies towards the development of efficient conditions began with the evaluation of various solvents (Scheme 6). The screening of a variety of solvents including toluene, dichloromethane (DCM) and diethyl ether (Et_2O) was performed (Table 2, entries 2–4). The reactions carried out in halogenated and aromatic solvents, such as dichloromethane and toluene, proceeded in good yields, but significantly lower enantioselectivity was observed. As shown in Table 2, the use of THF afforded superior results with respect to the yield and enantiomeric excess (entry 1) compared with the other solvents examined (entries 2–4). Subsequently, the catalyst loading amount and reaction temperature were also screened. Temperature was shown to have a significant effect on the yields and enantioselectivity (Table 2, entries 5–7). Decreasing the reaction temperature from 20 °C to −20 °C improved both the yield and stereoselectivity of the product **5aa** formation; however, it also resulted in a longer reaction time. Additionally, lowering the temperature to −78 °C caused a small decrease in yield and ee of product **5aa**. No significant changes in the stereoselectivities of the reaction were observed when a 20 mol % catalyst was used (Table 2, entry 8). Reducing the catalyst loading to 5 mol% caused a slight decrease in both the yield and enantiomeric excess of the product, and a longer reaction time was required. It revealed that 10 mol% of ligand is necessary for optimal yield and enantioselectivity. Further optimization aimed at improving catalytic activity revealed that 4 Å molecular sieves significantly influenced both the reactivity and stereoselectivity of the system. Such additives are known to play a crucial role in numerous zinc-catalyzed asymmetric transformations [35,36]. To our great delight, when 4 Å molecular sieves were added to the reaction, the conversion and enantioselectivity were enhanced; the desired product was obtained in 80% yield and 58% ee. Therefore, the optimized reaction conditions are summarized in entry 10 of Table 2.

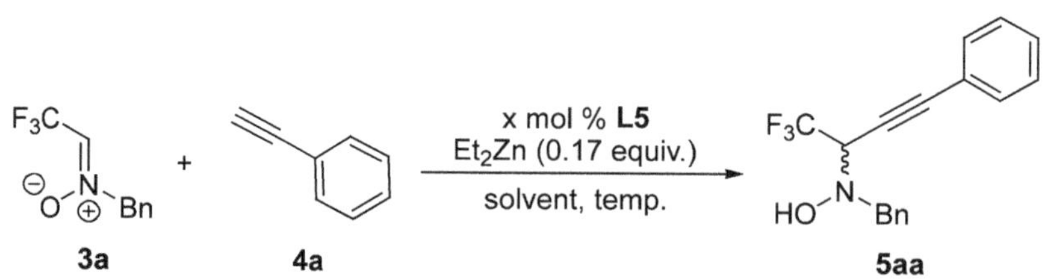

Scheme 6. Enantioselective additions of phenylacetylene (**4a**) to nitrone **3a** in diverse solvents.

Table 2. Screening Conditions for Reaction in the Presence of Ligand L5 [a].

Entry	Ligand (mol%)	Solvent	Time [h]	T [°C]	Yield [%] [c]	e.r. [%] [d]
1	10	THF	3	20	67	72:28
2	10	DCM	3	20	56	64:36
3	10	Toluene	3	20	49	59.5:40.5
4	10	Et_2O	3	20	23	55:45
5	10	THF	4	0	70	74:26
6	10	THF	4	−20	76	76.5:23.5
7	10	THF	4	−78	68	75:25
8	20	THF	4	−20	73	76:24
9	5	THF	5	−20	77	73.5:26.5
10 [b]	10	THF	4	−20	80	79:21

[a] Unless otherwise noted, all reactions were carried out with **3a** (0.1 mmol), **4a** (0.15 mmol), ligand (0.01 mmol) and Et_2Zn (0.17 mmol) in THF (2.0 mL) under nitrogen at for 3 h. [b] 40 mg of 4 Å MS were added. [c] Isolated yield after column chromatography. [d] Determined by chiral HPLC.

With the optimized reaction conditions in hand, we next examined the substrate scope of the developed protocol (Scheme 7). Regardless of the electronic nature of the substituents on the phenyl ring, the reactions afforded good yields, although variations in enantioselectivity were observed. Substrate **4b**, which contained an electron-donating substituent (-OMe) in the *para* position of the phenyl ring, was well tolerated and led to the desired product in the highest yield and enantioselectivity. It was noticed that in the case of the usage of acetylenes bearing electron-withdrawing substituents (Cl, CF_3) in the aryl ring as substrates, the products were isolated in lower yields and enantioselectivities. Surprisingly, in the presence of chiral ligands, the reaction product of acetylene **4e** with a sterically hindered *t*-Bu substituent was obtained in good yield but the decrease in yield and enantioselectivity was observed. However, changing of the CF_3 substituent to CHF_2 in the starting nitrone **3** resulted in a slight decrease in the yield, but the desired product was isolated with similar enantiomeric excess. In contrast, extending the reaction time to 12 h under the optimized conditions led to the formation of a complex reaction mixture, from which the cyclic products **6aa-6ad** were isolated in low yields and with poor enantioselectivity (<10%).

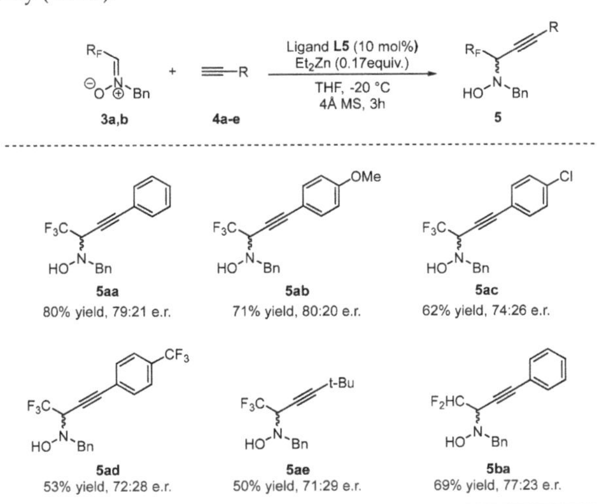

Scheme 7. Enantioselective additions of acetylenes **4a-e** to nitrones **3a-b** under the optimal conditions (All reactions were carried out with **3** (0.1 mmol), **4** (0.15 mmol), ligand (0.01 mmol) and Et_2Zn (0.17 mmol) in THF (2.0 mL) under nitrogen at −20 °C for 3 h; 20 mg of 4Å MS was added; Isolated yield after column chromatography; Determined by chiral HPLC).

3. Materials and Methods

3.1. General Information

Solvents and chemicals were purchased and used without further purification. Tetrahydrofuran (THF) and toluene (PhMe) were distilled over sodium/benzophenone(violet-colored solution prior to use). Dichloromethane (DCM) was distilled over sodium hydride. Zinc triflate ($Zn(OTf)_2$), diethyl zinc (Et_2Zn) (1M solution in hexane), triethylamine (Et_3N) and diverse acetylenes were purchased from Sigma-Aldrich (Merck, Poznań, Poland). Obtained products were purified by standard column chromatography on silica gel (230–400 mesh Merck, Poznań, Poland) or FLASH column chromatography using Grace Reveleris X2 apparatus with UV–Vis and ELSD detection (commercially available 12 g SiO_2 columns, pressure 20–25 psi, solvent flow rate 25–28 mL/min). Unless stated otherwise, yields refer to analytically pure samples. NMR spectra were recorded with Bruker Avance III 600 MHz (^{1}H NMR [600 MHz]; ^{13}C NMR [151 MHz]; ^{19}F NMR [565 MHz]) instrument. Chemical shifts are reported in ppm relative to solvent residual peaks (^{1}H NMR: $\delta = 7.26$ ppm [$CHCl_3$]; ^{13}C NMR: $\delta = 77.0$ ppm [$CDCl_3$]). For detailed peak assignments 2D (HMQC) spectra were measured. IR Spectra were measured with an Agilent Cary 630 FTIR spectrometer (in neat). MS Spectra were recorded on Varian 500 MS LS IonTrap spectrometer. The HPLC analysis were carried out using an Agilent 1260 Infinity system (Agilent, Waldbronn, Germany)at 25 °C using chiral columns Chiralcel AD and Chiralcel AD-H. Single-crystal XRD measurements were performed with a Rigaku XtaLAB Synergy, Pilatus 300K diffractometer. Melting points were determined in capillaries with a Stuart SMP30 apparatus (Stone, United Kingdom).

3.2. Synthesis of Nitrones Derived from Trifluoroacetaldehyde and Difluoroacetaldehyde

N-Benzyl *C*-(trifluoromethyl)nitrone (**3a**) and *N*-Benzyl *C*-(difluoromethyl)nitrone (**3b**) were prepared following the published protocol [14–16]. Chiral ligands (**L1-L5**) were also synthetized according to previously described methods [32–34]. More experimental data, e.g., NMR spectra of the products **5**,**6**, chromatograms and RTG structure details of nitrone **3a** and hydroxylamine **5aa** can be found in Supplementary Material.

3.3. Reactions of Nitrones 3 Derived from Fluorinated Aldehydes with Acetylenes 4

(a) *Reactions leading to racemic products*

Method A—A general procedure: Appropriate acetylene **4a-g** (3 mmol) and Et_3N (152 mg, 1.5 mmol) were added to a suspension of $Zn(OTf)_2$ (182 mg, 0.5 mmol) in anhydrous DCM (5 mL) placed in a dried round-bottom flask. The reaction was carried out under an inert gas atmosphere (argon). This mixture was magnetically stirred for 20 min at room temperature. Next, a solution of a corresponding nitrone **3a,b** (1 mmol) in DCM (1 mL). was added dropwise. The progress of the reaction was controlled by TLC (SiO_2; petroleum ether: ethyl acetate 3:2) and reaction completion time varied (1.5 h–12 h). Then a saturated NH_4Cl solution was added to the reaction mixture. The organic layer was separated, and water layer was extracted with DCM (3 × 15 mL). Combined organic layers were dried over anhydrous Na_2SO_4 and filtrated. Next, the solvent was evaporated under reduced pressure. Products were purified by automated flash chromatography (SiO_2 12 g cartridges; petroleum ether with increasing amount of DCM (0–60%) as an eluent).

Method B—A general procedure: Appropriate acetylene **4a-g** (1.4 mmol) was added to a solution of diethylzinc (1.4 mmol) in anhydrous THF placed in a dried round-bottom flask. The reaction was carried out under an inert gas atmosphere (argon). This mixture was magnetically stirred for 20 min at room temperature. Then the flask was placed in an ice bath (~0 °C) and next a solution of a corresponding nitrone **3a,b** (1 mmol) in THF (1 mL) was added dropwise. The progress of the reaction was controlled by TLC (SiO_2;

petroleum ether: ethyl acetate 3:2) and the time of completion of the reaction was about 12 h. Then a saturated NH_4Cl solution was added to the reaction mixture. The organic layer was separated, and water layer was extracted with DCM (3 × 15 mL). Combined organic layers were dried over anhydrous Na_2SO_4 and filtrated. Next, the solvent was evaporated under reduced pressure. Products were purified by automated flash chromatography (SiO_2 12 g cartridges; petroleum ether with increasing amount of DCM (0–60%) as an eluent). Analytically pure products were, in most cases, obtained by crystallization from an appropriate solvent.

N-Benzyl-N-(1,1,1-trifluoro-4-phenylbut-3-yn-2-yl)hydroxylamine (**5aa**). Yield: 198 mg (65%) (for method A), 134 mg (44%) (for method B); colorless crystals, m.p. 102–104 °C (DCM/petroleum ether). ^{1}H NMR (600 MHz, $CDCl_3$): δ 4.01, 4.27 (2d, 2H, $^2J_{H,H}$ = 12.84 Hz, $\mathbf{CH_2}$Ph), 4.33 (q, 1H, $^3J_{H,F}$ = 6.84 Hz, **CH**), 5.12 (s, 1H, **OH**), 7.32–7.43 (m, 8 arom. **H**), 7.56–7.59 (m, 2 arom. **H**) ppm. ^{13}C NMR (150 MHz, $CDCl_3$): δ 60.7 (q, $^2J_{C,F}$ = 31.5 Hz, CH), 62.2 ($\mathbf{CH_2}$Ph), 76.5 (q, $^3J_{C,F}$ = 3,0 Hz, **C**≡CPh), 89.5 (C≡**C**Ph), 121.6, 135.9 (2 arom. **C**), 123.1 (q, $^1J_{C,F}$ = 279.0 Hz, $\mathbf{CF_3}$), 128.0, 128.4, 128.6, 129.2, 129.4, 132.2 (10 arom. **C**H) ppm. ^{19}F NMR (565 MHz, $CDCl_3$): δ −71.47 (d, 3F, $^3J_{F,H}$ = 6.84 Hz, CF_3) ppm. IR: *v* 3275*m* (O-H), 2879*w*, 1491*w*, 1463*w*, 1359*m*, 1289*m*, 1187*s*, 1139*vs*, 1075*s*, 1030*m*, 818*s*, 751*vs*, 689*vs* cm^{-1}. ESI-MS (*m*/*z*): 306 (60, [M+H]$^+$), 328 (100, [M+Na]$^+$). Elemental analysis for $C_{17}H_{14}F_3NO$ (305.3) calculated: C 66.88, H 4.62, N 4.59; found: C 66.90, H 4.58, N 4.83.

N-Benzyl-N-(1,1-difluoro-4-phenylbut-3-yn-2-yl)hydroxylamine (**5ba**). Yield: 247 mg (86%) (for method A), 172 mg (60%) (method B); colorless crystals m.p. 85–87 °C (DCM/petroleum ether). ^{1}H NMR (600 MHz, $CDCl_3$): δ 3.99, 4.23 (2d, 2H, $^2J_{H,H}$ = 12.84 Hz, $\mathbf{CH_2}$Ph), 4.04–4.09 (m, 1H, **CH**), 5.06 (s, 1H, **OH**), 6.01 (dt, 1H, $^3J_{H,H}$ = 5.10 Hz, $^2J_{H,F}$ = 56.05 Hz, $\mathbf{CHF_2}$), 7.31–7.42 (m, 8 arom. **H**), 7.55–7.57 (m, 2 arom. **H**) ppm. ^{13}C NMR (150 MHz, $CDCl_3$): δ 61.6 (t, $^2J_{C,F}$ = 25.5 Hz, **CH**), 62.3 ($\mathbf{CH_2}$Ph), 78.6 (dd, $^3J_{C,F}$ = 3.0 Hz, **C**≡CPh), 89.4 (C≡**C**Ph), 113.7 (dd, $^1J_{C,F(1)}$ = 243.0 Hz, $^1J_{C,F(2)}$ = 244.5 Hz, $\mathbf{CHF_2}$), 121.9, 135.9 (2 arom. **C**) 128.0, 128.4, 128.6, 129.0, 129.6, 132.1 (10 arom. **C**H) ppm. ^{19}F NMR (565 MHz, $CDCl_3$): δ −125.22 (ddd, 1F, $^3J_{F(1),H}$ = 10.79 Hz, $^2J_{F(1),H}$ = 56.05 Hz, $^2J_{F,F}$ = 284.15 Hz, CHF_2), −121.55 (ddd, 1F, $^3J_{F(2),H}$ = 8.02 Hz, $^2J_{F(2),H}$ = 56.50 Hz, $^2J_{F,F}$ = 284.15 Hz, $CH\mathbf{F_2}$) ppm. IR: *v* 3237*m* (O-H), 2876*w*, 1491*w*, 1444*m*, 1384*m*, 1109*s*, 1087*s*, 1060*s*, 689*vs*, 540*vs* cm^{-1}. ESI-MS (*m*/*z*): 288 (45, [M+H]$^+$), 310 (100, [M+Na]$^+$). Elemental analysis for $C_{17}H_{15}F_2NO$ (287.3) calculated: C 71.07, H 5.26, N 4.88; found: C 71.26, H 5.02, N 5.17.

N-Benzyl-N-[1,1,1-trifluoro-4-(4′-methoxyphenyl)but-3-yn-2-yl]hydroxylamine (**5ab**). Yield: 101 mg (30%) (method A), 225 mg (67%) (method B); colorless crystals, m.p. 102–104 °C (Et_2O/petroleum ether). ^{1}H NMR (600 MHz, $CDCl_3$): δ 3.84 (s, 3H, $\mathbf{OCH_3}$), 3.99, 4.24 (2d, 2H, $^2J_{H,H}$ = 12.78 Hz, $\mathbf{CH_2}$Ph), 4.31 (q, 1H, $^3J_{H,F}$ = 6.84 Hz, CH), 5.22 (br.s, 1H, **OH**), 6.87–6.89 (m, 2 arom. **H**), 7.31–7.34 (m, 1 arom. **H**), 7.36–7.38 (m, 2 arom. **H**), 7.41–7.42 (m, 2 arom. **H**), 7.50–7.51 (m, 2 arom. **H**) ppm. ^{13}C NMR (150 MHz, $CDCl_3$): δ 55.3 ($\mathbf{OCH_3}$), 60.7 (q, $^2J_{C,F}$ = 31.5 Hz, CH), 62.2 (CH_2Ph), 75.0 (**C**≡CPh), 89.5 (C≡**C**Ph), 113.5, 135.8, 160.2 (3 arom. **C**), 114.0, 128.0, 128.7, 129.4, 133.7 (9 arom. **C**H) ppm. ^{19}F NMR (565 MHz, $CDCl_3$): δ −71.58 (d, 3F, $^3J_{F,H}$ = 6.84 Hz, CF_3) ppm. IR: *v* 3269*m* (O-H), 3026*w*, 2933*w*, 2885*w*, 2236*w*, 1610*m*, 1513*s*, 1454*m*, 1357*m*, 1271s, 1174*s*, 1133*vs*, 1000*s*, 972*w*, 827*vs*, 760*s* cm^{-1}. ESI-MS (*m*/*z*): 336 (100, [M+H]$^+$). Elemental analysis for $C_{18}H_{16}F_3NO$ (335.1) calculated: C 64.47, H 4.81, N 4.18; found: C 64.47, H 4.92, N 4.47.

2-Benzyl-5-(4′-methoxyphenyl)-3-(trifluoromethyl)-2,3-dihydroizoxazole (**6ab**). Yield: 107 mg (32%) (method A); 17 mg (5%) (method B); colorless crystals; m.p. 84–85 °C (Et_2O/petroleum ether). ^{1}H NMR (600 MHz, $CDCl_3$): δ 3.83 (s, 3H, CH_3), 4.02, 4.38 (2d, 2H, $^2J_{H,H}$ = 12.80 Hz, $\mathbf{CH_2}$Ph), 4.43 (dq, 1H, $^3J_{H,H}$ = 2.90 Hz, $^3J_{H,F}$ = 6.50 Hz, $\mathbf{CH}CF_3$), 5.01 (d, 1H, $^3J_{H,H}$ = 2.90 Hz, C=**CH**), 6.94–6.85 (m, 2 arom. **H**), 7.40–7.29 (m, 3 arom. **H**), 7.45–7.40 (m, 2 arom. **H**), 7.53–7.45 (m, 2 arom. **H**) ppm. ^{13}C NMR (150 MHz, $CDCl_3$): δ55.5 ($\mathbf{CH_3}$), 63.7 ($\mathbf{CH_2}$Ph),

70.8 (q, $^{2}J_{C,F}$ = 32.3 Hz, **C**HCF$_3$), 83.4 (q, C=**C**H), 114.1, 127.8, 128.2, 128.7, 129.8 (9 arom. **C**H), 124.2 (q, $^{1}J_{C,F}$ = 282.0 Hz, **C**F$_3$), 120.1, 134.9, 161.0 (3 arom. **C**), 157.4 (**C**=CH) ppm. ^{19}F NMR (565 MHz, CDCl$_3$): δ −77.78 (d, 3F, $^{3}J_{F,H}$ = 6.50 Hz, C**F**$_3$) ppm. **IR**: *v* 3034*w*, 2974*w*, 2940*w*, 2847*w*, 1662*m*, 1606*m*, 1513*m*, 1457*w*, 1428*w*, 1375*w*, 1256*s*, 1167*s*, 1126*vs*, 1047*m*, 1021*m*, 880*m*, 834*m* cm^{-1}. ESI-MS (*m/z*): 336 (100, [M+H]$^+$). Elemental analysis for $C_{18}H_{16}F_3NO$ (335.1) calculated: C 64.47, H 4.81, N 4.18; found: C 63.39, H 6.39, N 14.97.

N-Benzyl-N-[1,1,1-trifluoro-4-(4′-chlorophenyl)but-3-yn-2-yl]hydroxylamine (**5ac**). Yield: 14 mg (4%) (method A), 139 mg (41%) (method B); colorless crystals, m.p. 78–80 °C (Et$_2$O/petroleum ether) ^{1}H NMR (600 MHz, CDCl$_3$): δ 3.99, 4.24 (2d, 2H, $^{2}J_{H,H}$ = 12.78 Hz, C**H**$_2$Ph), 4.31 (q, 1H, $^{3}J_{H,F}$ = 6.84 Hz, C**H**), 5.21 (s, 1H, O**H**), 7.32–7.41 (m, 7 arom. **H**), 7.48–7.50 (m, 2 arom. **H**) ppm. ^{13}C NMR (150 MHz, CDCl$_3$): δ 60.6 (q, $^{2}J_{C,F}$ = 31.6 Hz, **C**H), 62.2 (**C**H$_2$Ph), 77.6 (**C**≡CPh), 88.2 (C≡**C**Ph), 120.0, 135.3, 135.7 (3 arom. **C**), 123.0 (q, $^{1}J_{C,F}$ = 279.2 Hz, **C**F$_3$), 128.1, 128.6, 128.7, 129.4, 133.4 (9 arom. **C**H) ppm. ^{19}F NMR (565 MHz, CDCl$_3$): δ −71.41 (d, 3F, $^{3}J_{F,H}$ = 6.84 Hz, C**F**$_3$) ppm. IR: *v* 3273*m* (O-H), 2877*w*, 2225*w*, 1490*s*, 1457*m*, 1394*m*, 1349*s*, 1263*s*, 1185*s*, 1133*vs*, 1085*vs*, 1006*s*, 972*m*, 920*m*, 979*m*, 820*s*, 753*s*, 700*s* cm^{-1}. ESI-MS (*m/z*): 340 (100, [M$_{Cl\text{-}35}$+H]$^+$), 342 (36, [M $_{Cl\text{-}37}$+H]$^+$). Elemental analysis for $C_{17}H_{13}ClF_3NO$ (339.7) calculated: C 60.10, H 3.86, N 4.12; found: C 60.00, H 3.78, N 4.42.

2-Benzyl-5-(4′-chlorophenyl)-3-(trifluoromethyl)-2,3-dihydroizoxazole (**6ac**). Yield: 302 mg (89%) (method A); colorless crystals; m.p. 116–117 °C (Et$_2$O/petroleum ether). ^{1}H NMR (600 MHz, CDCl$_3$): δ 4.04, 4.37 (2d, 2H, $^{2}J_{H,H}$ = 12.90 Hz, C**H**$_2$Ph), 4.47 (dq, 1H, $^{3}J_{H,H}$ = 2.90 Hz, $^{3}J_{H,F}$ = 6.50 Hz, C**H**CF$_3$), 5.14 (d, 1H, $^{3}J_{H,H}$ = 2.90 Hz, C=C**H**), 7.39–7.32 (m, 5 arom. **H**), 7.42–7.40 (m, 2 arom. **H**), 7.50–7.46 (m, 2 arom. **H**) ppm. ^{13}C NMR (150 MHz, CDCl$_3$): δ 63.8 (**C**H$_2$Ph), 70.8 (q, $^{2}J_{C,F}$ = 32.5 Hz, **C**HCF$_3$), 86.0 (q, $^{3}J_{C,F}$ = 1.5 Hz, C=**C**H), 124.0 (q, $^{1}J_{C,F}$ = 280.2 Hz, **C**F$_3$), 126.0, 134.6, 136.1 (3 arom. **C**), 127.5, 128.3, 128.7, 129.0, 129.8 (10 arom. **C**H), 156.6 (**C**=CH) ppm. ^{19}F NMR (565 MHz, CDCl$_3$): δ −71.32 (d, 3F, $^{3}J_{F,H}$ = 6.50 Hz, C**F**$_3$) ppm. IR: *v* 3127*w*, 3090*w*, 3034*w*, 2959*w*, 2896*w*, 2840*w*, 1651*w*, 1490*m*, 1453*w*, 1375*m*, 1341*m*, 1275*s*, 1222*m*, 1163*s*, 1129*vs*, 1085*m*, 1039*m*, 1010*m*, 879*m*, 834*m*, 798*m* cm^{-1}. ESI-MS (*m/z*): 340 (100, [M$_{Cl\text{-}35}$+H]$^+$), 342 (28, [M$_{Cl\text{-}37}$+H]$^+$). Elemental analysis for $C_{17}H_{13}ClF_3NO$ (339.7) calculated: C 60.10, H 3.86, N 4.12; found: C 60.06, H 3.87, N 4.28.

N-Benzyl-N-[1,1,1-trifluoro-4-(4′-trifluoromethylphenyl)but-3-yn-2-yl]hydroxylamine (**5ad**). Yield: 93 mg (25%) (method B); colorless crystals; m.p. 101–103 °C (Et$_2$O/petroleum ether). ^{1}H NMR (600 MHz, CDCl$_3$): δ 4.00, 4.26 (2d, 2H, $^{2}J_{H,H}$ = 12.84 Hz, C**H**$_2$Ph), 4.33 (q, 1H, $^{3}J_{H,F}$ = 6.84 Hz, C**H**), 5.16 (s, 1H, O**H**), 7.32–7.42 (m, 5 arom. **H**), 7.62–7.63 (m, 2 arom. **H**), 7.66–7.67 (m, 2 arom. **H**) ppm. ^{13}C NMR (150 MHz, CDCl$_3$): δ 60.5 (q, $^{2}J_{C,F}$ = 31.7 Hz, CH), 62.3 (**C**H$_2$Ph), 79.2 (q, $^{3}J_{C,F}$ = 1.5 Hz, **C**≡CPh), 87.8 (C≡**C**Ph), 123.0 (q, $^{1}J_{C,F}$ = 279.3 Hz, **C**F$_3$), 123.7 (q, $^{1}J_{C,F}$ = 270.5 Hz, **C**F$_3$), 125.3 (q, $^{3}J_{C,F}$ = 7.5 Hz, 2 arom. **C**H), 128.1, 128.7, 129.4, 132.5 (9 arom. **C**H), 130.9 (q $^{2}J_{C,F}$ = 7.5 Hz, 1 arom. **C**), 135.6 (1 arom. **C**) ppm. ^{19}F NMR (565 MHz, CDCl$_3$): δ −62.90 (s, 3F, $^{3}J_{F,H}$ = 6.84 Hz, C**F**$_3$), −77.53 (d, 3F, $^{3}J_{F,H}$ = 6.84 Hz, CHC**F**$_3$) ppm. IR: *v* 3422*m* (O-H), 3272*w*, 1618*m*, 1457*m*, 1405*m*, 1327*s*, 1275*s*, 1174*s*, 1126*vs*, 1003*vs*, 1050*s*, 1014*s*, 954*w*, 879*w*, 835*s*, 745*s* cm^{-1}. ESI-MS (*m/z*): 374 (100, [M+H]$^+$). Elemental analysis for $C_{18}H_{13}F_6NO$ (373.1) calculated: C 57.92, H 3.51, N 3.71; found: C 57.83, H 3.54, N 3.98.

2-Benzyl-5-(4′-trifluoromethylphenyl)-3-(trifluoromethyl)-2,3-dihydroizoxazole (**6ad**). Yield: 269 mg (72%) (method A); colorless crystals; m.p. 129–131 °C (Et$_2$O/petroleum ether). ^{1}H NMR (600 MHz, CDCl$_3$): δ 4.06 (d, 1H, $^{2}J_{H,H}$ = 12.60 Hz, C**H**$_2$Ph), 4.39 (d, 1H, $^{2}J_{H,H}$ = 12.60 Hz, C**H**$_2$Ph), 4.51 (dq, 1H, $^{3}J_{H,H}$ = 3.00 Hz, $^{3}J_{H,F}$ = 6.06 Hz, C**H**CF$_3$), 5.28 (d, 1H, $^{3}J_{H,H}$ =3.00 Hz, C=C**H**), 7.33–7.39 (m, 3 arom. **H**), 7.42–7.43 (m, 2 arom. **H**), 7.63–7.67 (m, 4 arom. **H**) ppm. ^{13}C NMR (150 MHz, CDCl$_3$): δ 63.7 (**C**H$_2$Ph), 70.6 (q, $^{2}J_{C,F}$ = 32.3 Hz,

$CHCF_3$), 87.5 (C=**CH**), 123.7 (q, $^1J_{C,F}$ = 270.5 Hz, CF_3), 123.8 (q, $^1J_{C,F}$ = 278.4 Hz, CF_3), 125.5 (q, $^3J_{C,F}$ = 3.8 Hz, 2 arom. **CH**) 126.4, 128.3, 128.6, 129.6 (7 arom. **CH**), 130.6, 134.3 (2 arom. **C**), 131.7 (q, $^2J_{C,F}$ = 27.6 Hz, 1 arom. **C**) 156.1 (**C**=CH) ppm. ^{19}F NMR (565 MHz, $CDCl_3$): δ −62.90 (s, 3F, CF_3), −77.53 (d, 3F, $^3J_{F,H}$ = 6.06 Hz, $CHCF_3$) ppm. IR: *v* 3127*w*, 3041*w*, 2903*w*, 1651*w*, 1414*w*, 1367*w*, 1330*m*, 1274*m*, 1159*s*, 1107*vs*, 1060*s*, 1017*m*, 850*s* cm^{-1}. ESI-MS (*m*/*z*): 374 (35, $[M+H]^+$), 396 (100, $[M+Na]^+$). Elemental analysis for $C_{18}H_{13}F_6NO$ (373.1) calculated: C 57.92, H 3.51, N 3.71; found: C 57.73, H 3.69, N 4.01.

N-Benzyl-N-(1,1,1-trifluoro-5,5-dimethylheks-3-yn-2-yl)hydroxyloamine (**5ae**). Yield: 151 mg (53%) (method A), 9 mg (3%) (method B); colorless crystals; m.p. 75–77 °C (DCM/petroleum ether). ^{1}H NMR (600 MHz, $CDCl_3$): δ 1.32 (s, 9H, 3CH_3), 3.87, 4.15 (2d, 2H, $^2J_{H,H}$ = 12.85 Hz, CH_2Ph), 4.07 (q, 1H, $^3J_{H,F}$ = 6.95 Hz, **CH**), 4.94 (s, 1H, **OH**), 7.30–7.33 (m, 2 arom. **H**). 7.34–7.38 (m, 3 arom. **H**) ppm. ^{13}C NMR (150 MHz, $CDCl_3$): δ 27.7 (**C**($CH_3)_3$), 30.8 (3CH_3), 60.2 (q, $^2J_{C,F}$ = 31.5 Hz, CH), 62.0 (CH_2Ph), 65.6 (q, $^3J_{C,F}$ = 3.0 Hz, **C**≡Ct-Bu), 99.2 (C≡**C**t-Bu), 123.2 (q, $^1J_{C,F}$ = 279.0 Hz, CF_3), 127.9, 128.6, 129.4 (5 arom. **CH**), 136.0 (1 arom. **CH**) ppm. ^{19}F NMR (565 MHz, $CDCl_3$): δ −72.02 (d, 3F, $^3J_{F,H}$ = 6.95 Hz, CF_3) ppm. IR: *v* 3269*m* (O-H), 2970*m*, 2922*m*, 2873*m*, 2240*w*, 1457*m*, 1360*s*, 1278*s*, 1203*w*, 1166*s*, 1130*vs*, 1092*m*, 1062*m*, 998*m*, 976*w*, 924*w*, 864*m*, 820*s*, 764*s*, 701*vs* cm^{-1}. ESI-MS (*m*/*z*): 286 (100, $[M+H]^+$), 308 (55, $[M+Na]^+$). Elemental analysis for $C_{15}H_{18}F_3NO$ (285.3) calculated: C 63.15, H 6.36, N 4.91; found: C 63.39, H 6.39, N 5.14.

Methyl 2-Benzyl-3-(trifluoromethyl)-4-isoxazoline-5-carboxylate (**6af**). Yield: 60 mg (21%) (method A); colorless crystals; m.p. 81–84 °C (DCM/petroleum ether). ^{1}H NMR (600 MHz, $CDCl_3$): δ 3.85 (s, 3H, CO_2CH_3) 4.0, 4.38 (2d, 2H, $^2J_{H,H}$ = 12.96 Hz, CH_2Ph), 4.46 (dq, 1H, $^4J_{H,H}$ = 3.06 Hz, $^3J_{H,F}$ = 6.40 Hz, **CH**), 5.65 (d, 1H, $^4J_{H,H}$ = 3.06 Hz, C=**CH**), 7.32–7.39 (m, 5 arom. **H**) ppm. ^{13}C NMR (150 MHz, $CDCl_3$): δ 52.7 (CO_2CH_3), 63.5 (CH_2Ph), 70.0 (q, $^2J_{C,F}$ = 31.5 Hz, CH), 99.6 (q, $^3J_{C,F}$ = 1.5 Hz, CH=CCO_2CH_3), 123.2 (q, $^1J_{C,F}$ = 279.0 Hz, CF_3), 128.4, 128.7, 129.7 (5 arom. CH), 133.5 (1 arom. **C**), 149.5 (CH=CCO_2CH_3), 158.2 (**C**=O) ppm. ^{19}F NMR (565 MHz, $CDCl_3$): δ −77.02 (d, 3F, $^3J_{F,H}$ = 6.40 Hz, CF_3) ppm. IR: *v* 3145*w*, 3090*w*, 3012*w*, 2959*w*, 2900*w*, 1733*vs* (C=O), 1647*m*, 1446*m*, 1334*m*, 1278*m*, 1226*s*, 1177*s*, 1121*vs*, 1067*m*, 1010*m*, 969*m*, 879*m*, 801*m*, 723*vs* cm^{-1}. ESI-MS (*m*/*z*): 288 (35, $[M+H]^+$), 310 (100, $[M+Na]^+$). Elemental analysis for $C_{13}H_{12}F_3NO_3$ (287.1) calculated: C 54.36, H 4.21, N 4.88; found: C 54.27, H 4.39, N 5.00.

N-Benzyl-N-(5,5-diethoxy-1,1,1-trifluoropent-3-yn-2-yl)hydroxylamine (**5ag**). Yield: 33 mg (10%) (method A); colorless oil. ^{1}H NMR (600 MHz, $CDCl_3$): δ 1.25–1.28 (m, 6H, 2OCH_2CH_3), 3.63–3.69 (m, 2H, OCH_2CH_3), 3.76–3.83 (m, 2H, CH_2CH_3), 3.91, 4.19 (2d, 2H, $^2J_{H,H}$ = 12.84 Hz, CH_2Ph), 4.16 (dq, 1H, $^5J_{H,H}$ = 1.17 Hz, $^3J_{H,F}$ = 6.42 Hz, **CH**), 5.20 (s, 1H, **OH**), 5.38 (d, 1H, $^5J_{H,H}$ = 1.17 Hz, **CH**($OEt)_2$), 7.30–7.38 (m, 5 arom. **H**) ppm. ^{13}C NMR (150 MHz, $CDCl_3$): δ 15.05, 15.06 (2OCH_2CH_3), 60.1 (q, $^2J_{C,F}$ = 31.5 Hz, CH), 61.1, 61.2 (2OCH_2CH_3), 62.2 (CH_2Ph), 73.0 (**C**≡CCH($OEt)_2$), 85.1 (C≡**C**CH($OEt)_2$), 91.1 (**CH**($OEt)_2$), 122.9 (q, $^1J_{C,F}$ = 279.0 Hz, CF_3), 128.0, 128.6, 129.4 (5 arom. CH), 135.7 (1 arom. **C**) ppm. ^{19}F NMR (565 MHz, $CDCl_3$): δ −71.46 (d, 3F, $^3J_{F,H}$ = 6.42 Hz, CF_3) ppm. IR: *v* 3331*w* (O-H), 2978*w*, 2926*w*, 2859*w*, 1673*w*, 1454*w*, 1353*w*, 1271*m*, 1384*m*, 1177*s*, 1140*vs*, 1051*s*, 928*w*, 883*w*, 827*w* cm^{-1}. HRMS (TOF AP$^+$) *m*/*z* $[M+H]^+$ calcd $C_{16}H_{21}NO_3F_3$: 332.1474, found: 332.1470; $[M+Na]^+$ calcd $C_{16}H_{20}NO_3F_3Na$: 354.1293, found: 354.1297.

(*b*) *Enantioselective reactions*

General procedure. Under an argon atmosphere, THF (0.5 mL) was syringed into a round-bottomed flask with a rubber septum containing the AziPhenol ligand (6 mg, 0.01 mmol). A solution of diethylzinc (200 μL of 1.0 M solution in hexane, 0.02 mmol) was added dropwise at room temperature. The mixture was stirred at room temperature until the solution became slightly cloudy. Then 40 mg 4Å MS was added to the mixture and cooled to −20 °C for an additional 10 min before Et_2Zn (1.5 mL of 1.0 M solution in hexane,

0.15 mmol) and the appropriate acetylene (0.15 mmol) in THF (0.5 mL) were added to the mixture. The nitrone (0.1 mmol) was dissolved in 1.0 mL of THF and added dropwise to the mixture and stirred at −20 °C. The progress of the reaction was controlled by TLC (SiO_2; hexane: ethyl acetate 3:2) and the time of completion of the reaction was about 3h. Then, a saturated NH_4Cl solution was added to the reaction mixture. The organic layer was separated, and the water layer was extracted with DCM (3 × 10 mL). Combined organic layers were dried over anhydrous Na_2SO_4 and filtrated. Next, the solvent was evaporated under reduced pressure. The crude mixture was purified by column chromatography (silica gel, hexane with ethyl acetate in gradient) to afford the corresponding products.

N-Benzyl-N-(1,1,1-trifluoro-4-phenylbut-3-yn-2-yl)hydroxylamine (**5aa**): 79:21 e.r.; $[\alpha]_D^{22} = -24.8$ (c 0.3, $CHCl_3$); HPLC analysis: Chiralcel AD-H, Hexanes: ^{i}PrOH = 90:10, flow = 1.0 mL/min, retention time: 6.59 (minor), 7.39 (major) min, wavelength = 250 nm.

N-Benzyl-N-[1,1,1-trifluoro-4-(4′-methoxyphenyl)but-3-yn-2-yl]hydroxylamine (**5ab**): 80:20 e.r.; $[\alpha]_D^{22} = -40.9$ (c 0.3, $CHCl_3$); HPLC analysis: Chiralcel AD-H, Hexanes: ^{i}PrOH = 95:5, flow = 1.0 mL/min, retention time: 18.54 (minor), 20.29 (major) min, wavelength = 250 nm.

N-Benzyl-N-[1,1,1-trifluoro-4-(4′-chlorophenyl)but-3-yn-2-yl]hydroxylamine (**5ac**): 74:26 e.r.; $[\alpha]_D^{22} = -33.4$ (c 0.3, $CHCl_3$); HPLC analysis: Chiralcel AD-H, Hexanes: ^{i}PrOH = 95:5, flow = 1.0 mL/min, retention time: 12.65 (minor), 17.19 (major) min, wavelength = 250 nm.

N-Benzyl-N-[1,1,1-trifluoro-4-(4′-trifluoromethylphenyl)but-3-yn-2-yl]hydroxylamine (**5ad**): 72:28 e.r.; $[\alpha]_D^{22} = -20.5$ (c 0.3, $CHCl_3$); HPLC analysis: Chiralcel AD-H, Hexanes: ^{i}PrOH = 95:5, flow = 1.0 mL/min, retention time: 13.38 (minor), 17.84 (major) min, wavelength = 250 nm.

N-Benzyl-N-(1,1,1-trifluoro-5,5-dimethylheks-3-yn-2-yl)hydroxyloamine (**5ae**): 71:29 e.r.; $[\alpha]_D^{22} = -9.9$ (c 0.3, $CHCl_3$); HPLC analysis: Chiralcel AD, Hexanes: ^{i}PrOH = 98:2, flow = 0.5 mL/min, retention time: 12.55 (minor), 14.23 (major) min, wavelength = 250 nm.

N-Benzyl-N-(1,1-difluoro-4-phenylbut-3-yn-2-yl)hydroxylamine (**5ba**): 77:23 e.r.; $[\alpha]_D^{22} = -19.2$ (c 0.3, $CHCl_3$); HPLC analysis: Chiralcel AD-H, Hexanes: ^{i}PrOH = 90:10, flow = 0.7 mL/min, retention time: 12.55 (minor), 14.23 (major) min, wavelength = 250 nm.

2-Benzyl-5-phenyl-3-(trifluoromethyl)-2,3-dihydroisoxazole (**6aa**): 50:50 e.r.; HPLC analysis: Chiralcel AD, Hexanes: ^{i}PrOH = 90:10, flow = 0.5 mL/min, retention time: 9.36, 10.14 min, wavelength = 250 nm.

2-Benzyl-5-(4′-methoxyphenyl)-3-(trifluoromethyl)-2,3-dihydroizoxazole (**6ab**): 51:49 e.r.; HPLC analysis: Chiralcel AD, Hexanes: ^{i}PrOH = 95:5, flow = 0.5 mL/min, retention time: 13.23, 14.15 min, wavelength = 250 nm.

2-Benzyl-5-(4′-chlorophenyl)-3-(trifluoromethyl)-2,3-dihydroizoxazole (**6ac**): 53:47 e.r.; HPLC analysis: Chiralcel AD, Hexanes: ^{i}PrOH = 98:2, flow = 0.3 mL/min, retention time: 10.65, 11.24 min, wavelength = 250 nm.

2-Benzyl-5-(4′-trifluoromethylphenyl)-3-(trifluoromethyl)-2,3-dihydroizoxazole (**6ad**): 51:49 e.r.; HPLC analysis: Chiralcel AD, Hexanes: ^{i}PrOH = 95:5, flow = 0.5 mL/min, retention time: 5.40, 5.78 min.

3.4. Computational Study

All calculations reported in this work were performed using the "Ares" cluster in the "Cyfronet" computational center in Cracow. The wb97xd functional and the 6-311+G(d) basis set included in the GAUSSIAN 09 package [37] were used. For all structures optimization the Berny algorithm was applied. Localized critical points were checked by vibrational frequency analyses to see whether they constituted minima or maxima on the potential energy surface (PES). All transition structures (TSs) showed a single imaginary frequency, whereas reactants, products and pre-reaction complexes had none. The intrinsic reaction coordinate (IRC) path was traced in order to check the energy profiles connecting each transition structure to the two associated minima of the proposed mechanism. GEDT indices were estimated according to Domingo equations [38]. The reaction environment polarity (THF) was simulated using polarizable continuum model (PCM) [39].

4. Conclusions

A method of synthesis of fluorinated hydroxylamines of type **5** was developed by adding acetylenes **4** to nitrones **3** derived from fluorinated aldehydes **1**. In these reactions, the formation of hydroxylamines **5** and/or oxazolines **6** was observed as products. However, the expected hydroxylamines **5** were the only or the main products in the reactions conducted in the presence of diethylzinc. Theoretical calculations clearly indicate that oxazolines **6** are the kinetic products and hydroxylamines **5** are the thermodynamic products of the studied process. In the extension of the studies the first enantioselective protocol of addition of acetylenes **4** to RF-nitrones **3** was elaborated. The desired products **5** were isolated in good yields and with moderate enantioselectivities. The obtained hydroxylamines **5** consist of an interesting class of building blocks for use in organic synthesis. In the literature, propargyl *N*-hydroxylamines have been used for acylaziridines [10], propargylamines by reduction [40–42], α,β-unsaturated ketones by the sequence of isomerization and hydrolysis reactions [7].

Supplementary Materials: The following supporting information can be downloaded at: https://www.mdpi.com/article/10.3390/molecules30234578/s1. Copies of ^{1}H and ^{13}C spectra of all new compounds, HPLC chromatograms and crystallographic data [43–46]. Figure S1: Molecular structure of **5aa** (left) and **3a** (right). Displacement ellipsoids are drawn at 50% probability level; Figure S2: Partial packing diagram of **5aa** (left) and **3a** (right). Displacement ellipsoids are drawn at 50% probability level; Figure S3: Selected interactions stabilizing molecular and supramolecular structure of **5aa**; Figure S4: Selected interactions stabilizing molecular and supramolecular structure of **3a**. Table S1: Selected structural data for 5aa and 3a; Table S2: Bond Lengths for **3a**_DEPO; Table S3: Bond Angles for **5aa**_DEPO; Table S4: Bond Angles for **3a**_DEPO; Table S5: Torsion Angles for 5aa_DEPO; Table S6: Torsion Angles for 3a_DEPO.

Author Contributions: Conceptualization and methodology, E.O., S.J. and R.J.; investigation, E.O., A.K. (part of bachelor's thesis), S.J. (preparative procedures), A.P. (crystallographic measurements, analysis and their description) and R.J. (DFT calculations); writing—original draft preparation, E.O., S.J. (preparative part), R.J. (calculations) and A.P. (crystallographic data description), writing—review and editing, E.O. and R.J.; supervision, E.O. and R.J.; project administration, E.O.; funding acquisition, E.O., S.J. and R.J.; All authors have read and agreed to the published version of the manuscript.

Funding: This research was funded by the University of Lodz in the framework of IDUB grant (S.J.; Grant No. 23/IDUB/MLOD/2021).

Institutional Review Board Statement: Not applicable.

Informed Consent Statement: Not applicable.

Data Availability Statement: Data are contained within the article.

Acknowledgments: We gratefully acknowledge the Polish high-performance computing infrastructure PLGrid (HPC Center: ACK Cyfronet AGH) for providing computer facilities and support within computational grant no. PLG/2025/018201.

Conflicts of Interest: The authors declare no conflicts of interest.

References

1. Maienfish, P.; Hall, R.G. The importance of fluorine in the life science industry. *Chimia* **2004**, *58*, 93–99. [CrossRef]
2. Kirsch, P. *Modern Fluoroorganic Chemistry*; Wiley-VCH: Weinheim, Germany, 2004.

3. Bégué, J.-P.; Bonnet-Delpon, D. *Bioorganic and Medicinal Chemistry of Fluorine*; Wiley: Hoboken, NJ, USA, 2008.
4. Nair, A.S.; Singh, A.K.; Kumar, A.; Kumar, S.; Sukumaran, S.; Koyiparambath, V.P.; Pappachen, L.K.; Rangarajan, T.M.; Kim, H.; Mathew, B. FDA-Approved Trifluoromethyl Group-Containing Drugs: A Review of 20 Years. *Processes* **2022**, *10*, 2054. [CrossRef]
5. Feuer, H. *Nitrile Oxides, Nitrones and Nitronates in Organic Synthesis: Novel Strategies in Synthesis*; John Wiley & Sons: Hoboken, NJ, USA, 2007.
6. Murahashi, S.-I.; Imada, Y. Synthesis and transformations of nitrones for organic synthesis. *Chem. Rev.* **2019**, *119*, 4684–4716. [CrossRef]
7. Das, P.; Hamme II, A.T. Zinc mediated direct transformation of propargyl *N*-hydroxylamines to α,β-unsaturated ketones and mechanistic insight. *Tetrahedron Lett.* **2017**, *58*, 1086–1089. [CrossRef] [PubMed]
8. Wang, Q.; Chit Tsui, G. Copper-mediated domino cyclization/trifluoromethylation of propargylic *N*-hydroxylamines: Synthesis of 4-trifluoromethyl-4-isoxazolines. *J. Org. Chem.* **2018**, *83*, 2971–2979. [CrossRef] [PubMed]
9. Ariyama, T.; Kusakabe, T.; Sato, K.; Funatogawa, M.; Lee, D.; Takahashi, K.; Kato, K. Pd(II)-Catalyzed ligand-controlled synthesis of 2,3-dihydroisoxazole-4-carboxylates and bis(2,3-dihydroisoxazol-4-yl)methanones. *Heterocycles* **2016**, *93*, 512–528. [CrossRef]
10. Miyamoto, Y.; Wada, N.; Soeta, T.; Fujinami, S.; Inomata, K.; Ukaji, Y. One-pot stereoselective synthesis of 2-acylaziridines and 2-acylpyrrolidines from *N*-(propargylic)hydroxylamines. *Chem. Asian J.* **2013**, *8*, 824–831. [CrossRef]
11. Norihiro, W.; Kentaro, K.; Yutaka, U.; Katsuhiko, I. Selective Transformation of *N*-(propargylic)hydroxylamines into 4-isoxazolines and acylaziridines promoted by metal salts. *Chem. Lett.* **2011**, *40*, 440–442. [CrossRef]
12. Reddy, C.R.; Vijaykumar, J.; Jithender, E.; Pavan, G.; Reddy, K.; Grée, R. One-pot synthesis of 3,5-disubstituted isoxazoles from propargylic alcohols through propargylic *N*-hydroxylamines. *Eur. J. Org. Chem.* **2012**, *2012*, 5767–5773. [CrossRef]
13. Gayon, E.; Szymczyk, M.; Gérard, H.; Vrancken, E.; Campagne, J.-M. Stereoselective and catalytic access to β-enaminones: An entry to pyrimidines. *J. Org. Chem.* **2012**, *77*, 9205–9220. [CrossRef]
14. Mlostoń, G.; Obijalska, E.; Celeda, M.; Mittermeier, V.; Linden, A.; Heimgartner, H. 1,3-Dipolar cycloadditions of fluorinated nitrones with thioketones. *J. Fluor. Chem.* **2014**, *165*, 27–32. [CrossRef]
15. Kowalski, M.K.; Mlostoń, G.; Obijalska, E.; Linden, A.; Heimgartner, H. First application of fluorinated nitrones for the synthesis of fluoroalkylated β-lactams via the Kinugasa reaction. *Tetrahedron* **2016**, *72*, 5305–5313. [CrossRef]
16. Jasiński, R. A new mechanistic insight on β-lactam systems formation from 5-nitroisoxazolidines. *RSC Adv.* **2015**, *5*, 50070–50072. [CrossRef]
17. Tanaka, K.; Ohsuga, M.; Sugimoto, Y.; Okafuji, Y.; Mitsuhashi, K. Applications of the fluorinated 1,3-dipolar compounds as the building blocks of the heterocycles with fluorine groups. Part XII. Synthesis of trifluoromethylisoxazolines and their rearrangement into trifluoromethylaziridines. *J. Fluor. Chem.* **1988**, *39*, 39–45. [CrossRef]
18. Tanaka, K.; Sugimoto, Y.; Okafuji, Y.; Tachikawa, M.; Mitsuhashi, K. Regio- and stereoselectivity of cycloadditions of C-(Trifluoromethyl)nitrone with olefins. *J. Heterocycl. Chem.* **1989**, *26*, 381–385. [CrossRef]
19. Milcent, T.; Hinks, N.; Bonnet-Delpon, D.; Crousse, B. Trifluoromethyl nitrones: From fluoral to optically active hydroxylamines. *Org. Biomol. Chem.* **2010**, *8*, 3025–3030. [CrossRef]
20. Zhang, Y.; Nie, J.; Zhang, F.-G.; Ma, J.-A. Zinc mediated enantioselective addition of terminal 3-en-1-ynes to cyclic trifluoromethyl ketimines. *J. Fluor. Chem.* **2018**, *208*, 1–9. [CrossRef]
21. Frantz, D.E.; Fässler, R.; Carreira, E.M. Catalytic in Situ Generation of Zn(II)-Alkynilides under Mild Conditions: A Novel C=N Addition Process Utilizing Terminal Acetylenes. *J. Am. Chem. Soc.* **1999**, *121*, 11245–11246. [CrossRef]
22. Aschwanden, P.; Frantz, D.E.; Carreira, E.M. Synthesis of 2,3-Dihydroisoxazoles from Propargylic N-Hydroxylamines via Zn(II)-Catalyzed Ring-Closure Reaction. *Org. Lett.* **2000**, *2*, 2331–2333. [CrossRef] [PubMed]
23. Karabuga, S.; Karakaya, I.; Ulukanli, S. 3-Aminoquinazolinones as chiral ligands in catalytic enantioselective diethylzinc and phenylacetylene addition to aldehydes. *Tetrahedron Asymmetry* **2014**, *25*, 851–855. [CrossRef]
24. Wróblewska, A.; Sadowski, M.; Jasiński, R. Selectivity and molecular mechanism of the Au(III)-catalyzed [3+2] cycloaddition reaction between (Z)-C,N-diphenylnitrone and nitroethene in the light of the molecular electron density theory computational study. *Chem. Heterocycl. Compd.* **2024**, *60*, 639–645. [CrossRef]
25. Domingo, L.R.; Ríos-Gutiérrez, M. A Useful Classification of Organic Reactions Based on the Flux of the Electron Density. *Sci. Radices* **2023**, *2*, 1–24. [CrossRef]
26. Mondal, A.; Mohammad-Salim, H.A.; Acharjee, N. Unveiling substituent effects in [3+2] cycloaddition reactions of benzonitrile N-oxide and benzylideneanilines from the molecular electron density theory perspective. *Sci. Radices* **2023**, *2*, 75–92. [CrossRef]
27. Kula, K.; Sadowski, M. Regio- and stereoselectivity of [3+2] cycloaddition reactions between (Z)-1-(anthracen-9-yl)-N-methyl nitrone and analogs of trans-β-nitrostyrene on the basis of MEDT computational study. *Chem. Heterocycl. Compd.* **2023**, *59*, 138–144. [CrossRef]
28. Chafaa, F.; Nacereddine, A.K. A molecular electron density theory study of mechanism and selectivity of the intramolecular [3+2] cycloaddition reaction of a nitrone–vinylphosphonate adduct. *Chem. Heterocycl. Compd.* **2023**, *59*, 171–178. [CrossRef]

29. Braga, A.L.; Paixão, M.W.; Westermann, B.; Schneider, P.H.; Wessjohann, L.A. Acceleration of Arylzinc Formation and Its Enantioselective Addition to Aldehydes by Microwave Irradiation and Aziridine-2-methanol Catalysts. *J. Org. Chem.* **2008**, *73*, 2879–2882. [CrossRef] [PubMed]
30. Carlos, A.M.M.; Contreira, M.E.; Martins, B.S.; Immich, M.F.; Moro, A.V.; Lüdtke, D.S. Catalytic asymmetric arylation of aliphatic aldehydes using a B/Zn exchange reaction. *Tetrahedron* **2015**, *71*, 1202–1206. [CrossRef]
31. Wang, M.-C.; Wang, Y.-H.; Li, G.-W.; Sun, P.-P.; Tian, J.-X.; Lu, H.-J. Applications of conformational design: Rational design of chiral ligands derived from a common chiral source for highly enantioselective preparations of (R)- and (S)-enantiomers of secondary alcohols. *Tetrahedron Asymmetry* **2011**, *22*, 761–768. [CrossRef]
32. Jarzyński, S.; Leśniak, S.; Pieczonka, A.M.; Rachwalski, M. *N*-Trityl-aziridinyl alcohols as highly efficient chiral catalysts in asymmetric additions of organozinc species to aldehydes. *Tetrahedron Asymmetry* **2015**, *26*, 35–40. [CrossRef]
33. Jarzyński, S.; Utecht, G.; Leśniak, S.; Rachwalski, M. Highly enantioselective asymmetric reactions involving zinc ions promoted by chiral aziridine alcohols. *Tetrahedron Asymmetry* **2017**, *28*, 1774–1779. [CrossRef]
34. Bonini, B.F.; Capitò, E.; Comes-Franchini, M.; Fochi, M.; Riccia, A.; Zwanenburg, B. Aziridin-2-yl methanols as organocatalysts in Diels–Alder reactions and Friedel–Crafts alkylations of N-methyl-pyrrole and N-methyl-indole. *Tetrahedron Asymmetry* **2006**, *17*, 3135–3143. [CrossRef]
35. Trost, B.M.; Hung, C.-I.J.; Koester, D.C.; Miller, Y. Development of Non-C2-symmetric ProPhenol Ligands. The Asymmetric Vinylation of *N*-Boc Imines. *Org. Lett.* **2015**, *17*, 3778–3781. [CrossRef]
36. Gao, Y.-Y.; Hua, Y.-Z.; Wang, M.-C. Asymmetric 1,6-Conjugate Addition of *para*-Quinone Methides for the Synthesis of Chiral β,β-Diaryl-α-Hydroxy Ketones. *Adv. Synth. Catal.* **2018**, *360*, 80–85. [CrossRef]
37. Frisch, M.J.; Trucks, G.W.; Schlegel, H.B.; Scuseria, G.E.; Robb, M.A.; Cheeseman, J.R.; Scalmani, G.; Barone, V.; Petersson, G.A.; Nakatsuji, H.; et al. *Gaussian 09. Revision D.01*; Gaussian, Inc.: Wallingford, CT, USA, 2013.
38. Domingo, L.R. A new C–C bond formation model based on the quantum chemical topology of electron density. *RSC Adv.* **2014**, *4*, 32415–32428. [CrossRef]
39. Scalmani, G.; Frisch, M.J.J. Continuous surface charge polarizable continuum models of solvation. I. General formalism. *Chem. Phys.* **2010**, *132*, 114110. [CrossRef]
40. Supranovich, V.I.; Levin, V.V.; Struchkova, M.I.; Dilman, A.D. Photocatalytic Reductive Fluoroalkylation of Nitrones. *Org. Lett.* **2018**, *20*, 840–843. [CrossRef] [PubMed]
41. Nelson, D.W.; Owens, J.; Hiraldo, D. α-(Trifluoromethyl)amine Derivatives via Nucleophilic Trifluoromethylation of Nitrones. *Org. Chem.* **2001**, *66*, 2572–2582. [CrossRef]
42. Konno, T.; Moriyasu, K.; Ishihara, T. A Remarkable Access to γ-Fluoroalkylated Propargylamine Derivatives or Fluoroalkylated Dihydroisoxazoles via the Reaction of Fluoroalkylated Acetylides with Various Nitrones. *Synthesis* **2009**, *7*, 1087–1094. [CrossRef]
43. Agilent. *CrysAlis PRO*; Agilent Technologies Ltd.: Yarnton, UK, 2014.
44. Sheldrick, G.M. SHELXT–Integrated space-group and crystal-structure determination. *Acta Cryst.* **2015**, *A71*, 3–8. [CrossRef] [PubMed]
45. Hübschle, C.B.; Sheldrick, G.M.; Dittrich, B. ShelXle: A Qt graphical user interface for SHELXL. *J. Appl. Cryst.* **2011**, *44*, 1281–1284. [CrossRef]
46. Sheldrick, G.M. Crystal structure refinement with SHELXL. *Acta Cryst.* **2015**, *C71*, 3–8. [CrossRef]

molecules

MDPI

Communication

Computational Analysis of the Asymmetric Hydrogenation of γ-Ketoacids: Weak Interactions and Kinetics

Ivan S. Golovanov * and Evgeny V. Pospelov *

N. D. Zelinsky Institute of Organic Chemistry, Russian Academy of Sciences, Leninsky Prospect 47, 119991 Moscow, Russia

* Correspondence: igolovanov@ioc.ac.ru (I.S.G.); evpospelov@ioc.ac.ru (E.V.P.)

Abstract

A computational study of the mechanism of asymmetric hydrogenation of γ-keto acids with the Ni(S,S)-QuinoxP* system was conducted. The main steps of the reaction mechanism were determined, including the formation of the NiH(S,S-QuinoxP*)$^+$ complex starting from a γ-keto acid molecule and the involvement of the hydrogen "metathesis" step. The rate-limiting and stereo-determining step of the reaction was identified as the transfer of a hydrogen atom from the catalytic particle to the carbonyl group of the substrate molecule. The stereochemical outcome of the process was calculated. The influence of weak interactions on the stereoselectivity of the process was demonstrated using NCI and sobEDAw analyses.

Keywords: asymmetric catalysis; non-bonding interactions; asymmetric hydrogenation; transition metal catalysis; DFT calculations; theoretical investigations into asymmetric catalysis and synthesis

1. Introduction

Asymmetric hydrogenation catalyzed by transition metal complexes (e.g., Ni, Cu, Fe, and Co) [1–6] is a rapidly developing area of modern chemistry. The possibility of replacing catalytic systems based on noble metals (Au, Ru, Rh, Ir, Pt, Pd) with relatively inexpensive and readily available systems based on transition metals opens a way to an inexpensive and highly stereoselective synthesis of various compounds, many of which are intermediates in the synthesis of pharmaceutically active compounds and natural products [7–9]. The synthesis of 5- and 6-membered lactones using γ/δ-keto acids, readily available from natural raw materials, has been actively studied in recent years [10]. Asymmetric hydrogenation of keto acids uses chiral catalysts (often Ir, Ru, Ni, or Co with specific ligands) to convert keto-functionalized acids into valuable chiral hydroxy acids or lactones, producing single enantiomers (optical isomers) crucial for pharmaceuticals, with high efficiency (TONs up to 100,000) and enantioselectivity (up to 99% ee) for γ- and δ-ketoacids, offering a green, scalable route to chiral building blocks and certain drugs (for example, Ezetimibe [11]). Catalytic systems based on Ir [11], Ru [12], and Pd [13] are used for synthesizing chiral lactones, as well as the recently proposed methods utilizing chiral Ni complexes [14,15]. Despite the availability of some experimental material, this topic has not yet been thoroughly studied. Moreover, this process lacks a sufficient explanation of its stereoselectivity and proposed mechanism. In this work, the mechanism of the Ni-catalyzed asymmetric hydrogenation reaction of γ-keto acids was studied using DFT calculations. In addition, the weak interactions contributing to the stereoselectivity of the studied reaction were visualized and quantified using NCI [16] and sobEDAw [17] analyses.

2. Results and Discussion

It can be noted that asymmetric hydrogenation reactions typically follow a specific pattern of elementary steps. The first step of any catalytic process is the formation of a catalytic particle. Considering that the starting compounds for the formation of the catalytic particle are nickel perchlorate and (S,S)-QuinoxP*, as well as the presence of hydrogen dissolved in the reaction mixture, it can be assumed that a number of nickel hydride complexes (including, but not limited to, mono- and dihydride complexes and their dimers) are formed during the reaction and can act as catalytic particles for the asymmetric hydrogenation process under consideration. Since in this case the catalytic system is of a "cocktail" type due to the possibility of interconversion of nickel hydride complexes into each other under these conditions, it is necessary to select a catalytic particle that, firstly, can exist under the reaction conditions, and secondly, effectively binds to the substrate molecules, namely a γ-keto acid. Thus, we decided to choose the hydride complex $NiH(S,S\text{-}QuinoxP^*)^+$ **1**, whose possible formation was demonstrated in one of our recent studies, as the catalytic particle [18].

The next step of the process is the interaction between the catalytic particle and the substrate molecule, leading to the formation of a prereaction complex. Note that although this stage itself is usually neither stereodetermining nor rate-limiting, the quality of the interaction between the catalytic particle and the substrate molecule at this step can significantly affect the subsequent steps. After the formation of the prereaction complex, hydride transfer occurs from the metal atom to the functional group of the substrate molecule. This step is often both rate-limiting and stereodetermining, since the hydrogen atom transferred to the substrate molecule typically does not participate further in the process, as shown by isotope labeling experiments in recent studies [1]. It is worth noting that the nature of the interactions between the substrate and catalytic particle molecules at this stage is of particular interest, as it can shed light on the possible influence of the interaction between the substrate and the catalyst (or part of it) on the stereoselectivity of the process.

Finally, a crucial step is the so-called hydrogen "metathesis," without which the regeneration of the catalytic particle cannot occur. This step typically occurs with a high reverse barrier but a low forward barrier, suggesting that this step is irreversible within the energy profile of the transformation under consideration. However, hydrogen "metathesis" requires the hydrogen molecule to successfully coordinate with the metal atom, which is sometimes difficult because the ligand environment creates certain steric hindrance.

However, the ligand structure and the steric hindrance it causes could be a double-edged sword. On the one hand, the typically "branched" ligand structure complicates the interaction of the catalytic particle with substrate molecules, solvent, or hydrogen. On the other hand, it is the ligand skeleton that is responsible for the majority of the weak interactions that influence the stereochemical outcome of asymmetric hydrogenation reactions. Therefore, it is important to maintain a balance between hindering the kinetics of the process and improving the interaction between the catalytic particle and the substrate molecule.

A reaction mechanism was proposed for the Ni-catalyzed asymmetric hydrogenation of γ-keto acids previously studied experimentally [14,15]. At the first step, complex **2** is formed from Ni catalyst **1** and a substrate molecule. The use of Ni-H species **1** as a catalyst has been substantiated in a number of previous works [1,6]. Then, the intramolecular addition of a H atom to the carbonyl carbon (hydrogen transfer step) occurs via **TS1** to

https://doi.org/10.3390/molecules31020385

give complex **3**. The process can proceed further through two pathways (pathway A and pathway B). Pathway A takes hydrogen, needed for product recovery and catalytic species regeneration, directly from the hydrogen molecule. Alternatively, pathway B utilizes a solvent, trifluoroethanol, as the hydrogen source, resulting in the formation of nickel alkoxide. In pathway A, at the next stage complex **4** is formed by the reaction of **3** with a hydrogen molecule (with a distance of 3.38 angstroms (pathway R) or 2.76 angstroms (pathway S) between the Ni atom and the hydrogen molecule). The subsequent "wedge-in" of the coordinated hydrogen molecule via **TS2** leads to species **5**. Finally, the "metathesis" stage of the hydrogen molecule in complex **5** via **TS3** results in particle **6**, which is actually a complex of the original catalytic particle and the γ-hydroxy acid. Their subsequent splitting completes the catalytic cycle and leads to the regeneration of the catalytic particle **1** (Scheme 1). The alternative pathway B involves the interaction of complex **3** with a trifluoromethanol molecule to give complex **4′**. After proton transfer from the alcohol molecule to the substrate molecule, the formation of the target hydroxy acid **7** and complex **5′** occurs. Then particle **5′** interacts with a hydrogen molecule, which leads to the formation of catalytic particle **1** and the closure of the catalytic cycle. Despite the fundamental possibility of the reaction occurring through either Pathway A or Pathway B, only Pathway A was calculated in this study, since previous studies [1] have demonstrated the possibility of hydrogen "metathesis" and shown that Pathway B requires overcoming a high activation barrier. Furthermore, regeneration of catalytic particle **1** still requires the interaction of **5′** with molecular hydrogen.

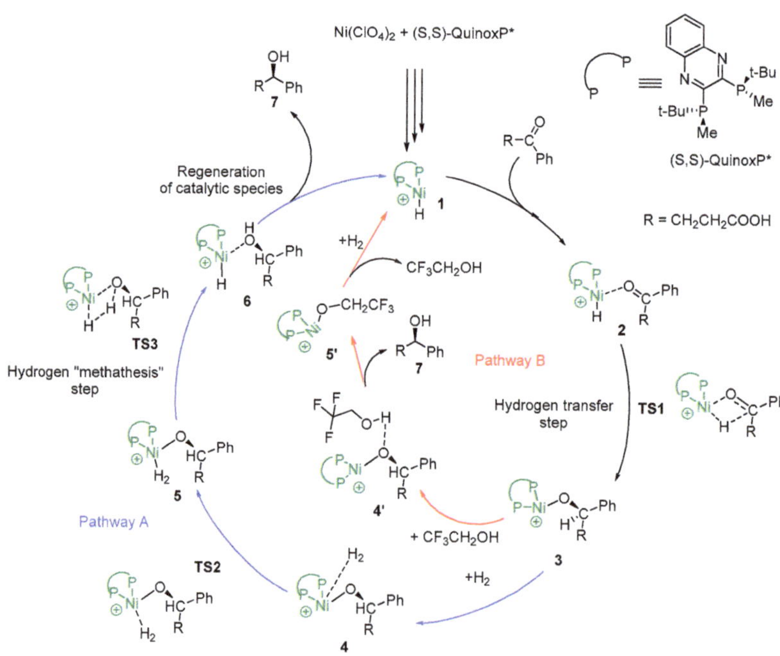

Scheme 1. Possible catalytic cycle of the Ni-catalyzed asymmetric hydrogenation of γ-keto acids.

We optimized both singlet and triplet states of Ni complex **1** and discovered that triplet-singlet gap $\Delta E_{T\text{-}S}$ = +10.9 kcal/mol. Furthermore, for all other calculated structures, stability analysis indicates a stable wavefunction, so the ground state is the singlet state. Thus we used multiplicity 1 throughout calculations. Calculated Gibbs free energy profile of the Ni-catalyzed asymmetric hydrogenation of a γ-keto acid (the "blue" pathway for R

https://doi.org/10.3390/molecules31020385

and the "red" pathway for S) is shown in Scheme 2. In fact, the formation of pre-reaction complexes **1-R** and **1-S** is an energetically favorable process (with a benefit of 8.4 kcal/mol for **1-R** and 11.2 kcal/mol for **1-S**). Their conversion into alcoholates **2-S** and **2-R** occurs via transition states **TS1-R** and **TS1-S** with similar energies, which are discussed in more detail below. The subsequent interaction with a hydrogen molecule leads (without a transition state) to complexes **3-R** and **3-S**, with complex **3-R** being 3.5 kcal/mol more favorable than the corresponding complex **3-S**. The distance between the hydrogen molecule and the nickel atom in structures **3-R** and **3-S** deserves special attention. Indeed, the difference in the corresponding distances in **3-R** and **3-S** is almost 0.5 angstroms (2.76 angstroms in **3-S** and 3.28 angstroms in **3-R**). This explains the difference in the activation barriers at the step of incorporation of a hydrogen molecule into the inner coordination sphere of the nickel atom (**TS2-R** and **TS2-S**). The subsequent step of hydrogen "metathesis" occurs via transition states with similar energies **TS3-R** (ΔG = 11.0 kcal/mol) and **TS3-S** (ΔG = 10.5 kcal/mol) and with a high barrier to the reverse transformation (19.7 kcal/mol for **TS3-R** and 19.0 kcal/mol for **TS3-S**). Then, everything ends with the regeneration of catalytic species **1** and the formation of γ-hydroxycarboxylic acids **6-S** and **6-R**.

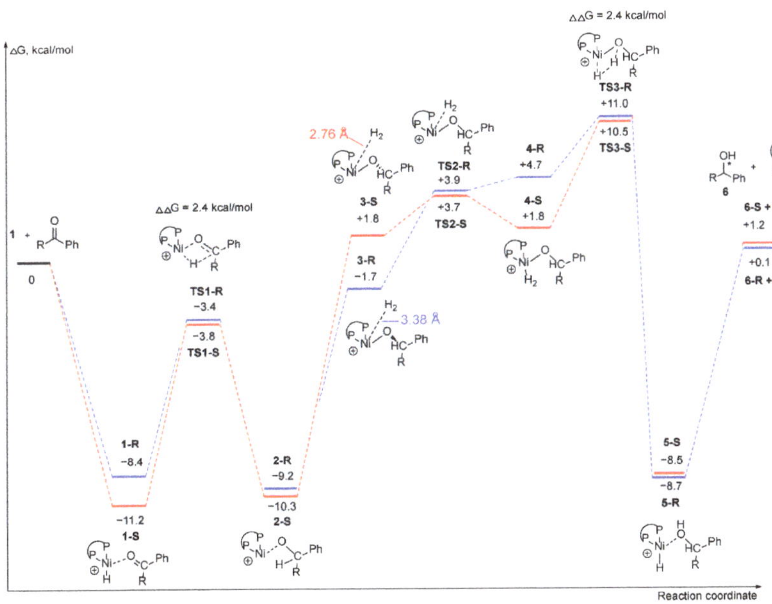

Scheme 2. Computed Gibbs free energy profiles for the asymmetric hydrogenation of γ-ketoacids catalyzed by the Ni complex of (S,S)-QuinoxP*. Calculations were performed at DFT level of theory with TPSSh functional, using the def2-TZVPP basis set on Ni and def2-SVP on other atoms, in 2,2,2-trifluoroethanol (SMD model).

As shown in the energy diagram, the main difference in the Gibbs free energy of activation is observed at the first step of the process, i.e., transfer of the hydrogen atom from the Ni atom to the C atom of the carbonyl group, with a barrier of 5.0 kcal/mol for **TS1-R** and 7.4 kcal/mol for **TS1-S**, $\Delta\Delta G_{act}$ = 2.4 kcal/mol (Scheme 3).

https://doi.org/10.3390/molecules31020385

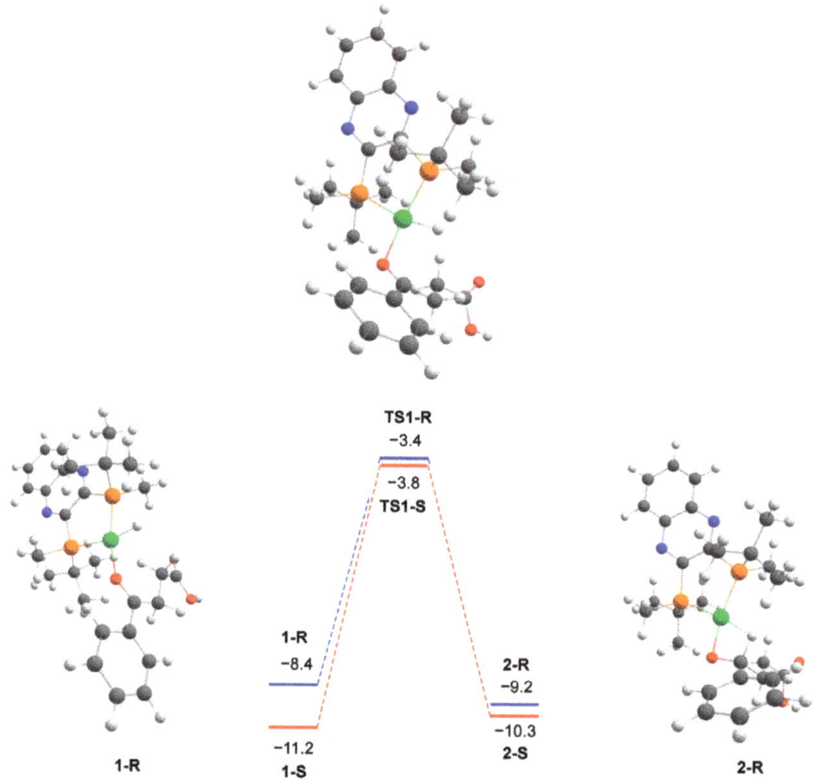

Scheme 3. The stereodetermining step of the reaction. The structures of the R-pathway are shown. The structures of the minor S-pathway are omitted for clarity.

To verify the reproducibility of the obtained result, this stage was recalculated using a different, suitable DFT functional, M11-L, yielded similar reaction barriers (4.5 kcal/mol for **TS1-R** and 7.2 kcal/mol for **TS1-S** ($\Delta\Delta G_{act}$ = 2.7 kcal/mol)).

The barrier of the "metathesis" of the hydrogen molecule is 6.3 kcal/mol for **TS3-R** and 8.7 kcal/mol for **TS3-S** ($\Delta\Delta G_{act}$ = 2.4 kcal/mol). Note that the barrier of the reverse reaction for the S-stereoisomer in step 1 is lower than that of the forward reaction. This should further promote the formation of the R-product.

To clarify the difference in the rates of the process under study for paths R and S, Eyring–Polanyi equation was used:

$$k(T) = \kappa \frac{k_B T}{h} e^{-\Delta G_0^{\neq}/RT}$$

$k(T)$—reaction rate constant; κ—transmission coefficient; k_B—Boltzmann constant; T—temperature; h—Planck constant; $\Delta G_0^{\neq}$—Gibbs free energy of activation; R—universal gas constant. Assuming that for the transformation of **1** into **2** via **TS1** the difference in transmission coefficients can be neglected, we obtain that the ratio k(R)/k(S) = 44.7.

Considering that the addition of the hydrogen atom, which determines the stereochemistry of the process, occurs at the **1** ⟶ **2** conversion step via **TS1**, NCI analysis was performed for **TS1-R** and **TS1-S** to visualize the weak interactions between the substrate molecule and the catalyst particle (Figure 1, see Supplementary Materials for color code for NCI).

https://doi.org/10.3390/molecules31020385

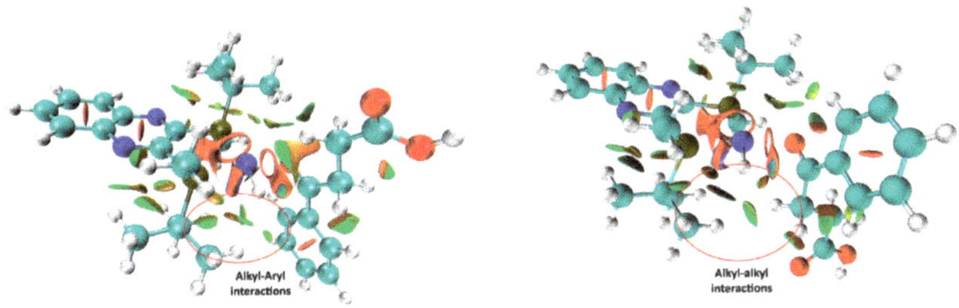

Figure 1. NCI analysis for **TS1-R** (**left**) and **TS1-S** (**right**).

NCI analysis reveals that in the case of **TS1-R**, interactions are observed between the π-orbitals of the phenyl group of the substrate and the tert-butyl group of the catalyst, leading to the stabilization of the transition state. Although such interactions are not observed in the case of **TS1-S**, they are replaced by interactions between the CH bonds of the substrate molecule and the tert-butyl moiety of the catalyst, which also leads to enhanced interactions between the substrate and the catalyst particle. To elucidate the nature of these interactions in **TS1-R** and **TS1-S** and to quantify them, energy decomposition analysis (sobEDAw) was also performed (Table 1).

Table 1. SobEDAw analysis for **TS1-R** and **TS1-S**.

E	TS1-R, kcal/mol	TS1-S, kcal/mol
E_{total}	−38.40	−37.64
E_{els}	−46.62	−47.89
E_{xrep}	67.74	66.88
E_{orb}	−35.91	−35.72
E_{disp}	−23.62	−20.92

Apparently, the main difference between **TS1-R** and **TS1-S** lies in the dispersion and electrostatic interactions. The electrostatic energy and exchange-repulsion energy were found to be higher in **TS1-R** than in **TS1-S**, while the dispersion interaction in **TS1-R** is significantly lower (by almost 3 kcal/mol) than in **TS1-S**. In summary, **TS1-R** has a lower interaction energy than **TS1-S**, which is consistent with experiments [14,15]. The large difference in dispersion energy may indicate that the ligand skeleton and the apparently non-reactive moiety of the substrate molecule may interact with each other and thereby improve the enantioselectivity of the reaction.

The kinetics of the "metathesis" step also predicts the preferential formation of the R-isomer of the γ-hydroxy acid (see discussion above). The formation of the R-isomer was also supported by NCI (Figure 2) and sobEDAw analyses of the corresponding transition states **TS3-R** and **TS3-S**.

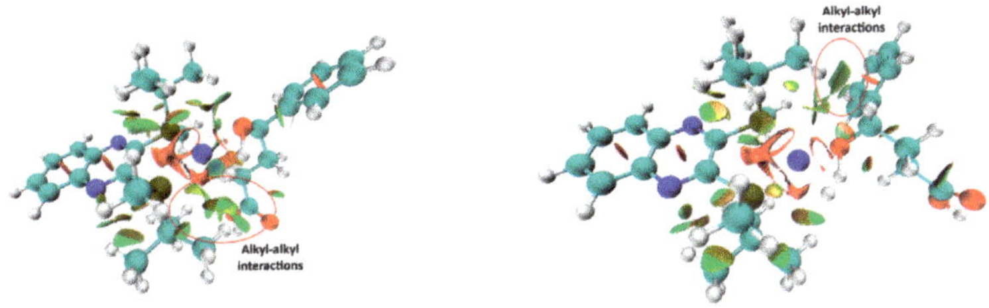

Figure 2. NCI analysis for **TS3-R** (**left**) and **TS3-S** (**right**).

https://doi.org/10.3390/molecules31020385

According to NCI for **TS3-R**, a strong interaction exists between the alkyl tail of the substrate molecule and the tert-butyl group of the ligand moiety. Furthermore, an interaction is also observed between the alpha-aromatic hydrogen atom of the substrate molecule and the methyl group of the catalytic particle backbone. It is worth noting that such interactions are almost nonexistent in **TS3-S**. In turn, **TS3-S** exhibits the interaction between the tert-butyl group of the ligand and the alpha-aromatic hydrogen atom of the substrate molecule, as well as the weak interaction between the hydrogen atoms of the phenyl group and the methyl group of the ligand. No other weak interactions leading to improved binding between the substrate and catalyst molecules are observed.

To quantify the different components of the binding energy between the substrate and catalyst molecules in **TS3-R** and **TS3-S**, energy decomposition analysis (sobEDAw) was also performed (Table 2).

Table 2. SobEDAw analysis for **TS3-R** and **TS3-S**.

E	TS3-R, kcal/mol	TS3-S, kcal/mol
E_{total}	−254.70	−250.33
E_{els}	−273.02	−260.69
E_{xrep}	217.82	193.24
E_{orb}	−163.49	−155.20
E_{disp}	−36.02	−27.68

Overall, the interaction between the molecules of the catalytic particle and the substrate is stronger in **TS3-R** than in **TS3-S**. Thus, the difference in the total interaction energy between **TS3-R** and **TS3-S** is 4.37 kcal/mol. The largest difference in its constituent energies lies in the electrostatic interaction (12.33 kcal/mol), as well as in the orbital (8.29 kcal/mol) and dispersion (8.34 kcal/mol) interactions. We suggest that the main contribution to the latter interactions is made by the weak interactions between the ligand skeleton and various moieties of the initial γ-keto acid molecule. It is worth mentioning separately that the exchange-repulsion energy is also higher in **TS3-R**; however, the total contribution of all components still leads to better binding of the substrate and catalyst molecules than in **TS3-S**.

To get insights into the electron distribution in catalytic species **1** and transition states of hydrogen transfer step (**TS1**), we examined them using the Extended Transition State—Natural Orbitals for Chemical Valence method (ETS-NOCV). In **1** the main part of contribution to ΔEorb stems from the sum of the first and second NOCV pairs ($\Delta E = -101.2$ kcal/mol). As follows from the visualization of NOCV pair density isosurface, it is clear that the sum of two first NOCV pairs corresponds to electron donation from both phosphorus and nitrogen atoms of the ligand to the nickel and hydrogen atoms (Figure 3).

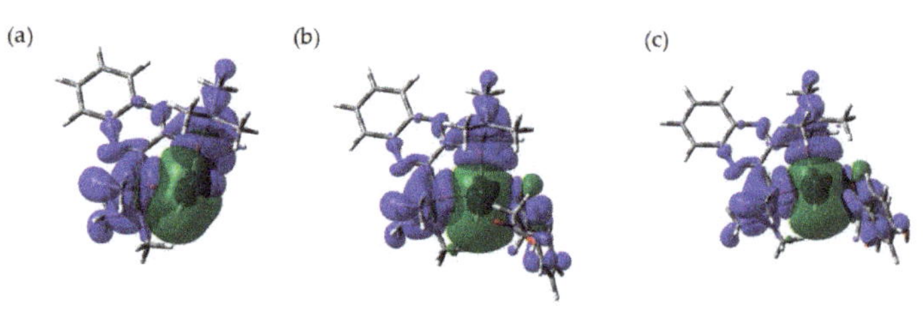

Figure 3. (**a**) Visualization of NOCV pair density isosurfaces for **1**. (**b**) Visualization of NOCV pair density isosurfaces for **TS1-R**. (**c**) Visualization of NOCV pair density isosurfaces for **TS1-S**. The blue/green colored isodensity surface defines electron density depletion/accumulation.

In **TS1** unlike in **1**, the π orbitals of the aromatic moiety of the substrate molecule also participate in the electron distribution in the transition state. This may indicate the influence

https://doi.org/10.3390/molecules31020385

of the substrate's structural features on the reaction and may indirectly explain the different stereochemical outcomes of substrates that differ only in the substitution pattern of the aromatic moiety. It should be noted that the first two NOCV pairs also make a significant contribution to ΔE_{orb} ($\Delta E = -116.8$ kcal/mol for **TS1-R** and $\Delta E = -115.9$ kcal/mol for **TS1-S**).

3. Materials and Methods

DFT calculations were performed using Gaussian 16 Rev.C01 [19]. The TPSSh DFT functional with the def2-TZVPP basis set on Ni and def2-SVP on the other atoms was used for geometry optimization, thermodynamic and kinetic calculations. All calculations were performed in 2,2,2-trifluoroethanol (SMD model). Noncovalent interaction (NCI) analysis [16] and ETS-NOCV [20] were performed using Multiwfn ver. 3.8. [21], while sobEDAw analysis was performed according to the reported procedure [17]. See Supplementary Materials for details.

4. Conclusions

A mechanism for the asymmetric hydrogenation of γ-keto acids was proposed and modeled by DFT calculations. The stereoselectivity of the process was shown to be influenced by both the reaction kinetics and weak interactions in the transition states of the stereodetermining step–hydrogen atom transfer stage. DFT computations have shown that the difference in the activation barriers of this stage is 2.4 kcal/mol, which determines the stereoselectivity of the formation of the R-product. The difference in Gibbs free energy between the R and S pathways predicts 96% ee for the hydrogenation product, in good agreement with the experimental data [14,15]. Using the Eyring–Polanyi equation, the ratio of the rate constants for pathways R and S can be estimated as k(R)/k(S) = 44.7. According to NCI and sobEDAw analyses, weak interactions in **TS1** between π-orbitals of the phenyl group of the substrate and the tert-butyl group of the catalyst have an impact on preferential formation of the R-product.

Supplementary Materials: The following supporting information can be downloaded at: https://www.mdpi.com/article/10.3390/molecules31020385/s1, Details of DFT calculations, Cartesian coordinates and energies for all calculated structures.

Author Contributions: Conceptualization, E.V.P.; methodology, I.S.G. and E.V.P.; investigation, I.S.G. and E.V.P.; calculations, I.S.G.; writing—original draft preparation, E.V.P.; writing—review and editing, I.S.G. and E.V.P.; supervision, E.V.P.; project administration, I.S.G. and E.V.P. All authors have read and agreed to the published version of the manuscript.

Funding: This research received no external funding.

Institutional Review Board Statement: Not applicable.

Data Availability Statement: All data are included in the manuscript and the Supplementary Materials.

Conflicts of Interest: The authors declare no conflicts of interest.

References

1. Wei, H.; Luo, Y.; Li, J.; Chen, J.; Gridnev, I.D.; Zhang, W. Enantioselective Synthesis of Chiral β^2-Amino Phosphorus Derivatives via Nickel-Catalyzed Asymmetric Hydrogenation. *J. Am. Chem. Soc.* **2025**, *147*, 342–352. [CrossRef] [PubMed]
2. Guan, J.; Luo, Y.; Wang, Q.; Chen, J.; Zhang, W. Copper-Catalyzed Asymmetric Hydrogenation of Unsymmetrical *Ortho*-Br Substituted Benzophenones. *Angew. Chem. Int.* **2025**, *64*, e202416313. [CrossRef] [PubMed]
3. Li, Y.; Yu, S.; Wu, X.; Xiao, J.; Shen, W.; Dong, Z.; Gao, J. Iron Catalyzed Asymmetric Hydrogenation of Ketones. *J. Am. Chem. Soc.* **2014**, *136*, 4031–4039. [CrossRef]
4. Wang, Z.; Li, M.; Zuo, W. Cobalt-Catalyzed Asymmetric Hydrogenation of Ketones Enabled by the Synergism of an N–H Functionality and a Redox-Active Ligand. *J. Am. Chem. Soc.* **2024**, *146*, 26416–26426. [CrossRef]

https://doi.org/10.3390/molecules31020385

5. Gridnev, I.D. Co-Catalyzed Asymmetric Hydrogenation. The Same Enantioselection Pattern for Different Mechanisms. *Int. J. Mol. Sci.* **2023**, *24*, 5568. [CrossRef] [PubMed]
6. Li, B.; Chen, J.; Liu, D.; Gridnev, I.D.; Zhang, W. Nickel-Catalysed Asymmetric Hydrogenation of Oximes. *Nat. Chem.* **2022**, *14*, 920–927. [CrossRef]
7. Bellido, M.; Riego-Mejías, C.; Diaz-Moreno, A.; Verdaguer, X.; Riera, A. Enantioselective Ir-Catalyzed Hydrogenation of Terminal Homoallyl Sulfones: Total Synthesis of (−)-Curcumene. *Org. Lett.* **2023**, *25*, 1453–1457. [CrossRef]
8. Zhang, W.; Chen, X.; An, Y.; Wang, J.; Zhuang, C.; Tang, P.; Chen, F. Enantioselective Total Syntheses of (−)-20-*Epi*-Vincamine and (−)-20-*Epi*-Eburnamonine by Ir-Catalyzed Asymmetric Imine Hydrogenation/Lactamization Cascade. *Chem. Euro. J.* **2020**, *26*, 10439–10443. [CrossRef] [PubMed]
9. Guo, C.; Sun, D.-W.; Yang, S.; Mao, S.-J.; Xu, X.-H.; Zhu, S.-F.; Zhou, Q.-L. Iridium-Catalyzed Asymmetric Hydrogenation of 2-Pyridyl Cyclic Imines: A Highly Enantioselective Approach to Nicotine Derivatives. *J. Am. Chem. Soc.* **2015**, *137*, 90–93. [CrossRef] [PubMed]
10. Sharma, R.; Solanki, A.K.; Jayamurugan, G. One-Pot Synthesis of Lactones from Ketoacids Involving Microwave Heating and Sodium Borohydride: Application in Biomass Conversion. *Chem. Commun.* **2025**, *61*, 93–96. [CrossRef] [PubMed]
11. Hua, Y.-Y.; Bin, H.-Y.; Wei, T.; Cheng, H.-A.; Lin, Z.-P.; Fu, X.-F.; Li, Y.-Q.; Xie, J.-H.; Yan, P.-C.; Zhou, Q.-L. Iridium-Catalyzed Asymmetric Hydrogenation of γ- and δ-Ketoacids for Enantioselective Synthesis of γ- and δ-Lactones. *Org. Lett.* **2020**, *22*, 818–822. [CrossRef] [PubMed]
12. Li, J.; Ma, Y.; Lu, Y.; Liu, Y.; Liu, D.; Zhang, W. Synthesis of Enantiopure γ-Lactones via a RuPHOX-Ru Catalyzed Asymmetric Hydrogenation of γ-Keto Acids. *Adv. Synth. Catal.* **2019**, *361*, 1146–1153. [CrossRef]
13. Zhang, K.; Zhang, X.; Chen, J.; Liu, Z.; Pan, C.; Zhu, Y.; Wu, S.; Fan, B. Palladium/Zinc Co-Catalyzed Asymmetric Hydrogenation of γ-Keto Carboxylic Acids. *Chem. Asian J.* **2021**, *16*, 1229–1232. [CrossRef] [PubMed]
14. Deng, C.-Q.; Deng, J. Ni-Catalyzed Asymmetric Hydrogenation of Aromatic Ketoacids for the Synthesis of Chiral Lactones. *Org. Lett.* **2022**, *24*, 2494–2498. [CrossRef] [PubMed]
15. Xiao, G.; Xie, C.; Guo, Q.; Zi, G.; Hou, G.; Huang, Y. Nickel-Catalyzed Asymmetric Hydrogenation of γ-Keto Acids, Esters, and Amides to Chiral γ-Lactones and γ-Hydroxy Acid Derivatives. *Org. Lett.* **2022**, *24*, 2722–2727. [CrossRef] [PubMed]
16. Johnson, E.R.; Keinan, S.; Mori-Sánchez, P.; Contreras-García, J.; Cohen, A.J.; Yang, W. Revealing Noncovalent Interactions. *J. Am. Chem. Soc.* **2010**, *132*, 6498–6506. [CrossRef] [PubMed]
17. Lu, T.; Chen, Q. Simple, Efficient, and Universal Energy Decomposition Analysis Method Based on Dispersion-Corrected Density Functional Theory. *J. Phys. Chem. A* **2023**, *127*, 7023–7035. [CrossRef]
18. Pospelov, E.V.; Golovanov, I.S.; Sukhorukov, A.Y.; Chen, J.; Zhang, W.; Gridnev, I.D. Mechanism of formation of active catalytic species in nickel-catalysed asymmetric hydrogenation. *Mend. Comm.* **2025**, *35*, 708–710. [CrossRef]
19. Frisch, M.J.; Trucks, G.W.; Schlegel, H.B.; Scuseria, G.E.; Robb, M.A.; Cheeseman, J.R.; Scalmani, G.; Barone, V.; Petersson, G.A.; Nakatsuji, H.; et al. *Gaussian 16*; Revision C.01; Gaussian, Inc.: Wallingford, CT, USA, 2016.
20. Mitoraj, M.P.; Michalak, A.; Ziegler, T. A Combined Charge and Energy Decomposition Scheme for Bond Analysis. *J. Chem. Theory Comput.* **2009**, *5*, 962–975. [CrossRef]
21. Lu, T.; Chen, F. Multiwfn: A Multifunctional Wavefunction Analyzer. *J. Comput. Chem.* **2012**, *33*, 580–592. [CrossRef] [PubMed]

https://doi.org/10.3390/molecules31020385

MDPI AG
Grosspeteranlage 5
4052 Basel
Switzerland
Tel.: +41 61 683 77 34

Molecules Editorial Office
E-mail: molecules@mdpi.com
www.mdpi.com/journal/molecules

www.ingramcontent.com/pod-product-compliance
Lightning Source LLC
Chambersburg PA
CBHW061456210926
51850CB00008B/1733
* 9 7 8 3 7 2 5 8 6 9 1 0 7 *